333.79097 SOVAC
Sovacool, Benjamin K.
The dirty energy
dilemma :

WRIGHT LIBRARY

Advance Praise for *The Dirty Energy Dilemma*

This brilliant volume by Benjamin Sovacool will provide new insights and stimulate new debates. Its appearance could not be timelier. It should be read carefully by all, from academics to policymakers, from businessmen to journalists.
Kishore Mahbubani
Dean of the Lee Kuan Yew School of Public Policy at the National University of Singapore, former Ambassador to the United Nations, and author of *The New Asian Hemisphere*

In this timely work, Benjamin Sovacool explains both the alternative energy possibilities now available and the political, economic, and social impediments that the United States must overcome to create a clean energy society.
Dr. David E. Nye
Professor of American Studies at the University of Southern Denmark and author of *Consuming Power: A Social History of American Energies*

The Dirty Energy Dilemma *is a thoughtful and thoroughly researched look into what's stopping the use of renewable energy in the U.S. An important and useful read for advocates of clean power.*
Dr. Godfrey Boyle
Director of the Energy & Environment Research Unit at The Open University and author of *Renewable Energy: Power for a Sustainable Future*

An extensively researched, highly accessible, authoritative explanation of the challenges facing the American electric utility sector. The Dirty Energy Dilemma *will stimulate debate about its content, and offers inspiration for those who seek to promote more progressive energy policy.*
Dr. Art Rosenfeld
California Energy Commission, and Recipient of the 2006 Enrico Fermi Award

In an era of rising fossil fuel prices and possibly catastrophic climate change, finding ways to promote clean power sources such as onshore and offshore wind have never been more important. The Dirty Energy Dilemma *clearly describes the challenges facing such clean power systems in the United States, but it also provides sound advice for regulators wishing to overcome social obstacles to clean energy all over the world.*
Steve Sawyer
Secretary General of the Global Wind Energy Council

This is a book the President and his staff should read. They should give copies to every member of the House and Senate if they are to intelligently vote on energy policy. Insightful, clear, and forceful, Sovacool explains how America's renewable energy policy has been hijacked for 30 years and what we need to do to reclaim it for a safer, cleaner, and more secure country.
Paul Gipe
renewable energy advocate and author of *Wind Energy Comes of Age*

This is a bold and brilliant work that examines why wind, solar, and other clean power technologies have not penetrated the American electricity market, but it also provides a much needed blueprint for how they could.
The value of this book is that it provides a fresh look at a long-standing issue: despite all of its clear advantages, how can alternative clean power (including energy efficiency) compete with the U.S. electric power system of today? By viewing the electric utility system as a set of social, cultural, economic, and political interests, Sovacool provides a holistic and comprehensive perspective in addressing this question. His policy recommendations for achieving the widespread adoption of clean energy should provide a good start for those working in a new Administration in 2009.
Dr. Ed Vine
Staff Scientist at the Lawrence Berkeley National Laboratory and Research Coordinator for the California Institute for Energy and Environment

The Dirty Energy Dilemma *offers a new insight into America's energy issues. Sovacool takes us past the oft-visited and necessary, but insufficient questions of hardware and finance and into new terrain. By viewing the electric utility industry as a set of human interests and perceptions, Sovacool adds a vital element: cultural analysis. With a blend of sensitivity and bluntness, recognizing that "we have met the enemy and they are us," Sovacool outlines the perceptions—the paradigms—that have tied our minds to old and dirty energy choices. Shining a light on the ways that we have limited our thinking, he opens the chance for us, together, to go beyond our past and into a future that we consciously make far, far, better. We all—as analysts, as technologists, as economists, or as* humans—*owe Sovacool a thanks for showing us the limits of our thoughts, so we can transcend them in our future.*
Michael Dworkin
Professor of Law & Director, Institute for Energy and the Environment, Vermont Law School

This book thoroughly, cogently, forcefully, and colorfully lays out the political, financial, and technical obstacles to meeting the need for reliable and clean electricity at a cost we can afford. But its unique value is in its description of the surprising institutional and cultural barriers that keep electricity dirty, expensive, and prone to failure. Policymakers

and advocates should pay particular attention to the optimism and the proposals with which the book concludes.
Jerrold Oppenheim and Theo MacGregor
coauthors of *Democracy And Regulation: How the Public Can Govern Essential Services*
Dr. Hermann Scheer
member of the German Bundestag, President of the European Association for Renewable Energy, General Chairman of the World Council for Renewable Energy, and the Recipient of the 1999 Right Livelihood Award

The Dirty Energy Dilemma *is an urgent and engrossing dissection of the American way of energy. In vivid and incisive detail Benjamin Sovacool lays bare the social, political, and cultural matrix that shapes and distorts energy choices, stifling the clean and fostering the dirty. Are the obstacles surmountable? Sovacool says yes, and shows how. Informative, provocative, readable, and re-readable.*
Walt Patterson
Senior Research Fellow at the Royal Institute of International Affairs in London and author of *Keeping The Lights On: Towards Sustainable Electricity*

The Dirty Energy Dilemma *creatively combines traditional policy analysis with sociology, history, and philosophy to create a rich account of the interactions among electricity, culture, and technology. It offers policymakers a clear and logical playbook for overcoming the regulatory challenges that confront clean-power systems. It really is a first-rate book.*
Dick Munson
Senior Vice President of Recycled Energy Development and author of *From Edison to Enron*

Sovacool's work is the perfect antidote to the sophistry that underpins the notion of a nuclear renaissance. The Dirty Energy Dilemma *is a sobering analysis of the forces arrayed against the clean energy revolution in the United States and the obstacles that must overcome if we are to achieve it.*
Jim Riccio
Nuclear Policy Analyst for Greenpeace

The Dirty Energy Dilemma *provides a bold analysis of the social, cultural, economic, and political forces that impede development of a clean electric energy system. Arguing that many clean energy technologies are now available and viable if given the right policy milieu, Benjamin Sovacool provides a strong vision of and policy prescriptions for a transition from a highly centralized, capital intensive, vulnerable, and environmentally damaging system to one marked by high efficiency, resilience, locally distributed modular generation, and sustainable renewable resources.*
Rodney Sobin
Director of the Office of Small Business Assistance for the Virginia Department of Environmental Quality

A comprehensive and insightful exploration of the barriers facing the adoption of clean energy and a clear portrayal of the role that renewable energy and energy efficiency can play in the multifaceted effort needed to address climate change. An exceptional and extremely timely book for policymakers or anyone interested in the subject of energy or climate policy.
Janet Peace
Director of Market and Business Strategy at the Pew Center on Global Climate Change

An important book. Sovacool's The Dirty Energy Dilemma *should be required reading for all who believe that a cleaner, more secure energy system is not only possible but essential, and for anyone who has ever doubted the potential for renewable energy and efficiency to rapidly meet a growing share of our energy needs if we have the political will and the right policies.*
Dr. Janet L. Sawin
Senior Researcher, Worldwatch Institute

This comprehensive overview of the history and current reality of energy systems within the United States presents an insightful analysis of the reasons why keeping the lights on costs people so much more than they assume. The environmental and social challenges that plague energy systems, not just within the United States but the world over, reflect an unacceptable complacency about the long-term stakes at hand. Sovacool makes a compelling case for a new set of parameters and processes to define a new energy future: citizens and civil society have a crucial role to play—and indeed a responsibility—to ensure a more sustainable trajectory for the choices made.
Smita Nakhooda
Institutions and Governance Program–The Electricity Governance Initiative, World Resources Institute

Sovacool tells a story that policymakers urgently need to hear. The twenty-first century needs a wiser and more caring energy policy. This book provides the way forward in a clear and accessible way.
Miguel Mendonça
Research Coordinator for the World Future Council and author of *Feed-In Tariffs: Accelerating the Deployment of Renewable Energy*

With remarkable poignancy, Sovacool examines why Americans have yet to embrace readily available alternatives to our dirty fossil fuels. His work taps a dizzying amount of research to reveal the stark reality that many of his contemporaries are afraid to admit: the enemy of clean energy is not technology, politics, or even markets. The enemy is us!
Christopher Cooper
Principal, Oomph Consulting, LLC

The Dirty Energy Dilemma *offers us all a thoughtful strategy for addressing the daunting energy and climate challenges that we currently face.*
Wilson Rickerson
Principal, Energy Strategies, LLC

A fresh and innovative look at energy policy in America, bringing out with remarkable discernment the social implications of renewable, fossil fuel, and nuclear power technologies. Here is a book that makes most other books in its field seem outdated and obsolete.
Kyle Rabin
Director, Network for New Energy Choices

THE DIRTY ENERGY DILEMMA

What's Blocking Clean Power in the United States

Benjamin K. Sovacool

Foreword by
Marilyn A. Brown

Westport, Connecticut
London

Library of Congress Cataloging-in-Publication Data

Sovacool, Benjamin K.
The dirty energy dilemma : what's blocking clean power in the United States / Benjamin K. Sovacool ; foreword by Marilyn A. Brown.
p. cm.
Includes bibliographical references and index.
ISBN 978-0-313-35540-0 (alk. paper)
1. Electric utilities–Environmental aspects–United States. 2. Clean energy–United States. 3. Energy policy–United States. I. Title.
HD9685.U5S697 2008
333.790973–dc22 2008023813

British Library Cataloguing in Publication Data is available.

Copyright © 2008 by Benjamin K. Sovacool

All rights reserved. No portion of this book may be reproduced, by any process or technique, without the express written consent of the publisher.

Library of Congress Catalog Card Number: 2008023813
ISBN: 978-0-313-35540-0

First published in 2008

Praeger Publishers, 88 Post Road West, Westport, CT 06881
An imprint of Greenwood Publishing Group, Inc.
www.praeger.com

Printed in the United States of America

The paper used in this book complies with the Permanent Paper Standard issued by the National Information Standards Organization (Z39.48–1984).

10 9 8 7 6 5 4 3 2 1

To Kelly. Without you, this book (and my perpetual state of happiness) would never have happened.

We must bear in mind, then, that there is nothing more difficult and dangerous, or more doubtful of success, than an attempt to introduce a new order of things in any state. For the innovator has for enemies all those who derived advantages from the old order of things, whilst those who expect to be benefited by the new institutions will be but lukewarm defenders. This indifference arises in part from fear of their adversaries who were favored by the existing laws, and partly from the incredulity of men who have no faith in anything new that is not the result of well-established experience. Hence it is that, whenever the opponents of the new order of things have the opportunity to attack it, they will do it with the zeal of partisans, whilst the others defend it but feebly, so that it is dangerous to rely upon the latter.

—Nicolo Machiavelli (1469–1527), Chapter Six of *The Prince*
(New York: Random House, 1922), p. 20

The difficulty lies not with the new ideas, but in escaping the old ones.

—John Maynard Keynes (1883–1946), Preface to
The General Theory of Employment, Interest and Money
(New York: Harcourt Publishers, 1935), p. iv

Contents

List of Illustrations

Figures

Maps

Photographs

Tables

Foreword

As the United States and the world face a crisis of energy availability and security, many must wonder why electricity continues to be delivered to peoples' homes, apartments, businesses, and factories much as it always has. Overhead transmission lines transport electrons from remote coal, natural gas, and nuclear power plants to substations where they are converted to lower voltages suitable for use in an array of appliances and gadgets. By the time the power lights an incandescent bulb, 97 percent of the energy embodied in the original lump of coal burned to drive a turbine in a conventional power plant has been lost.

Why has the U.S. electric system not been transformed to take advantage of the new electrical engineering accomplishments in the past half century? Why does more than half of the total power produced in Denmark come from combined heat and power (CHP) facilities, while only a fraction of U.S. power is drawn from CHP? Why is about 8 percent of total installed electricity capacity in the European Union from renewable sources other than hydropower and biofuels, while the United States boasts only 2 percent? Why have roof-mounted solar photovoltaic (PV) panels become a common feature in Germany and Japan, while many states in the United States do not even have "one-stop-shop" providers of PV products?

The Dirty Energy Dilemma: What's Blocking Clean Power in the United States answers these questions and many more. First written by Benjamin Sovacool as a doctoral dissertation in Virginia Tech's Department of Science and Technology Studies, this book has been transformed into an encyclopedic, yet highly readable, explanation of the underachievement of clean energy technologies in this country. Expanding on an initial list of three types of clean energy (distributed generation, combined heat and power, and renewable resources), the book goes beyond the dissertation's scope by incorporating an important additional clean resource—energy efficiency. For each of these four clean energy types, Sovacool presents a wealth of information about their market potential and the barriers that prevent this potential from being realized.

Not to be satisfied by simply identifying obstacles to the success of clean energy technologies, the book also offers a platform of four policy remedies: make clean power mandatory, eliminate subsidies, get the price right, and inform the public and protect the poor. It wisely calls for federal action to help implement these four policies for reasons of distributive justice, transaction costs, and historical precedence (among other arguments). Based on the sobering historical evidence presented by Sovacool, it is clear that the large-scale market uptake of clean energy sources in the United States will not be done in any "business-as-usual" policy scenario. Placed in the context of urgent energy and climate challenges, stronger federal engagement is long overdue.

The Dirty Energy Dilemma: What's Blocking Clean Power in the United States is a major resource for clean technology advocates, and an asset for our political representatives and their staff as the United States enters an active season of debate about the nation's energy and climate policies. I strongly recommend that stakeholders and policy analysts take the time to read it.

Marilyn A. Brown
Professor of Energy Policy, Georgia Institute of Technology
and
Commissioner, National Commission on Energy Policy
and
Distinguished Visiting Scientist, Oak Ridge National Laboratory

Preface

William Blake (1757 to 1827) once wrote that "energy is eternal delight." In fact, the very word "energy" first appeared in English in the sixteenth century, and then it had no scientific meaning. It simply referred to forceful or vigorous language, and it was not until the 1800s that the concept of "energy" encompassed anything resembling its modern form, when natural philosophers began to use it to describe phenomena such as the motion of the planets, transfer of heat, and operation of machinery. The concept continued to evolve into today's common scientific definition that energy is the capacity to do work, or the ability to move an object against a resisting force.[1]

Writing this book met both modern connotations: it was a great deal of work and felt like it was constantly moving against resisting forces. Because I believe writing is like jazz (both can require a curious mix of improvisation and repetition), the book also reiterated many ideas from previously published works.[2]

Walt Patterson correctly points out that when everyone speaks about energy production and consumption, their statements are inaccurate since "energy" itself can be neither created nor destroyed.[3] We do not consume energy; we use it. When we discuss "energy efficiency" or "energy conservation" we are really talking about "fuel efficiency" or "energy performance." When we talk about "energy production," we really mean "the production of energy carriers" and "energy consumption" is really "energy use." While I agree with Patterson that talking about energy as if it could be produced and consumed tends to distort our understanding of what we actually "do" with energy, I still adhere to conventional nomenclature because it offers useful shorthand widely used by analysts currently talking about energy and electricity policy. I fully admit, however, that its use should be contextualized and bracketed by Patterson's concerns, which is why I mention them here.

I had the distinct pleasure of traveling across the United States and to Belgium, Denmark, Germany, Indonesia, Japan, Malaysia, the Philippines, Singapore, Spain,

Thailand, and the United Kingdom to learn more about clean power technologies and electricity policy. To discover what impedes clean power and what can truly promote it, I was privileged to conduct 181 formal, semistructured interviews at more than 82 institutions (including electric utilities, regulatory agencies, interest groups, energy systems manufacturers, nonprofit organizations, consulting firms, universities, national laboratories, and state institutions) from 2005 to 2008. The interviews represent a broad overview of the interests and academic training of those connected to the electric utility sector and clean power, and each of those interviewed are to be commended for taking the time out of their busy schedules to speak with me about this project. Those seeking full transcripts for some of these interviews should consult Appendix 2 of my dissertation.[4]

I am grateful to the U.S. National Science Foundation for grants SES-0522653, ECS-0323344, and SES-0522653, which have supported elements of the work reported here, along with an extremely generous fellowship from the Centre on Asia and Globalisation, part of the distinguished Lee Kuan Yew School of Public Policy at the National University of Singapore. I must state that any opinions, findings, and conclusions or recommendations expressed in this material are my own and do not necessarily reflect the views of the National Science Foundation or the Centre on Asia and Globalisation.

I am also appreciative to the Centre on Asia and Globalisation, Oak Ridge National Laboratory, Virginia Center for Coal and Energy Research, Society for the Social Studies of Science, the National Science Foundation's Electric Power Networks Efficiency and Security Program, Virginia Tech Department of Science and Technology Studies, Virginia Tech Department of Philosophy, Virginia Tech Department of History, Virginia Tech School of Public and International Affairs, and the Virginia Tech Graduate Student Assembly for travel support to conduct research interviews for this project.

There are so many other people to thank that I almost considered not trying. At the top of the list have to be the members of my dissertation committee, Professors Richard Hirsh, Daniel Breslau, Saul Halfon, Eileen Crist, and Timothy Luke. Each of them, especially Richard and Daniel, spent countless hours reading over my thoughts and providing outstanding advice.

A special thank you goes out to Marilyn Brown, for writing the book's foreword and mentoring me, and to Christopher Cooper, one of the sharpest minds I have ever met. Janet Sawin, Wilson Rickerson, and Skip Laitner also get special thanks for providing comments on earlier drafts and helping me think through some of the policy recommendations.

Elizabeth Siman, my mother-in-law, gets a very genuine "thank you!" for keeping me updated with recent events in the energy sector by sending me highlighted copies of magazine and newspaper articles.

Ann Florini, Toby Carroll, and all of my associates at the Centre on Asia and Globalisation have been exemplary colleagues, and I thank them for putting up with me.

Even though my mother's reply to my idea for this book was "it sounds completely boring," she, my father, and my brother all have my deepest gratitude.

Abbreviations

List of Abbreviations, Acronyms, and Key Terms

ACEEE	American Council for an Energy-Efficient Economy
ASE	Alliance to Save Energy
ASME	American Society of Mechanical Engineers
AWEA	American Wind Energy Association
Btu	British thermal unit
CEC	California Energy Commission
¢/kWh	cents per kWh
CFLs	compact fluorescent light bulbs
CHP	combined heat and power
CO_2	carbon dioxide
DG	distributed generation
DOD	U.S. Department of Defense
DOE	U.S. Department of Energy
DSM	demand-side management
€¢/kWh	Eurocents per kWh
EIA	U.S. Energy Information Administration
ELCC	effective load carrying capability
EPA	U.S. Environmental Protection Agency
EPRI	Electric Power Research Institute
EU	European Union
FERC	Federal Energy Regulatory Commission
FIT	feed-in tariff
FPL	Florida Power and Light
gCO_2e/kWh	grams of CO_2 equivalent per kWh
GAO	U.S. Government Accountability Office
GDP	gross domestic product

Gg	gigagrams
GNEP	Global Nuclear Energy Partnership
GW	gigawatt (one thousand MW)
GWh	gigawatt-hour
Hg	mercury
IAEA	International Atomic Energy Agency
IEA	International Energy Agency
IEEE	Institute of Electrical and Electronics Engineers
IGCC	integrated gasification combined cycle
IPCC	Intergovernmental Panel on Climate Change
ISO	Independent Systems Operator
ITER	International Thermonuclear Experimental Reactor
kW	kilowatt
kWh	kilowatt-hour
LBNL	Lawrence Berkeley National Laboratory
LCOE	levelized cost of electricity
LNG	liquefied natural gas
MBtu	million Btu
MW	megawatt (one thousand kW)
MWh	megawatt-hour
NARUC	National Association of Regulatory Utility Commissioners
NERC	North American Electric Reliability Corporation
NO_x	nitrogen oxides
NRC	Nuclear Regulatory Commission
NRDC	Natural Resource Defense Council
NREL	National Renewable Energy Laboratory
NYSERDA	New York State Energy Research and Development Authority
NWPA	Nuclear Waste Policy Act of 1982
OECD	Organization for Economic Cooperation and Development
OPEC	Organization of Petroleum Exporting Countries
ORNL	Oak Ridge National Laboratory
OPA	Ontario Power Authority
OTA	Office of Technology Assessment
OTEC	Ocean Thermal Energy Conversion
PG&E	Pacific Gas & Electric Company
PM	particulate matter
PSE	Puget Sound Energy
PUC	Public Utility Commission
PUHCA	Public Utility Holding Company Act of 1935
PURPA	Public Utility Regulatory Policy Act of 1978
PV	photovoltaics
R&D	research and development
RPS	renewable portfolio standard
SBCs	system benefits funds
SO_2	sulfur dioxide
T&D	transmission and distribution
TLR	Transmission Loading Relief

TVA	Tennessee Valley Authority
TW	terrawatt (one thousand GW)
UCS	Union of Concerned Scientists
UNEP	United Nations Environment Program
U.S.	United States
USGS	U.S. Geologic Survey
WPPSS	Washington Public Power Supply System

Introduction

Some advocates of the conventional electricity industry send the message to the American public that fossil fuel and nuclear technologies are time-tested, reliable, and essential to continued low electricity prices and stable economic growth. Alternative clean power technologies, by contrast, they ruefully dismiss as immature, untried, intermittent, costly, prone to accident and failure, and essential only for niche applications such as lighting an off-grid parking lot or powering a remote weather station.

But the story that the American public hears is not the whole story. In fact, as this book will demonstrate, the central message that clean power is not ready for prime time is simply wrong.

The quest for the true story about alternative energy technologies begins with a few hard facts about the U.S. electric power system of today:

- The average coal-fired generator can emit 913 tons of carbon dioxide (CO_2), or more than its physical weight, every single hour;[1]
- The power consumed for one day in the average American home needs 775 gallons of water to produce it;[2]
- Every month, unexpected blackouts cost businesses and industries more than three times what the federal government spends on education over the same period;[3]
- The United States wastes more energy from the inefficient production of electricity each year than Japan harnesses for its entire economy;[4]
- Air pollution from just nine power plants currently operating in the Midwest will directly and incontrovertibly cause 1,500 premature deaths, 70,000 asthma attacks, and more than 2 million daily incidents of upper respiratory symptoms among the 33 million people living within close proximity of the plants every five years;[5]
- Every decade the American electricity industry wastes about $689 billion in heat and power inefficiencies;[6]
- When priced to include all costs and benefits (or positive and negative externalities) using the best methods available, clean power technologies are 2 to 15 times cheaper than conventional sources;[7]

- Relying only on published, peer-reviewed studies, commercially available renewable power generators could provide 3.7 times more electricity than the country currently uses.[8]

Put succinctly, the current power system has you using more water for electricity than for bathing, showering, drinking, and watering your lawn. It deteriorates human health and spews hazardous material into our air, land, and water. It is the single biggest contributor to climate change. It costs customers $206 billion every year for power they needed but never got, and overcharges customers $68.9 billion for the power that they did get. It utilizes the most expensive generators when all social costs and benefits are included, and ignores those technologies that could cleanly meet almost four times the country's electricity needs.

The Dirty Energy Dilemma

This book presents a sobering assessment of fossil-fueled, nuclear, and alternative technologies in the electricity sector. Once taken for granted as a stable and secure consortium of publicly regulated and efficiently run monopolies, the electric utility industry in the United States has over the past three decades become increasingly unstable, fragmented, unreliable, insecure, inefficient, expensive, and harmful to our environment and public health. This alarming trend is being driven by steadily rising prices and periodic disruptions in fossil fuel supply; burgeoning levels of pollution; growing inefficiency in long-distance transmission and centralized distribution networks; and increasing vulnerability to corporate boardroom manipulations, cascading blackouts, accidents, sabotage, and natural disasters.

The fix for this ugly array of problems is relatively simple: clean power in the form of energy efficiency, distributed generation (DG), combined heat and power (CHP), and renewable energy. Energy efficiency refers to improving the physical performance of specific end uses such as lighting, heating, and motors, achieved by replacing, upgrading, substituting, or maintaining existing equipment. DG encompasses a set of small-scale energy systems that produce power in modular increments situated close to where it is consumed. CHP systems produce thermal energy and electricity from a single fuel source, recycling low-grade energy that would otherwise be wasted. Renewable energy generators create electricity from sunlight, wind, falling water, biomass, waste, and/or geothermal heat.

In contrast to the belching of pollution by centralized fossil fuel plants that currently generate almost 70 percent of American electricity, these clean power systems produce few harmful by-products, relieve congestion on the transmission grid, require less maintenance, are not subject to price volatility, and enhance the security of the national energy system from natural catastrophe, terrorist attack, and dependence on supply from hostile and unstable regions of the world. Some renewable energy technologies using wind and landfill gas even produce power at lower prices than the most advanced coal, natural gas, and nuclear plants.

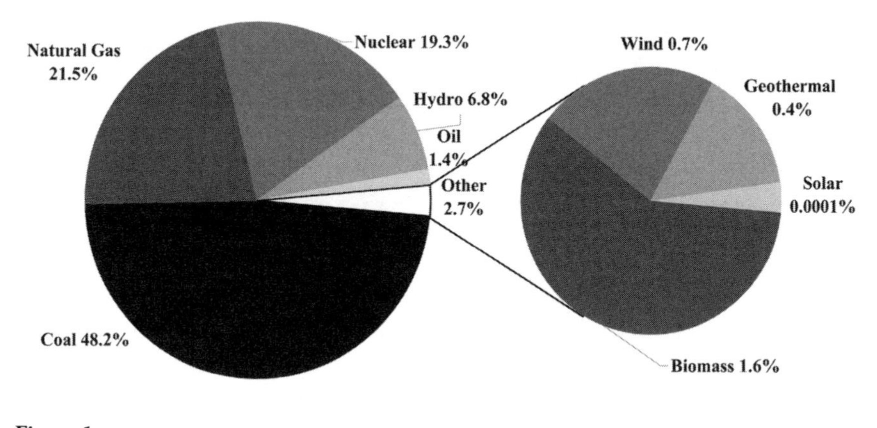

Figure 1
U.S. Electricity Generation by Source, 2008

Solar, wind, and geothermal sources together account for only 1.1 percent of 2008 U.S. electricity generation.

Source: Figure 4 of U.S. Energy Information Administration, *Annual Energy Outlook 2008.*

Despite their clear advantages, solar, wind, and geothermal sources together account for little more than 1 percent of U.S. electricity generation—90 times less than fossil fuels and nuclear plants do. If distributed equally to every home in America, renewable resources would not provide enough juice at their current utilization to power even two 60-watt light bulbs for ten hours every day.[9] They would not produce enough energy to run the hot-water faucet for five minutes.[10] Though ballyhooed and lauded as inexhaustible, clean, modular, and universally accessible, solar power still accounts for less than one-thousandth of 1 percent of national electricity output.

This relatively minor use of renewable energy systems has created a general attitude among energy analysts, scholars, laboratory directors, and corporate executives that renewable electricity technologies are not viable sources of electricity supply. For example, while Rodney Sobin, former Innovative Technology Manager for the Virginia Department of Environmental Quality, is a strong advocate of renewables, he does concede that "in many ways, renewable energy systems were the technology of the future, and today they still are."[11] Ralph D. Badinelli, a professor of Business Information Technology at Virginia Tech, explains that renewable energy technologies do not contribute significantly to U.S. generation capacity because "such sources have not yet proven themselves . . . Until they do, they will be considered scientific experiments as opposed to new technologies."[12] Similarly, Mark Levine, the Environmental Energy Technologies Division Director at the Lawrence Berkeley National Laboratory (LBNL), comments that despite all of the hype surrounding renewable energy, such systems are still only "excellent for niche applications, but the niches aren't large."[13] Lee Raymond, the former chief executive officer of Exxon

Mobil, told the *World Petroleum Congress* that "non-petroleum sources of energy" were merely "fashionable," and that "with no readily available economic alternatives on the horizon, fossil fuels will continue to supply most of the world's energy needs for the foreseeable future." [14]

DG and CHP technologies fare even worse. Less than 1 percent of industrial DG systems operate continuously, and overall the U.S. Department of Energy (DOE) projects that DG and CHP systems will grow to just 3,000 megawatts (MW) of installed utility power in 2025, or 0.25 percent of total estimated capacity.[15] Tom Casten, chief executive officer of a manufacturer of fuel processing cogeneration steam plants, notes that even though DG technologies can reduce energy costs for industrial firms by more than 40 percent, such generators remain "the exception instead of the rule." [16] These technologies have so much untapped promise that if they were deployed to their fullest cost effective potential in just three energy-intensive industries—petroleum refining, chemicals manufacturing, and paper and pulp processing—their estimated market value could exceed $1 trillion (in 2007 dollars).[17]

Energy efficiency holds equal promise. Almost two decades ago, researchers calculated that cost-effective energy efficiency measures could cut the nation's electricity use by 30 to 75 percent without significant life-style changes and without reducing the growth of U.S. gross domestic product (GDP).[18] Nine years later, a second study estimated that if businesses made minor mechanical alterations to their industrial processes they could cut their electricity consumption in half—with net savings of $110 billion a year.[19] Four years after that, the National Commission on Energy Policy warned that electricity consumption could be 40 percent higher than cost-minimizing levels.[20] In 2005, economists concluded that the United States could cost-effectively reduce energy use 25 percent or more during the next 15 years in ways that increase overall productivity.[21] Law professor John Dernbach projected in 2007 that 60 percent of the country's 70 million owner-occupied residences were still not well-insulated and 70 percent of the country's 4.6 million commercial buildings entirely lacked roof or wall insulation.[22] Researchers at the American Council for an Energy-Efficient Economy (ACEEE), a nonprofit think tank, estimated in 2008 that the country could have $700 billion worth of potential investments in energy efficiency by 2030.[23] Using their most recently compiled data, the U.S. Energy Information Administration (EIA) reports utilities across the nation used energy efficiency and demand-side management programs to reduce electricity consumption by only about 1.4 percent.[24]

Given these statistics, a number of questions arise: If clean power systems deliver such impressive benefits, why are they languishing in neglect at the margins of the American energy portfolio? Why do the clean power technologies with the most promise to ease the problems of the American electric utility industry have the least use? And why does the United States lag so far behind Europe and other industrialized economies, where conversion to clean power systems has already taken off in a big way?

The Conventional View

The conventional view, of course, is that we cannot do better, and why should we try? For most of the industry's history since Thomas Edison flipped the switch on his first industrial generator in 1882, the electric utility system has coupled declining electricity prices with steadily rising profits. Providing an almost textbook example of "economies of scale" in action, the capacity of large generators doubled every 6.5 years from 1930 to 1970 while, at the same time, electricity prices in nominal terms dropped from almost $1 to less than 7 cents per kilowatt-hour (¢/kWh).[25]

The success of the conventional power system became associated with a number of related assumptions.[26] Planners believed that the U.S. power system should consist of relatively few but large units of supply and distribution, and that those units should be composed of large, monolithic apparatuses rather than small, redundant models. Managers clustered units geographically near oil fields, coal mines, sources of water, demand centers, and each other, interconnecting units sparsely and making them heavily dependent on a few critical nodes and links. System builders wove interconnected units into a synchronous system in such a way that it is difficult for a section to operate in isolation. Operators provided little storage to buffer successive stages of energy conversion and distribution, meaning that failures tend to be abrupt and unexpected rather than gradual and predictable. Utilities located generators remotely from users, so that supply-chain links had to be long and the overall system lacked qualities of user controllability, comprehensibility, and interdependence.

Within this conventional view, clean power sources have failed to achieve widespread use in the United States because of seemingly insurmountable technical barriers. Brian O'Shaughnessy, chief executive officer of Revere Copper, told senators that "since the evolution of renewable power is at a very early stage in its development, mandating renewable power with today's technology is like trying to go to the moon in the 1950s."[27] The Electric Power Research Institute's *Electricity Technology Roadmap* concluded that technical problems relating to energy capture, storage, and manufacturing meant that "the market penetration of renewable technologies" was "limited."[28] An article in *Newsweek* proclaimed, "wind and solar are nice and clean—but the sun doesn't work 24/7 and the wind is fickle."[29] The George W. Bush administration has officially opposed mandating the use of clean power sources on the grounds that they create "winners" and "losers" among regions of the country and increase electricity prices in places where renewable resources are less abundant or harder to cultivate.[30]

This Book's View

An alternative view is that, in an era of rapid climate change and rising electricity costs, we have to do better. We need to pursue a smarter energy strategy, one that seeks to perform tasks with the least possible energy; one that recognizes that energy varies in its type, quality, unit scale, and geographic source; and one that promotes technologies that are simple enough to have construction lead times measured in days, weeks, and months instead of decades. A strategy is needed that adheres to ecological limits, protects future generations, encourages public participation and control, creates more autonomy, consumer choice, social stability, and equity, and decreases the conflict and concentration of knowledge and materials.

According to this view, there is nothing wrong with clean power systems as they are currently available. Rather, the conventional system's ability to continue along a given path results from the actions of numerous stakeholders (such as educational and regulatory institutions), the investment of billions of dollars in equipment, and the work and culture of people employed in the electricity industry. In concert, these elements promote "business as usual" and tend to culminate in dedicated constituencies that link commercial success with public welfare.[31] Managers of the system obviously prefer to maintain their domain and, while they may seek increased efficiencies and profits, they do not want to see the introduction of new and disruptive "radical" technologies that may alter momentum and reduce their control over the system.[32]

For most of their history, electric utilities have been able to lead the virtual "quiet life" of a monopolist without the threat of competition. Knowing that their markets were secure and future growth virtually guaranteed, electric utilities became less innovative and cost-conscious than other industries.[33] Utilities managed risks to ensure their own exposure was limited relative only to that of their limited number of competitors. The result is an electric utility system where stakeholders levy for *comparative* advantage, rather than improvement in any *absolute* sense.[34] While current technologies may be less than optimally efficient, they enable a highly integrated network of industry players to produce almost certain profits in highly uncertain environments.[35]

What Else Makes This Book Different?

In viewing the electric utility system in this manner—as a set of social, cultural, economic, and political interests, rather than a "black box" of power technologies—this book differs from most scholarship on electricity and energy.

First, it contrasts energy reports such as those from the U.S. Energy Information Administration (EIA) and International Energy Agency (IEA), which focus primarily on estimating generation capacities, projecting fuel costs, and predicting the environmental impacts of particular energy technologies. For example, the paragon of excellence among these types of reports, the EIA's *Annual Energy Outlook,* projects current trends of energy consumption to provide perspective about future incomes and prices, but it does not anticipate future policy changes or provide policy recommendations. The report assumes the existing configuration of the industry, and thus restricts consideration to a very narrow range of alternatives.

Second, those studies that do attempt to provide a rich, contextualized approach tracing social, historical, and institutional factors in the acceptance of energy technologies have not tended to focus on the availability of cost-effective clean power resources. Two of the best investigations of the social nature of electricity technologies—Thomas Hughes's *Networks of Power* and David Nye's *Electrifying America*—limit their analysis from the 1880s to the 1940s, and focus exclusively on either generation (Hughes) or consumption (Nye).[36] David Nye's other influential book, *Consuming Power: A Social History of American Energy Technologies,* dedicates only a chapter to electricity and only a few paragraphs to renewable energy systems and DG technologies.[37] The work of historian Richard F. Hirsh on the managerial practices and technological choices facing the American electric utility industry provides excellent insight into how large-scale and centralized fossil fuel generators lost both technical and social momentum throughout the 1960s, 1970s, and 1980s. While Hirsh does talk about renewable energy and energy efficiency in the 1990s, he does not emphasize the importance of social factors and their relationship to clean power technologies in the 2000s. Martin V. Melosi and Vaclav Smil provide well-written and thorough cultural histories of energy systems in the United States and the world, but conclude their investigation with the oil crises of the 1970s.[38] In other words, none of these outstanding works focus on changes affecting energy efficiency, DG, CHP, and renewables in the electric utility sector in the past 10 to 20 years.

When some recent studies do investigate particular energy technologies, they tend to focus on fossil fuels (especially since such fuels and generators have historically dominated American electricity generation) and cover either supply or demand (rather than both). Additionally, the majority of these works have been written well before the creation of the Energy Policy Act of 1992, the consequent electricity industry restructuring throughout most of the 1990s, the occurrence of the California electricity crisis of 2000–2001, the subsequent blackouts in the northeast (New York) and south (Hurricane Katrina), the enactment of the Energy Policy Act of 2005, and the very recent passage of the Energy Independence and Security Act of 2007. Each of these events drastically altered the trajectory of the American electric utility system and existing studies suffer from their exclusion.

Third and finally, one cannot fully understand the failure to adopt clean power technologies without studying electricity as a heterogeneous system. Most recent policy briefs and books that focus on electricity in America attack the problem within narrow disciplinary boundaries. For example, the Edison Electric Institute (EEI) and Electric Power Research Institute (EPRI) tend to center their studies purely on the economics of electricity generation, transmission, and distribution, while reports from groups like the Pew Foundation and Natural Resource Defense Council (NRDC) emphasize the environmental dimensions of electricity generation and consumption. The National Academies of Science and Union of Concerned Scientists (UCS) have produced insightful analysis of the security and infrastructure challenges facing the sector, while groups like the Alliance to Save Energy (ASE) and ACEEE remain mostly concerned with conservation and industrial efficiency (with some notable exceptions). Those groups that have even attempted a social analysis of the impediments facing particular technologies—such as the American Wind Energy Association (AWEA), the American Solar Energy Society, or the Combined

Heat and Power Association—limit their analyses to individual technologies, rather than how such technologies operate in the electric utility system as a whole.

Put concisely, very few of these organizations has produced an interdisciplinary or holistic analysis of energy technologies that takes into account *all* of these factors in an investigation of the social influences impeding the use of clean power resources. Virtually none have connected research on cultural and social attitudes with patterns of electricity generation and consumption. And few have focused on how clean power systems can be considered a class of technologies (rather than individual artifacts) that emerge in a sophisticated social and technical network.

In contrast, for the first time, this book demonstrates that the only barrier blocking the conversion of a significant proportion of the U.S. energy portfolio to cleaner sources is not technological—we have the technology—but institutional. The impediments to clean power in the United States are not technical but social, political, regulatory, and cultural. Extensive interviews of public utility commissioners, utility managers, system operators, manufacturers, researchers, business owners, and ordinary consumers reveal that they are hobbled by organizational conservatism and self-interest, paralyzed by market failures, mired in legal inertia, limited by weak and inconsistent political incentives, and confined by ill-founded prejudices and institutionalized apathy.

Utility operators reject distributed resources because they are trained to think only in terms of big, centralized power plants. Consumers ignore renewable power systems because they are not given the economic option of choosing them. Intentional market distortions (such as subsidies), and unintentional market distortions (such as split incentives), prevent consumers from becoming invested in their electricity choices. As a result, new twists on dirty energy technologies that are compatible with the technical, economic, political, and cultural structure of the existing electrical generation system stand a much better chance of being adopted than do new clean technologies that may offer superior transmission efficiencies and social benefits but are not consistent with the dominant paradigm.

Preview of Chapters

To explain the dirty energy dilemma, Chapter 1 assesses the four big challenges facing the American electric utility sector: meeting demand projections, finding clean and abundant sources of energy supply, maintaining the infrastructure needed to distribute electricity, and minimizing the destruction of the environment. It begins by offering a brief primer on the modern electricity industry and argues that hydrogen, clean coal, fusion, advanced nuclear, and conventional units all suffer from the same difficulties: They are capital intense, must adhere to a brittle transmission and distribution infrastructure, are prone to accidents, are subject to fuel price spikes and interruptions, require excessive water usage, emit greenhouse gases and other air pollutants, destroy the land, and pollute our food chain.

Chapter 2 argues that clean power systems can meet the big four energy problems better than fossil-fueled systems: They satisfy growth in demand while giving communities greater control over energy technologies. They use domestically available energy sources in ways that increase the stability of the transmission grid and minimize transmission losses. They insulate the industry from price spikes and fuel interruptions. They operate more efficiently and more cleanly than their conventional counterparts. They are already the cheapest, most cost-effective ways to save or generate electricity when all costs and benefits are considered. They are less prone to cost overruns. And they consume less water, pollute less air, and use less land than conventional units.

Chapter 3 discusses the financial impediments, market barriers, and market failures associated with clean power. Most property owners, business leaders, utility managers, and electrical system operators interviewed for this book believe that clean power technologies entail prohibitively high capital costs. Split incentives between builders and homeowners, landlords and tenants, and utilities and consumers further block investments in clean power. Utilities and system operators use their market power (along with vestiges of monopoly legislation in federal and state regulation of utilities) to prevent new firms from entering the industry or connecting to their grid.

Chapter 4 exposes the political and regulatory barriers to clean power technologies. Exaggerated projections and hyperbolic claims in the 1970s backfired by turning industry away from clean power technologies that could have seen widespread adoption in the 1980s. Weak and inconsistent political incentives for renewable and distributed technologies, varying state standards, competition among utilities, and both public and private underfunding of clean power research and development (R&D) deprived public utility commissioners and system operators of any serious motivation to adopt alternatives to fossil fuels.

Chapter 5 analyzes the cultural and behavioral impediments to adoption of clean power technologies. Public apathy and misunderstanding, psychological resistance, and historical patterns of production and consumption all play a role in impeding the use of clean power sources.

Chapter 6 examines the aesthetic and environmental impediments to clean power technologies stemming from the mistaken belief that these systems have unacceptably large and destructive footprints. This chapter illustrates how, even though clean power technologies are much better than alternatives, their symbolic nature and the conflicting values and interests of consumers prevents them from becoming more widely accepted.

Chapter 7 outlines how policymakers can overcome these economic, regulatory, political, cultural, behavioral, environmental, and aesthetic obstacles. The chapter proposes four big policy mechanisms for achieving widespread adoption of clean energy: make clean power mandatory, eliminate energy subsidies for conventional electricity technologies, price electricity accurately, and establish a national systems benefits charge to inform the public, provide low-income assistance, and fund energy efficiency.

Chapter 8 concludes the book by discussing how, even while the country faces an impending electricity crisis, clean power technologies offer an optimal solution.

Even though these technologies are impeded by a seamless tangle of obstacles, history tells us that all energy systems have had to overcome similar challenges, and that comprehensive, consistent, and sustained policy action can eliminate the barriers to more widespread use of clean energy.

Four Caveats

As is obvious by now, this book has chosen to focus exclusively on the United States and its electric utility system instead of other countries. This is because, by any standard, the United States is the largest consumer of energy. In 2007, it absorbed roughly one-fourth of the world's total primary energy consumption even though it has less than one-twentieth of the global population. The country constitutes the world's largest market for power equipment. It has more nuclear reactors, pipelines, refineries, power producers, and electricity customers than any other country, and its citizens consumed 23 percent of the world's electricity.[39] Since the United States remains the world leader in the per capita extraction, consumption, and use of oil, uranium, coal, and natural gas, its decisions affect the global energy marketplace. The country also has the technical and financial resources needed to accelerate the development of clean power sources on a global scale.

The book also could have focused on clean or alternative energy broadly to include transportation fuels such as ethanol, biodiesel, and oil shale, but it did not. Apart from the obvious limits of space, electricity fuels were determined to be more important than transportation fuels for four reasons. Less than 80 percent of the American population own their own vehicles, but almost 100 percent of the population own devices that consume electricity.[40] People are generally less aware about electricity, with the DOE reporting that only about 12 percent can pass a "basic" electricity-literacy test. The opposite is true for transportation, since every few days most individuals put gasoline into their vehicles (with their own hands) and experience the effects of volatile prices.[41] From a greenhouse gas emissions standpoint, the electricity sector emits roughly 38 percent of the country's carbon dioxide, while the transportation sector emits only 32 percent.[42] Efforts to promote coal-to-liquids, hydrogen, and plug-in hybrid electric vehicles tend to merge the two sectors, so that focusing on electricity here still has the potential to address transportation problems in the future. In short, technological progress in the transportation industry means that electricity could soon constitute a significant transportation fuel.

Third, the book may come across as "one sided" to some, and it may sound "anti-progress" or "anti-electrification" to others. While this book is critical of the electric utility industry, it is not against electrification, and it recognizes that electricity brings manifold benefits to society too numerous to list. Utility executives and power providers have incredibly challenging jobs trying to transform "dirty" fuels into "clean" forms of energy.

In identifying the "risks" involved with the current electric utility system, this book says nothing about whether they are "worth it." Rather, this book maintains that the problem is that the current system makes many of those risks invisible and

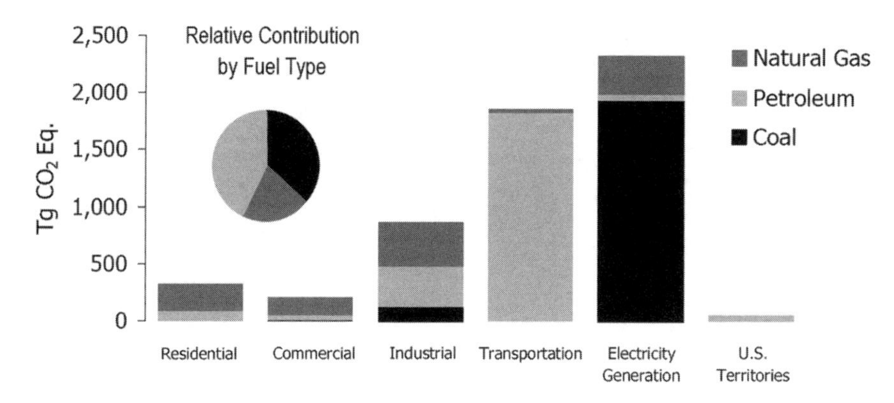

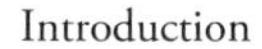

Figure 2
Fossil Fuel Emissions of CO_2 by Sector and Fuel Type, 2005

In 2005, the electric utility sector was by far the largest source of greenhouse gas emissions, emitting more carbon dioxide than the transportation sector or the residential, commercial, and industrial sectors combined.

Source: U.S. Environmental Protection Agency, *Human-Related Sources and Sinks of Carbon Dioxide* (2007).

involuntary. Rock climbers, construction workers, soldiers, and prostitutes all take a somewhat active and voluntary role in their risky behavior. Those suffering the impact of particle inhalation, nuclear meltdowns, exploding gas clouds, and petroleum-contaminated water do not. And it is the author's belief that *all* costs and benefits must be made visible before informed decisions about energy policy can be made. Furthermore, the responsibility for the problems facing the electricity industry does not lie with utilities alone. Many of the faults within the existing system transcend any individual or organization; no single "evil mastermind" carefully manipulates the course of energy technologies for its own benefit. Instead, a complex amalgam of mutually reinforcing political, social, economic, and technical interests shape the course of our energy system.

Finally, in laying out the book's chapters, the author did not intend to suppose that demarcations between "financial and market," "political and regulatory," "cultural and behavioral," and "environmental and aesthetic" impediments really exist in distinct, separate classes. For instance, the repeal of the Public Utility Holding Company Act of 1935 (PUHCA) removed incentives for utilities to engage in collaborative R&D and oriented them to focus more on short-term goals and rapid profits. Whether this is an example of an economic or political barrier is unclear. The Reagan administration's reduction of federal subsidies for clean power in the 1980s caused a large number of firms to go bankrupt, creating a social stigma against renewable technologies such as wind and solar. Is this obstacle social, economic, or political? Dividing the "social" from the "technical," or even the "economic" from the "political," is counterproductive and dangerous, since it misses the point that such impediments exist in an integrated nexus, and it is done here only to make such obstacles easier to identify.

CHAPTER 1

THE BIG FOUR ENERGY CHALLENGES

When asked about investment advice, Mark Twain reputedly responded, "Buy land, they ain't making any more of it." His remark recognized that natural resources are finite and that as they are depleted their value increases. In essence, Twain's remarks summarized the situation with fossil fuels and uranium: God is not making any more of them anytime soon, and the fundamental premise that they will become more valuable (i.e., costly) is more a shortcoming than a measure of their economic strength. From now until the point that they truly become depleted, coal, natural gas, oil, and uranium will always be subject to perceived scarcity, market speculation, manipulation, and price gouging.[1]

This chapter demonstrates that conventional coal, natural gas, and nuclear generators are prone to insolvable infrastructural, economic, social, and environmental problems. Coal and natural gas plants require complex, brittle, congested, and expensive infrastructure to transform them into useful forms of energy and to deliver that energy to market; they are susceptible to escalating and volatile prices; and they are quickly becoming too dirty and dangerous for most utilities to effectively utilize. Nuclear facilities face expiring licenses, immense capital costs, rising uranium fuel prices, and irresolvable problems with reactor safety, waste storage, proliferation, and vulnerability to attack. Finally, this chapter shows why hydrogen, clean coal, fusion, and advanced nuclear technologies are not mature enough to respond to any of the current challenges facing the industry.

To fully appreciate these concerns, it is useful to briefly explore how the electric utility system functions.

A Tale of Some Toast

Consider what it takes for you to transform a dull-looking slice of white bread into a delicious piece of crunchy toast. Now, even before you flip a switch, press a button, or turn a knob, six sets of energy losses typically will occur. To remind readers of just a little basic science, the first law of thermodynamics stipulates that energy is a constant that can neither be created nor destroyed. It can and does change form, but as it does the amount of useful energy decreases.

First, all the way back at the power plant, the chemical energy from combusted fuel is converted into thermal energy to produce steam, with a majority of the energy discharged into the surrounding environment as low-temperature waste heat and exhaust. Second and third, similar losses occur as the thermal energy is converted to mechanical energy to spin turbines, and mechanical energy to electrical energy to produce an electric current. These first three sets of energy losses are staggering: power plants convert the potential chemical energy of coal and other fuels into thermal energy at 100 percent efficiency, but typical usable energy is just 33 percent of that, with the remaining two-thirds simply wasted. Two Oak Ridge National Laboratory (ORNL) researchers have suggested that the poor thermal efficiency of most power plants translates into enough potential power lost every year to meet the needs of five to ten cities the size of Manhattan.[2]

A fourth loss occurs when some of this remaining energy is used by the plant itself in order to operate, run its pollution controls, conduct emissions monitoring, and power other equipment. Conventional coal plants with air pollution controls use 10 percent of their own electricity within the facility; nuclear plants use around 9 percent of their energy to run fuel and cooling cycles.

A fifth loss occurs as electricity is transmitted and distributed. Transmission and distribution lines do not conduct electricity with perfect efficiency. The average transmission line is about 90 percent efficient, with performance worsening in hotter weather or over exceptionally long distances.

Finally, at your home, a sixth loss occurs as the toaster converts the electrical energy coming out of your socket back into heat to toast your slice of bread.

Accounting for all six sets of losses, what you are left with as you set the toaster is just 27.6 units of usable energy out of every 100 units you started with. In terms of making toast, it would have been nearly four times more efficient just to burn a lump of coal and place your bread over the flame.[3] Each time energy is converted from fuel to steam to mechanical energy to electricity to end use, a substantial amount of energy is lost. For lighting, it is even worse: one study found that converting coal at a power plant into the incandescent light used in your home wastes nearly 97 percent of the original energy potential![4]

It gets worse. The six sets of losses described above do not include inefficiencies that occur between the extraction of fuels and their arrival at the power plant. The goliath-size draglines, bulldozers, dump trucks, barges, and trains that extract and transport coal are not calculated in this exercise, but they represent significant losses nonetheless. The net energy loss of decommissioning the energy-intensive materials and processes used to construct and then take apart a conventional power plant after its useful operating life also is not included. The energy use associated with cleaning up air pollutants that have already been emitted and treating polluted water (both energy-intensive activities) also is excluded. So our toast example likely significantly *underestimates* the amount of total energy lost within the current system.

We tend to forget that energy is a means and not an ends, and that energy is useful only insofar as it performs specific tasks relative to those ends. We do not consume electricity or oil directly, but rather the comfort, cooked food, heating, cooling, refrigeration, motive power, showering, illumination and lighting, mobility, entertainment, information, communication, and a host of other services these fuels provide. The current energy system expends enormous amounts of energy to mine or drill a dwindling supply of materials from the earth, transport them, process them into fuels, burn them to heat water to thousands of degrees, divert the steam to turn a turbine to generate electricity, and then transmit that electricity over hundreds of miles of wire only so that a consumer can use it to burn toast at a few hundred degrees.[5] The system is equivalent to "cutting butter with a chainsaw—inelegant, expensive, messy, and dangerous."[6]

A Primer on the Electric Utility System

At the turn of the millennium, the National Academy of Engineering decided to rank the greatest technological achievements of the twentieth century. Contenders included automobiles, television, the mechanization of agriculture, airplanes, water systems, microelectronics, radios, computers and the Internet, the telephone, highways, medical technologies, spacecraft, lasers, and improved materials. Do you know what won?

The electric utility system.

Capital-intensive, complicated, and expansive, the modern electricity grid consists of 16,924 conventional power plants providing about 1,000 gigawatts (GW) of installed capacity that send electricity through 351,000 miles of high-voltage transmission lines and 21,688 substations.[7] Given that the average human body produces between 60 to 90 Watts (W) of equivalent energy per hour, it would take 13.3 billion people—more than twice as many as exist on Earth—to naturally produce as much energy as America's electricity grid does in 60 minutes.[8] How do they do it?

From the beginning of the twentieth century until the 1970s, the electric utility industry took advantage of incrementally improving large-scale technology and managerial innovations to produce growing amounts of centralized power at declining costs. Driven by resource availability, economies of scale, and lower staffing levels per facility, the predominately fossil-fueled and nuclear-fueled units produced immense amounts of electricity, often between 300 and 1,300 MW. Power *plants,* made up of several units, generate multiples of these amounts.

To understand the basics about how power systems function, you need to remember at least two characteristics: instantaneousness and variability. The first and most important principle of electric power systems is that they must be planned and operated to follow instantaneously customer demand, often called "load." The second important principle of power systems is that customer demands change continuously and exhibit daily, weekly, and seasonal load cycles.

At each moment the supply of power must meet the demand of customers and utilities must maintain power frequency and voltages within appropriate limits across the entire transmission system.[9] Daily load variance occurs as routine practices reinforce the effects of changing from day to night, such as turning lights

Table 1
Existing Electricity Capacity by Energy Source, 2006

The bulk of the power flowing through the American electric utility system comes from 10,916 fossil-fueled and nuclear plants. Hydroelectric resources and a small number of renewable generators make up the rest.

Fuel Source	Number of Generators	Installed Capacity (MW)
Coal[a]	1,493	335,830
Petroleum[b]	3,744	64,318
Natural Gas[c]	5,470	442,945
Other Gases[d]	105	2,563
Nuclear	104	105,585
Hydroelectric Conventional[e]	3,988	77,419
Other Renewables[f]	1,823	26,470
Pumped Storage	150	19,569
Other[g]	47	976
Total	**16,924**	**1,075,677**

[a] Anthracite, bituminous coal, subbituminous coal, lignite, waste coal, and synthetic coal.
[b] Distillate fuel oil (all diesel and No. 1, No. 2, and No. 4 fuel oils), residual fuel oil (No. 5 and No. 6 fuel oils and bunker C fuel oil), jet fuel, kerosene, petroleum coke (converted to liquid petroleum), and waste oil.
[c] Includes a small number of generators for which waste heat is the primary energy source.
[d] Blast furnace gas, propane gas, and other manufactured and waste gases derived from fossil fuels.
[e] The net summer capacity and/or the net winter capacity may exceed nameplate capacity due to upgrades to and overload capability of hydroelectric generators.
[f] Wood, black liquor, other wood waste, municipal solid waste, landfill gas, sludge waste, tires, agriculture by-products, other biomass, geothermal, solar thermal, photovoltaic energy, and wind.
[g] Batteries, chemicals, hydrogen, pitch, purchased steam, sulfur, tire-derived fuels and miscellaneous technologies.

Source: Energy Information Administration Form EIA-860 *Annual Electric Generator Report.* Totals may not equal sum of components because of independent rounding.

on, raising indoor temperature when waking up, taking showers before breakfast, cooking in the dinner hour, and washing dishes. Over the course of a week, energy use changes as the weekend approaches and, throughout the year, as seasonal changes in temperature and climate occur. To match these loads, utility companies employ a series of practices that includes bringing generators with different cycles and corresponding cost structures online at different times. Base-load plants have the longest cycling times and lowest average costs. Peaking plants have the shortest cycling times but the highest average costs. Intermediate load plants fall somewhere in the middle.[10]

The modern electric utility industry is highly integrated and complex—and by some metrics, more important than religion, sex, and fast food (even though the industry's recent growth has been stunted by escalating fuel prices, changing consumer preferences, and technological stasis). In 2007, the U.S. electricity industry possessed more than $863 billion of embedded investment, making it the largest investment sector of the American economy. Annual sales of electricity among all firms for the same year were approximately $360 billion, or 27 times more than the country spent on pornography and almost 4 times more than it gave to religious organizations.[11] Electricity is regulated by 53 federal, state, and city public service commissions and more than 44,000 different state and local codes. About 240 investor-owned utilities (such as Exelon, Dominion, and American Electric Power) operated three-quarters of the country's total electrical capacity. In addition to these investor owned utilities, more than 3,187 other private utilities provided power along with 900 cooperatives, 2,012 public utilities, 400 power marketers, 2,168 nonutility generating entities, and 9 federal utilities (such as the Tennessee Valley Authority and Bonneville Power Administration).[12] Put another way, the country has more power providers than Burger King Restaurants.[13]

Despite its amazing scope and complexity, the electric utility industry faces four urgent and intractable problems. Demand for (and consumption of) electricity continues to grow, but utilities are finding it more difficult to provide it to customers due to aging generation technology and the capital resources needed to build large generators. The transmission grid and associated infrastructure with fuel cycles connected to conventional generators is becoming more insecure and fragile. Coal, natural gas, and uranium must pass through many layers of vertically and horizontally managed subsystems before they can be converted into electricity, making them susceptible to interruptions in the supply chain and volatile prices. The human health and environmental costs of the conventional paradigm actually surpass the industry's entire revenue, according to some estimates.

No. 1: The Demand Challenge

In their most recent analysis, the EIA projects that total electricity consumption will grow at an annual rate of 1.3 percent per year, or from 3,821 billion kWh (kilowatt-hours) in 2006 to 5,149 billion kWh by 2030.[14] This may not sound like much, but if the projection is accurate it means that nationwide electricity demand will actually *double* before 2040.[15]

As Chapter 2 will explore in detail, it is true that remarkable improvements in energy efficiency have contributed greatly to reducing the growth rate of electricity consumption. Sadly, however, the historical record suggests that energy efficiency practices and demand-side reduction programs seem unable to offset steady increases in demand by themselves. For example, on-site electricity consumption per household in the United States dropped 27 percent between 1978 and 1997, yet the number of households grew by 33 percent. Over the same period, electricity's share of household energy consumption actually increased from 23 percent to 35 percent.[16] John Wilson, a staff member for the California Energy Commission (CEC) put it this way:

> New Californian homes use one sixth of the energy per square foot for air conditioning than they did 20 years ago. The bad news is that homes are almost twice as big as they were 30 years ago. We are making things more efficient, but making more of them.[17]

Population growth and changing tastes and preferences have thus outpaced gains in energy efficiency. Air conditioning did not exist in large quantities 50 years ago, and it now consumes 35 percent of all nationwide electricity on hot days. In some states, such as Texas, it can consume three-quarters of all electricity. Therefore, despite the significant gains made by energy efficiency improvements, demand historically has surpassed them.

Many industries, for example, have shifted to digital controls and electronic communications and computing equipment, placing greater stress on the power grid. Joe Loper, former vice president for research at the ASE, comments that "electric generation and electric-using equipment are far more efficient than they used to be, but demand for electricity continues to soar due to new uses, bigger homes to heat, and increased population." [18] Hans Blix, the former director of the International Atomic Energy Agency (IAEA), stated that "the more efficient use of energy will only partially slow down the expanding use of energy. Although our light bulbs will save electricity, we shall have more lights." [19] The EIA's *International Energy Outlook* also noted that "rapid additions to commercial floorspace, the continuing penetration of new telecommunications technologies and medical imaging equipment, and increased use of office equipment are projected to offset efficiency gains for electric equipment in the sector." [20]

It appears the industry is trying as hard as possible to meet the EIA's growth projections with conventional sources (see Figures 3 and 4). Utilities currently plan to construct more than 120 conventional coal-fired power plants, equating to about 6,000 MW of installed coal capacity per year, from 2008 to 2018.[21] Sherri K. Stuewer, a senior vice president at ExxonMobil, recently made the controversial declaration that "it is irresponsible for utilities and energy companies to stop using fossil fuels." [22] John Hoffmeister, president of Shell Oil Company, argued that it is his *job* to "constantly look for more fossil fuels, more fossil fuels, and more fossil fuels to try to meet demand." [23] Using one of the most rigorous methodological tools yet invented to estimate future renewable energy deployment—the National Energy Modeling System (NEMS)[24]—the EIA estimates that 90.4 percent of future capacity additions from 2007 to 2011 will be met by fossil fuels.[25] System operators and utilities have announced plans to build more than 150 conventional coal-burning electricity plants (representing 85 GW of capacity) in 42 states by 2025.[26]

EIA notes that poor financing, comparatively higher capital costs for renewable power, and the need to build or upgrade transmission capacity from remote resource areas will likely discourage significant investments in clean energy. As the EIA put it:

> Despite the rapid growth projected for biofuels and other non-hydroelectric renewable energy sources . . . oil, coal, and natural gas still are projected to provide roughly the same 86 percent share of the total U.S. primary energy supply in 2030 that they did in 2005.[27]

What is depressing about this statement, obviously, is that it illustrates the country will remain wedded to fossil fuels for the foreseeable future. Interesting, however,

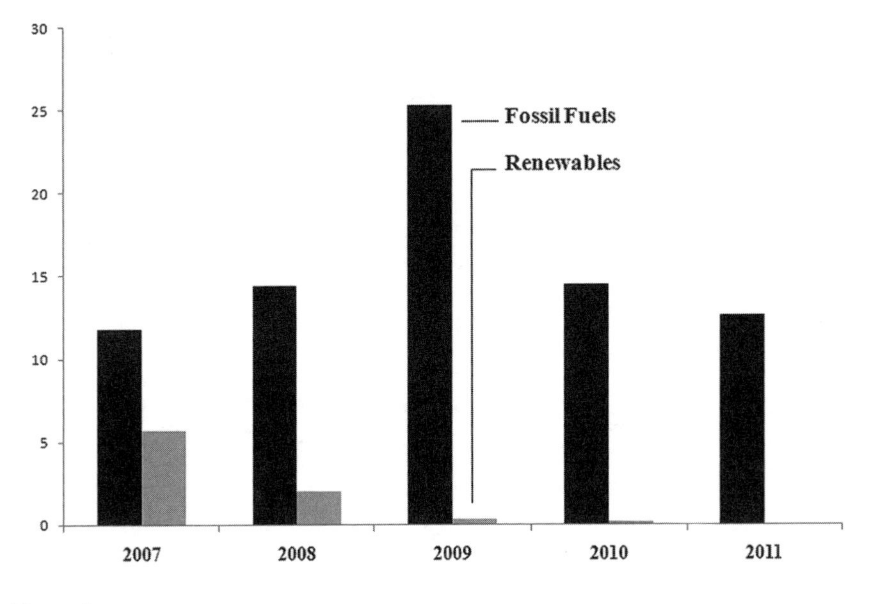

Figure 3
Projected Electricity Capacity Additions by Fuel Type, 2007–2011

Looking at the next four years, power plant operators expect to add a significant amount of fossil-fueled capacity, especially coal and natural plants.

Source: U.S. Energy Information Administration's *Electric Power Annual 2006.* The term "fossil fuels" includes coal, petroleum, natural gas, and other gases. "Renewables" includes conventional hydroelectric, wind, solar, biomass, and geothermal.

is that the statement never mentions what role hydrogen, clean coal technologies, fusion, and advanced nuclear plants are expected to play in the next few decades. Why is it that the EIA projections exclude these four technologies?

Hydrogen

To help promote hydrogen powered fuel cells, President George W. Bush announced a $1.2 billion "Hydrogen Fuel Initiative" in his 2003 State of the Union Address. The program attempts to harness the energy potential of hydrogen as a way to power cars, trucks, homes, and businesses. As the President remarked a few days after his address, "Hydrogen fuel cells represent one of the most encouraging, innovative technologies of our era . . . One of the greatest results of using hydrogen power, of course, will be energy independence for this nation."

Nonetheless, a transition to the much-touted "hydrogen economy" faces tenacious infrastructural challenges: inability to manufacture cost effective fuel cells, as well as problems extracting, compressing, storing, and distributing hydrogen-based fuels. Engineers and scientists researching hydrogen fuel cells continually talk about the difficulty in overcoming a "chicken and egg problem," as policymakers see little incentive to invest in fuel cells without a well-developed distribution infrastructure.

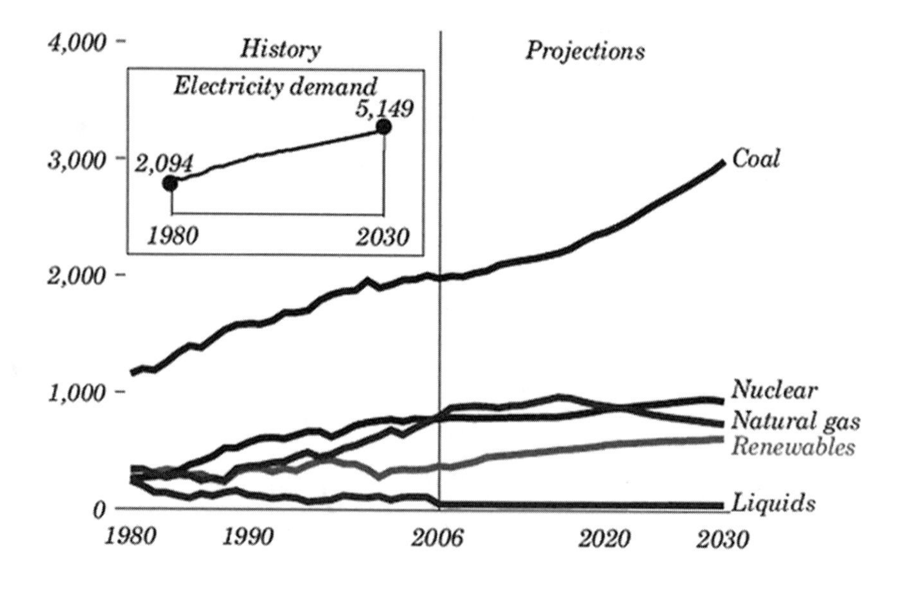

Figure 4
Projected Future Electricity Portfolio, 2006–2030 (billion kWh)

Utilities in the United States have told the DOE that they expect to significantly increase their use of coal and nuclear fuels over the next 25 years. Renewable generators of all types, by contrast, utilities intend to use less.

Source: Energy Information Administration Form EIA-860 *Annual Electric Generator Report.*

And yet, industry has no incentive to build the infrastructure without operating fuel cells on the market.

Hydrogen is a source of energy only if it can be taken in its pure form and reacted with another chemical, such as oxygen. Natural forces have already oxidized all the hydrogen on earth, with the exception of hydrocarbons, so that none of it is available as usable fuel. The rest has to be "made." Oil refineries use hydrogen to purify fuels, and chemical manufacturers employ it to make ammonia and other compounds. Both industries obtain a vast majority from high temperature processing of natural gas and petroleum. The method, however, sacrifices 30 percent of the energy content in the gas to obtain 70 percent of the energy in hydrogen, and emits prodigious amounts of CO_2 in the process—about 32 kilograms of CO_2 for one kilogram of hydrogen.[28] This small amount of hydrogen production already accounts for a shocking 2 percent of global energy demand.[29]

The second method, electrolysis, presents its own problems. Scientists at the Lawrence Livermore National Laboratory calculated that to produce enough hydrogen through electrolysis to displace the country's oil needs would require 230,000 tons of hydrogen to be produced every day—enough to fill 13,000 Hindenburg blimps—and the country would need to *double* its electricity generating capacity.[30] Methods for producing hydrogen are exceptionally expensive, and researchers in the European Union (EU), where electricity prices are already much more expensive than in the United States, concluded that energy prices would have to jump *ten times* today's rates before the hydrogen production techniques would be considered cost-effective.[31]

In terms of distribution and storage, long-distance hydrogen pipelines would need to be immense because gaseous hydrogen is so diffuse and takes up far more space than oil or gas. Such pipelines would have to be specially constructed, as hydrogen cannot be used with current petroleum pipelines due to the brittleness of their material components, the inadequacy of seals, and the incompatibility of pumping lubricants.[32] Hydrogen storage is equally daunting. Storing hydrogen in its gaseous state requires large, high-pressure cylinders, taking up significant amounts of space. Although liquid hydrogen can take up much less space, scientists must supercool it to temperatures below −423 degrees Fahrenheit (about as close to absolute zero as you can get). At these temperatures, around 40 percent of the energy in the hydrogen must then be expended to liquefy it. The American Physical Society went even further and concluded that a new material must be *invented* to solve the hydrogen economy storage problem.[33] "Given these technical difficulties," notes one engineer, "the implementation of an economically viable method of retail hydrogen distribution from large-scale central production factories is essentially impossible."[34]

The National Research Council recommends that the DOE should halt many components of its hydrogen strategy, such as research on high-pressure tanks and cryogenic liquid storage, because they have little promise of long-term applicability to the country's energy needs. As they concluded:

> In no prior case has the government attempted to promote the replacement of an entire, mature, networked energy infrastructure before market forces did the job. The magnitude of change required . . . exceeds by a wide margin that of previous transitions in which the government has intervened.[35]

The Congressional Research Service has also warned legislators that "it is doubtful that a widespread system of hydrogen distribution will emerge."[36] Even under a best-case scenario when the government throws an unlimited amount of money into hydrogen research and each of the above barriers are miraculously overcome, the Pacific Northwest National Laboratory concluded that "in the advanced technology case with a carbon constraint, hydrogen doesn't penetrate . . . in a major way until after 2035."[37]

Clean Coal

In order to make the combustion of coal more efficient and environmentally friendly, the DOE initiated the FutureGen project in 2003 and the Clean Coal Power Initiative in 2004. These programs aim to develop a diverse array of technologies including supercritical pulverized coal plants (that boost thermal efficiency by operating at higher temperatures), integrated gasification combined cycle (IGCC) plants (that use chemical processes that gasify coal to remove sulfur and mercury), pressurized fluid bed combustion plants (that use elevated pressure to capture sulfur dioxide and nitrogen oxides), and carbon sequestration/management techniques (deep underground geologic formations that are engineered to capture and store excess CO_2).

Hyperefficient IGCC power plants are widely held to be much less efficient than conventional coal-fired plants. While defending TXU Energy's plan to build 11 new coal-fired units in Texas, one TXU vice president noted:

> IGCC is a promising technology, but is not yet viable on a large-scale commercial basis for the types of coal available in Texas. There are only two IGCC units in operation today in the U.S.—both are small, were heavily subsidized, and actually have dirtier emissions profiles than the supercritical plants we have proposed. Further, both these plants continue to operate at low reliability levels more than five years after coming on line.[38]

The reason behind such pessimism is that almost all clean coal technologies rely on a four-step process of capturing, storing, transporting, and sequestering CO_2.

The "capture" stage requires separating CO_2 from emissions and exhaust into pure waste streams, then pressurizing it for transport.[39] While effective at preventing carbon from escaping directly into the atmosphere, capturing creates serious trade-offs in operating efficiency. The easiest way to capture CO_2 is to "scrub" it from the flue or exhaust gas using a chemical-like amine to extract the greenhouse gas. But this process typically entails a 30 percent energy penalty (meaning it consumes one-third of the plant's output). Alternatively, power plants can operate on pure oxygen rather than air, but doing so requires a 24 to 40 percent energy penalty.[40]

The most serious challenge concerns storage and sequestration, which entails injecting CO_2 into deep reservoirs, such as depleted oil and gas fields, saline reservoirs, and unmineable coal seams. Engineers prevent CO_2 from escaping by laying impermeable caprock over the site, drilling capillaries, and dissolving CO_2 into aquifer fluids. Identifying sites that have the needed permanence, cost, and safety requirements is an enormously complicated issue, and considerable underground testing must occur before injection begins.

The environmental and liability issues associated with sequestration are intimidating. Groundwater contamination can occur if stored CO_2 leaks or unexpectedly migrates. CO_2 injection can result in pressure buildup and an increase in seismic activity. Operational leakage to the surface can create a public health risk since high concentrations of CO_2 can induce fatal asphyxia. Slow, sudden, or chronic releases of CO_2 can occur to the surface, accelerating climate change. Sequestration can contaminate underground assets such as brines and natural gas fields, exacting significant property damage. Environmental degradation can occur as sequestered CO_2 leaks to the surface and impacts vegetation, trees, and soil composition.[41]

Perhaps even more discouraging are the infrastructural and economic challenges associated with clean coal. A 2007 interdisciplinary study from the Massachusetts Institute of Technology (MIT) concluded that carbon capture and storage technologies were still very expensive, costing about $25 per ton to capture and pressurize

CO_2 and another $5 per ton to transport and store it. If implemented widely, the MIT team calculated the carbon capture and storage technologies would almost double the cost of coal-fired power. A modern subcritical pulverized coal plant produced power at about 4.84 ¢/kWh in 2006, but the cost of generation jumps 70 percent to 8.16 ¢/kWh with the use of carbon capture and sequestration. The MIT study also found that if all of the CO_2 emitted from power plants in 2006 were transported for sequestration, the quantity would be three times the weight and one-third the annual volume of natural gas transported by the entire U.S. natural gas pipeline system. If just 60 percent of the CO_2 emissions were to be captured and compressed in liquid form for geologic sequestration, the volume of liquid carbon would be equivalent to all of the oil currently consumed in the United States (equal to about 20 million barrels of oil per day). The researchers argued that no CO_2 storage project currently operating has the necessary modeling, monitoring, and verification capability to overcome the technical issues with such a task, and that every large-scale clean coal system would have unique complex characteristics, traits that make them hard to mass produce.[42]

Consequently, Eileen Claussen, president of the Pew Center on Global Climate Change, concluded that:

> Carbon capture and storage and IGCC are nowhere near prime time. Right now, to stretch the analogy further, they are far enough from prime time to be on the air around 3 a.m. with a bunch of annoying infomercials. And they won't get any closer to prime time without substantial investment in R&D, as well as a major policy commitment to these technologies.[43]

To make matters worse, in many cases two to three tons of CO_2 must be sequestered to offset every ton emitted. William L. Sigmon, a senior vice president at American Electric Power, estimates that, given the energy-intensive nature of carbon capture, his company had to sequester two tons of carbon for every one ton emitted (at a cost of $40 to $45 for every ton of carbon displaced).[44] Perhaps as a result of these challenges, the DOE cancelled the FutureGen project in January 2008 after the costs of running and operating their demonstration clean coal plants had risen from $830 million to $1.8 billion.

Fusion

Others have promoted the promise of nuclear fusion, where energy is produced by fusing together deuterium and tritium to form heavier atoms, releasing enormous amounts of energy in the process. The United States is contributing $1.1 billion to the International Thermonuclear Experimental Reactor (ITER), a venture that has been dubbed the most expensive joint scientific project after the International Space Station. Current Secretary of Energy Samuel W. Bodman explained that "as partners in ITER, we are pursuing the promise of unlimited, clean, safe, renewable, and commercially available energy from nuclear fusion, which has the potential to significantly strengthen energy security, at home and abroad."

Despite such promise, ITER remains only an experiment, and the facility is not scheduled to become operational until 2016. Even if it produces electricity, ITER would not be a viable source, since its immense capital cost means that it would produce power at around 36 ¢/kWh, or more than three times current average electricity rates. Jack Barkenbus, a senior research associate at Vanderbilt University, wryly remarks that "people have been saying for the past 40 years that fusion will be on a near commercial basis in 30 years. It is still 30 years away from commercialization today."[45]

Advanced Nuclear

To encourage R&D in advanced nuclear reactors, the DOE announced its "Generation IV" initiative in 2002 (funding R&D on the "next generation" of nuclear power systems) and launched a Global Nuclear Energy Partnership (GNEP) in 2006 (promoting nuclear energy abroad to create export opportunities for American technology firms). Both initiatives, however, face serious hurdles and are unlikely to commercialize advanced nuclear technology within the next decade.

First, consider the cost of Generation IV reactors. While each of the advanced designs being studied are still experimental, nuclear plants are the most expensive and capital-intensive structures to build in an industry that is already the most capital intense in the economy.[46] The Nuclear Energy Agency reports that close to 60 percent of the investment needed for a new nuclear project goes toward initial construction (compared to about 40 percent for coal and 15 percent for natural gas projects).[47] Even assuming the low end of industry averages, new reactors would cost a minimum of $2,000 per installed kW—meaning a 4 GW plant would cost $8 billion. And the estimate of $2,000 per installed kW reported by the industry is extremely conservative and woefully out of date. Researchers from the Keystone Center, a nonpartisan think tank, consulted with 27 nuclear power companies and contractors, and concluded in June 2007 that the cost for building new reactors would be between $3,600 and $4,000 per installed kW (with interest). They also projected that the operating costs for these plants would be remarkably expensive: 30 ¢/kWh for the first 13 years until construction costs are paid followed by 18 ¢/kWh over the remaining lifetime of the plant. Just a few months later, in October 2007, Moody's Investor Service projected even higher operating costs due to the quickly escalating price of metals, forgings, other materials, and labor needed to construct reactors. They estimated total costs for new plants, including interest, at between $5,000 and $6,000 per installed kW. Florida Power & Light informed the Florida Public Service Commission in December 2007 that their estimated cost for building two new nuclear units at Turkey Point in South Florida was $8,000 per installed kW, or a shocking $24 billion. In 2008 Progress Energy pegged its cost estimates for two new units in Florida to be about $14 billion plus an additional $3 billion for T&D.[48] Georgia Power, which is currently looking into building two new 1,100 MW units at their Vogtle facility near Waynesboro, similarly expects them to cost $14 billion to construct (about $6,300 per installed kW at a site that already has an existing nuclear plant). It is no surprise, then, that a MIT study concluded that only by implementing a carbon tax of $200 per ton on conventional power plants could advanced nuclear reactors become cost competitive with existing technologies.[49]

If EIA's projections were met solely by Generation IV nuclear plants in 2040, the nation would need to build thousands of new nuclear reactors over the next 20 years. One lobbyist even estimated that a new nuclear plant would need to be built roughly every 10 days to meet forecasted demand.[50] In order for advanced nuclear plants to even maintain the 19 percent share of power generation held by conventional nuclear units, an additional 190 GW of new capacity would have to be built. Taking an average reactor size of 1,000 MW, this equates to about six plants per year, every year until 2040, at a capital cost that could surpass $1.5 trillion. The historical record suggests that the task is insurmountable. France, which currently generates 76 percent of its electricity from nuclear units, has the fastest record for deploying nuclear plants in history: 58 between 1977 and 1993, or an average of 3.4 reactors per year, close to half the 6 per year needed in the United States. In addition, 190 new nuclear plants would require the additional construction of four large enrichment plants, five fuel fabrication plants, and three waste disposal sites, each the size of Yucca Mountain.[51]

Furthermore, researchers at Georgetown University, University of California Berkeley, and Lawrence Berkeley National Laboratory (LBNL) assessed financial risks for advanced nuclear power plants utilizing a three-decade historical database of delivered costs from each of 99 conventional nuclear reactors operating in the United States.[52] Their assessment found a significant group of plants with extremely high costs: 16 percent in the more than 8 ¢/kWh category. The study pointed out two unique attributes of advanced nuclear power plants that make them prone to unexpected increases in cost: (a) their dependence on operational learning, a feature not well suited to rapidly changing technology and market environments subject to local variability in supplies, labor, technology, public opinion, and the risks of capital cost escalation; and (b) difficulty in standardizing new nuclear units, or the idiosyncratic problems of relying on large generators whose specific site requirements do not allow for mass production. As one example, the Generation IV program envisions six different reactor designs. Past technology development patterns suggest that many high-cost surprises will occur in the planning and deployment process for new nuclear units. These "hidden" but inevitable cost overruns may be one explanation for why people are not investing in Generation IV reactors.

Arms control advocates, the National Research Council, and the Congressional Budget Office have also attacked the GNEP program. In October 2007, leading arms control experts signed a letter urging the Senate Appropriations Committee to eliminate all funding for the GNEP on the grounds that it would encourage weapons proliferation.[53] The National Research Council of the National Academies issued a highly critical assessment of GNEP and argued that the program's rapid deployment schedule entailed considerable financial and technical risks and prematurely narrowed the selection of acceptable reactor designs. The report also faulted the DOE for not seeking sufficient independent peer reviewers for projects and for failing to adequately address waste management challenges.[54] The following month, the nonpartisan Congressional Budget Office warned that GNEP's plan to reprocess spent fuel would cost 25 percent more than a wide range of other storage and reprocessing options.[55]

Moreover, a study commissioned by the Office of Science and Innovation in the United Kingdom also concluded that R&D on "all of the [Generation IV] systems" face "several key challenges" that will require considerable expense and ingenuity to overcome. The report identified significant gaps in materials technology, manufacturing processes, modeling and simulation capabilities, and waste storage.[56] Analogously, researchers at North Carolina State University concluded that the materials used in conventional reactors will not be suitable for Generation IV technology. Zirconium alloys, for instance, are currently utilized as fuel cladding in light and heavy water reactors but will not work under the higher temperature environments envisioned by Generation IV proponents. Other core components, such as pressure vessels and pipes, will no longer be able to be made from low alloy ferritic steels, and pressure vessels needed to handle expected temperatures from Generation IV reactors would likely double the size of existing reactors. The researchers noted that a lack of fast spectrum irradiation facilities and high temperature testing facilities greatly restrict the ability for scientists and engineers to even design, test, and evaluate the necessary structural materials for advanced reactors, implying that the Generation IV strategy relies, as does hydrogen, on discovering new materials that have yet to be invented.[57] Researchers for the European Commission agreed and stated in 2007 that an unexpected technological breakthrough must occur before either Generation IV technology would become feasible, stating that they found it "inconceivable" that the "long-term objective of sustainable development of nuclear fission energy" could be met with existing technology.[58] This could be why one analyst referred to Generation IV systems as "paper reactors" since they are likely never to be built and will exist only on paper.

Conventional Problems with Conventional Units

Sticking with conventional units, if one assumed that the EIA's projections would be met entirely by coal or natural gas, the country would need to build nearly 4,950 new midsized coal plants[59] or close to 10,000 new midsized natural gas plants.[60] Researchers from ORNL projected that to meet expected electricity demand by 2020, one power plant would need to be built every 4.9 days.[61] This level of conventional generation could create logistical nightmares.

The volume of coal being transported (coal accounted for 41 percent of all freight moved by U.S. rail carriers in 1999) already creates frequent bottlenecks, contributing to congestion and higher transportation costs. MIT economist Wendell H. Wiser estimates that in the typical transportation of coal by rail, an individual freight car spends as much as 50 percent of its time in a switchyard and 40 percent in customer yards and sidings. This means that an average ton of coal shipped by rail spends as little as 10 percent of its time actually moving toward its destination.[62] The University of Wyoming estimates that up to 80 percent of the cost of coal for ratepayers in Illinois is to cover railway costs. Coal at the mouth of a mine in Wyoming, for example, costs about $5 per ton, but by the time it reaches a power plant outside of Chicago, that same coal costs about $30 a ton.[63]

The cumulative costs to transport new natural gas may be even higher. Significant investments in distribution infrastructure will be needed to ensure natural gas supplies can be adequately allocated to power providers and consumers. Matching this projected growth with adequate supply would require vast transportation networks and import agreements. Transportation and distribution already account for 41 percent of the residential price of natural gas. Since the construction of natural gas pipelines can cost as much as $420,000 per mile, fully constructing the natural gas infrastructure recommended by the George W. Bush administration's National Energy Plan (which calls for over 301,000 miles of new natural gas transmission and distribution pipelines) could cost ratepayers as much as $126.4 billion.[64]

Amazingly, planned retirements threaten to exacerbate the situation even more. Forty percent of the U.S. nuclear capacity is scheduled to retire by 2020 as their licenses expire, and utilities retired 180 conventional units totaling 3,804 MW in 2006.[65] Through 2017, more than half of all nonfederal hydroelectric capacity totaling 32,000 MW of power will be forced to go through a federal relicensing process or be retired. Many of the plants, which are already aging and provoking protest from environmental groups in the Pacific Northwest, are not likely to receive renewed licenses. Moreover, it will cost about $3.2 billion to process license applications for 307 hydropower projects in 39 states. Even if every plant were to receive federal approval for a new license, the relicensing process typically takes years to complete. Since most plants cannot operate without a license, utilities will desperately need to dispatch new generation supplies in hydroelectric regions just to maintain current levels of production, let alone expand them.[66]

No. 2: The Infrastructure Challenge

Conventional electricity systems and their associated fuel cycles also manifest vulnerability and insecurity at multiple points, from the T&D system to the power plant it supports to associated pipelines and storage facilities.

Transmission and Distribution Reliability

One of the most highly stressed energy infrastructures in the nation is its high voltage transmission grid. Scott Sklar, the president of an energy consulting firm, used the following analogy to describe this strained network:

> The electric utility sector is like the Space Shuttle. When it works, it works fantastically well. But when it fails, it screws up devastatingly. We spend billions of dollars on surge protectors, which means our electric utility system doesn't have the quality it needs. Instead, issues of quality (and when it fails) are shoved onto the consumer. And it may only get worse: twenty years ago the country wasn't moving towards a digitized economy. Now our manufacturing controls, computers, and telecommunications

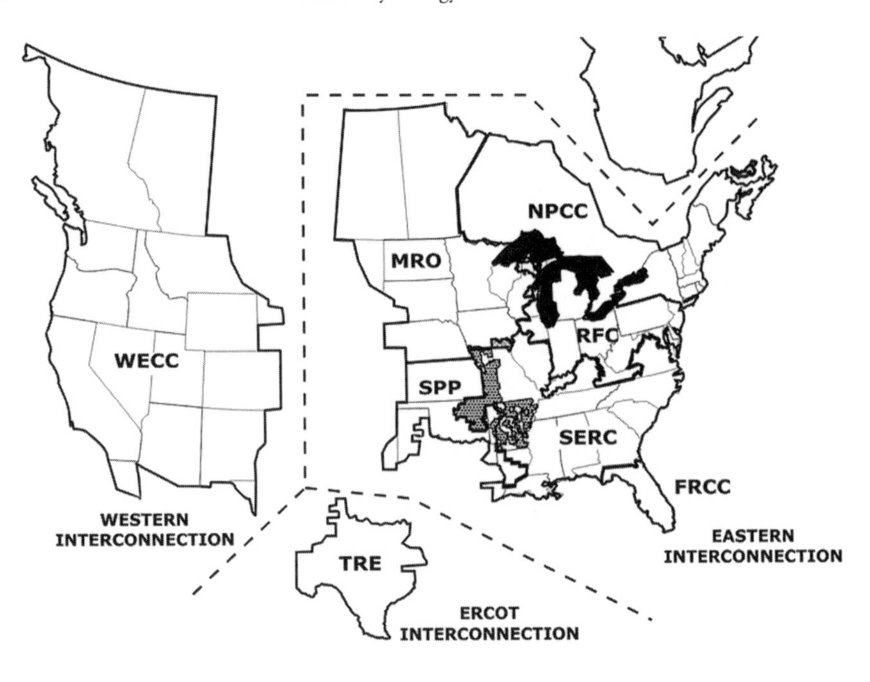

Map 1
The North American High Voltage Transmission Grid

The American high voltage transmission network is fragmented and balkanized into eight regional zones that must siphon their electricity through three interconnection points.
Source: North American Electricity Reliability Council. Figure shows transmission lines for 230 kV and above.

> devices (even our appliances) are starting to move towards complete digitization, which will place even more stress on the existing power grid.[67]

David Hill, deputy laboratory director at the Idaho National Laboratory, explains that, despite its immense size, the electric power grid was not centrally planned in this country. Instead, the system grew over time, full of "inefficiencies, instabilities, and piecemeal privatization of utilities," resulting in a tangle of regional systems stitched together like a quilt, instead of one fluid, nationwide system (see Map 1).[68]

To put the grid's complexity in perspective, Ralph Loomis, an executive vice president for Exelon (the largest electric utility in the United States), has noted that the grid Exelon uses to distribute power to its customers is enormously complicated, spanning 13 states and millions of homes. Loomis notes that "it is one machine extending over hundreds of thousands of miles. It is amazing it works as well as it does."[69]

Consumers still pay to have the network work as well as it does. Transmission and distribution (T&D) often accounts for more than twice the cost of generation on consumer's electricity bills. For most of the 1970s and 1980s, the T&D network accounted for about 70 percent of the average cost of delivered electricity to residential customers. During hot weather, when power lines stretch and conductivity decreases, T&D losses can exceed 25 percent. Imagine purchasing a dozen eggs at your local grocery store, but having between one and three eggs break *every* time you transported them to your home, year after year. To give an idea of how significant these overhead costs still are, the Public Service Commission of Utah estimates that only 3 cents out of the 6.3 ¢/kWh Utah Power charges its customers are related to the actual generation of electricity.[70]

Perhaps incongruously, growth in electricity production did not accompany a proportionate investment in transmission infrastructure during most of the 1990s. Like prisons, high school cafeterias, and public housing, transmission lines face perpetual underinvestment. U.S. expenditures in transmission infrastructure peaked at almost $10 billion in 1970 and declined to less than $4 billion in 2000. The Energy Policy Act of 1992 further complicated the problem by increasing grid utilization while further removing incentives to invest in transmission. Shelly Strand, an energy consultant, put it this way:

> The [current electric utility] model assumes that energy is like sugar: you can put sugar into X and get it out at Y and it will be the same sugar regardless of which terminus you get it from. Energy doesn't work that way. You cannot put energy into the grid in Texas and take it out in California and have it be the same energy because of voltage drop and other factors. The whole deregulation model is set up on the false assumption that the grid system is national and reaches everywhere equally in scope. We don't have a national grid system in this country that works like that.[71]

Strand suggests that the current, balkanized power grid is not responsive enough to meet the needs of a restructured and competitive electricity market. And it is precisely this market that utilities argued would better enable them to lower electricity prices through competition. Instead, such utilities have apparently used the rubric of "restructuring" to avoid making needed improvements to the system. For instance, Adam Serchuk, a senior program manager for the Energy Trust of Oregon, argues that "the current transmission systems used to deliver electricity were designed for a different institutional structure, without the flexibility needed for an efficient trading-based energy system." [72]

As a result, the ratio of transmission capacity to peak electricity demand has decreased—by 16 percent between 1989 and 1998. In fact, transmission investment in the United States declined by about 1.5 percent per year between 1975 and 1998, even while electricity demand more than doubled.[73] In New York, for example, the aging underground electricity grid is so degraded that it costs local utilities over $1 billion a year just to replace corroded cable needed to make the system function at current demand levels. James Gallagher, the Director of the Office of Electricity and Environment for the New York State Public Service Commission, comments that the billion dollars "doesn't even allow us to get proactive to get ahead of the problem. The scale of the challenge is enormous." [74]

The 2003 blackout of the Northeastern seaboard prompted calls for up to $100 billion in new transmission investments to prevent bottlenecks and relieve strained power lines. Nevertheless, investment continues to lag woefully. While electricity demand is forecast to grow by 20 percent between 1998 and 2008, transmission capacity is set to grow by only 5 percent.[75] Eric Hirst, a corporate fellow at ORNL, estimated that normalized transmission capacity declined by almost 19 percent between 1992 and 2002, and that it is expected to continue to drop 11 percent from 2002 to 2012.[76] One estimate concluded that maintaining transmission adequacy *at current levels* for the next six years will require a minimal additional investment of 26,600 miles of new transmission costing $65.8 billion.[77] Compare this staggering figure with estimated planned construction of only 6,200 miles and the investment shortfall becomes almost stupefying.[78]

The result of transmission underinvestment has been reduced power reliability and increased congestion in many regions of the United States, with grid components being operated closer to their technical limits and beyond their originally planned lifetimes. The ultimate consequence is an increased incidence of severe blackouts and power outages. Between 1964 and 2005, for instance, the Institute of Electrical and Electronics Engineers estimates that no less than 17 major blackouts have affected more than 195 million residential, commercial, and industrial customers in the United States, with 7 of these major blackouts occurring in the past ten years. Sixty-six smaller blackouts (affecting between 50,000 and 600,000 customers) occurred from 1991 to 1995 and 76 occurred from 1996 to 2000.[79] The costs of these blackouts are monumental: the DOE estimates that power outages and power quality disturbances cost customers as much as $206 billion annually, or more than the entire nation's electricity bill for 1990.

It is important to note that the frequency of outages determines their costs rather than their duration.[80] In the modern digital economy, even short power outages can shut down communication devices, refrigerators, water and process pumping, and the use of credit card, fax, and electronic cash registers. One chief executive from a global microchip company commented that:

> My local utility tells me they only had 20 minutes of outages all year. I remind them that these four 5-minute episodes interrupted my process, shut down and burnt out some of my controls, idled my work force. I had to call my control service firm, call in my computer repair firm, direct my employees to test the system. They cost me eight days and millions of dollars.[81]

The IEA estimated that the average cost of a one-hour power outage was $7.5 million for brokerage operations and $3.1 million for credit card companies.[82] A 2005 survey found that one minute of downtime cost automotive businesses an average of $24,000.[83] The figures grow dramatically for the semiconductor industry, where a two-hour power outage can cost close to $54 million.[84]

As consumers demand more electricity than the system can deliver, U.S. ratepayers could soon face serious congestion-driven rate increases. The North American Electric Reliability Corporation (NERC) warns that grid congestion will continue to increase and in some situations "lead to supply shortages and involuntary customer interruptions" and that "some portions of the grid will not be able to support all desired electricity market transactions."[85] Consider a metric known as calls for transmission loading relief (TLR)—instances where market sales cannot be executed because of transmission constraints, which forces operators to use more expensive and less efficient equipment (or in severe cases halt transmission altogether).[86] The number of times system operators in the Eastern Interconnection called for Level 2 or TLR increased from around 300 in 1998 and 1999 to more than 1,000 in 2000, 1,500 in 2002, and 2,000 in 2003.[87] Level 5 TLRs, the most serious when reliability is completely threatened, jumped 26 percent from 2002 to 2003.[88] Looking at just the PJM network in the Northeast, one can see that calls for TLR have jumped incredibly from just 53 in 1999 and 132 in 2000 to 750 in 2004 and 2,090 in 2005—an increase of almost a factor of 40 over the course of six years.[89]

Why is the T&D problem not being addressed by the industry? Under normal market conditions, some utilities actually benefit from limited transmission resources. When the transmission system is saturated, less supply is available to meet existing demand, and prices increase. In some cases, market forces create perverse incentives for utilities to delay transmission upgrades unless or until they risk catastrophic system failure. Owners of congested T&D lines will see payments decline when lines are upgraded, thus giving them an incentive to keep lines congested and operating poorly. Those with strategically located generation units can also affect T&D rates, leading to situations where generators collude with T&D operators to keep prices high. T&D investors can also play "preemption games" where early investors attempt to use T&D infrastructure to keep other participants out of the market.[90] Even the Federal Energy Regulatory Commission (FERC) has observed:

> Companies that own both transmission and generation under-invest in transmission because the resulting competitive entry often decreases the value of their generation assets.[91]

Market dynamics can create situations where congestion prices benefit some electricity generators at the expense of customers, who not only pay higher prices, but suffer costs from the increased risk of blackouts.[92] Sadly, little incentive exists for utilities and system operators to fix the problem. The National Commission on Energy Policy recently documented a general "failure to advance projects" among the electric utility industry that would "relieve electricity transmission bottlenecks in the Northeast, Southeast, and Northwest, despite the well-documented benefits that reduced grid congestion would bring in terms of cost, environmental impact, and the improved reliability of regional electricity systems." [93]

The current structure of the U.S. transmission system also encourages some utilities to load limited transmission lines to crowd out other generators. Florida Power and Light (FPL) accused TXU of intentionally flooding West Texas transmission lines with high-cost power to prevent FPL's wind power from reaching customers across the state. The state's independent electricity market monitor found TXU guilty of similar market manipulations during the summer of 2005 and the state's Public Utility Commission recommended that TXU be fined $210 million for that offense.[94]

The National Council on Electric Policy—a joint task force created by the DOE and U.S. Environmental Protection Agency (EPA)—noted that many system operators have an extra incentive to promote congestion through a technique known as pancaked rates.[95] These rates come into play when power under contract traverses more than one power system. Each operator can then charge full rates for transmission service as long as more than 70 percent of their system is currently under use.

Leonard S. Hyman, an economist specializing in utility industry finance, concluded that "the evidence suggests that investor owned utilities have reduced transmission and distribution spending to bare-bones levels, that spending will have to rise significantly in the near future in order to meet the needs of customers, and that the higher level of spending will trigger rate hike filings in order to cover the costs of the new capital."[96] One MIT economist, Paul Joskow, recently predicted that transmission congestion in the PJM network, New York, New England, California, Texas, and the Midwest will only continue to grow. Joscow noted that in 2003 transmission congestion already cost consumers in the PJM territory $499 million and $688 million in the state of New York alone.[97]

T&D SECURITY

A comprehensive, three-year Department of Defense (DOD) study concluded that relying on centralized plants to transmit and distribute electric power created unavoidable (and costly) vulnerabilities. The study noted that T&D systems constituted "brittle infrastructure" that could be easily disrupted, curtailed, or attacked.[98] One of the authors, physicist Amory Lovins, who is currently helping the U.S. military streamline energy-intensive sectors, has long argued that if you build an inefficient, centralized electricity system, major failures, whether by accident or malice, become inevitable by design. In Britain during the coal miner strikes of 1976, a leader of the power engineers famously told Lovins that "the miners brought the country to its knees in 8 weeks, but we could do it in 8 minutes." This is because the massive, complex, and interconnected infrastructure needed to distribute power from a centralized generation source is brittle and subject to cascading failures easily induced by severe weather, human error, sabotage, or even the interference of small animals. "Electrical grids," notes Lovins, "distribute a form of energy that cannot readily be stored in bulk, supplied by hundreds of large and precise machines rotating in exact synchrony across a continent, and strung together by a frail network of aerial arteries that can be severed by a rifleman or disconnected by a few strikers."[99]

DOD's conclusions complement a similar study undertaken by the IEA, which noted that centralized energy facilities create tempting targets for terrorists because they would need to attack only a few, poorly guarded facilities to cause large, catastrophic power outages.[100] Thomas Homer-Dixon, chair of Peace and Conflict Studies at the University of Toronto, cautions that it would take merely a few motivated people with minivans, a limited number of mortars, and a few dozen standard balloons to strafe substations, disrupt transmission lines, and cause a "cascade of power failures across the country," costing billions of dollars in direct and indirect damage.[101] A deliberate, aggressive, well-coordinated assault on the electric power grid could devastate the electricity sector and leave critical sectors of the economy without reliable sources of energy for a long time. Paul Gilman, former executive assistant to the Secretary of Energy, has argued that the time needed to replace affected infrastructure would be "on the order of Iraq," not "on the order of a lineman putting things up a pole." [102]

The security issues facing the modern electric utility grid are almost as serious as they are invisible. In 1975, the New World Liberation Front bombed assets of the Pacific Gas and Electric Company more than 10 times, and members of the Ku Klux Klan and San Joaquin Militia have been convicted of attempting to attack electricity infrastructure.[103] Internationally, organized paramilitaries such as the Farabundo Marti National Liberation Front were able to interrupt more than 90 percent of electric service in El Salvador and penned manuals for successfully attacking power systems.[104] A natural gas pipeline in Columbia has been shot so many times that operators humorously refer to it as "the flute."

The vulnerabilities of centralized generation systems to accidental or intentional disaster has never been so apparent as in Iraq, where determined insurgents destroy critical infrastructure faster than American contractors can rebuild it. James Robb, a former "black ops" agent and expert in counterterrorism, warns that a terrorist-criminal symbiosis is developing out of the situation in Iraq. There, terrorists have learned to fight nation-states strategically, without weapons of mass destruction using a new method of "systems disruption," a simple way of attacking electricity and natural gas networks that require centralized coordination.[105] In the past three years of the U.S. occupation of Iraq, relatively simple attacks on oil and electricity networks reduced or held delivery of these services to prewar levels, with a disastrous affect on the country's infant democracy and economy.

Insurgents were not the first to use such tactics. In its initial wave of precision air strikes in January 1991 and March 2003 in Iraq and during the Kosovo air bombing campaign in 1998, the U.S. military targeted energy infrastructure, including three nuclear plants, both to disrupt military systems and to enhance the overall psychological and economic impact of the attacks. Similarly, under its unilateral and multilateral sanctions regime, the United States has barred entry of materials used to build or repair electricity generators, knowing full well how essential such technologies are to a country's economic well-being.[106] Such disruptions are designed to erode the target state's legitimacy by keeping it from providing the services it must deliver to command the allegiance of its citizens.

Several recent trends in the electric utility industry, however, have only increased the vulnerability of T&D infrastructure. To improve their operational efficiency, many utilities and system operators have increased their reliance on automation and computerization. Low margins and various competitive priorities have encouraged industry consolidation, with fewer and bigger facilities and intensive use of assets centralized in one geographical area. As the National Research Council noted, "[power control systems are] more centralized, spare parts inventories have been reduced, and subsystems are highly integrated across the entire business." [107] Restructuring and consolidation has resulted in lower investment in security in recent years, as cash-strapped utilities seek to minimize costs and maximize revenue available for other areas.

Natural Gas and LNG Security

The facilities for natural gas and liquefied natural gas (LNG), a highly pressurized form of natural gas, exhibit unique vulnerabilities as well. Because natural gas and LNG are highly toxic and flammable, natural gas refineries, processing centers, storage facilities, and pipelines represent an attractive target for terrorists. Of special concern are LNG regasification terminals, of which there are six currently in the United States and its territories: an export terminal in Kenai, Alaska; and five import terminals in Everett, Massachusetts; Cove Point, Maryland; Elba Island, Georgia; Lake Charles, Louisiana; and Penuelas, Puerto Rico. FERC recently admitted that a leak from an LNG tanker at one of these facilities could catch fire and endanger people up to a mile away, and that a loss of an entire tanker could produce a fire more than one mile wide, resulting in second-degree burns to inhabitants across the region.[108] The Congressional Research Service has noted that "a consensus" exists among experts that an intentional or accidental LNG spill could result in

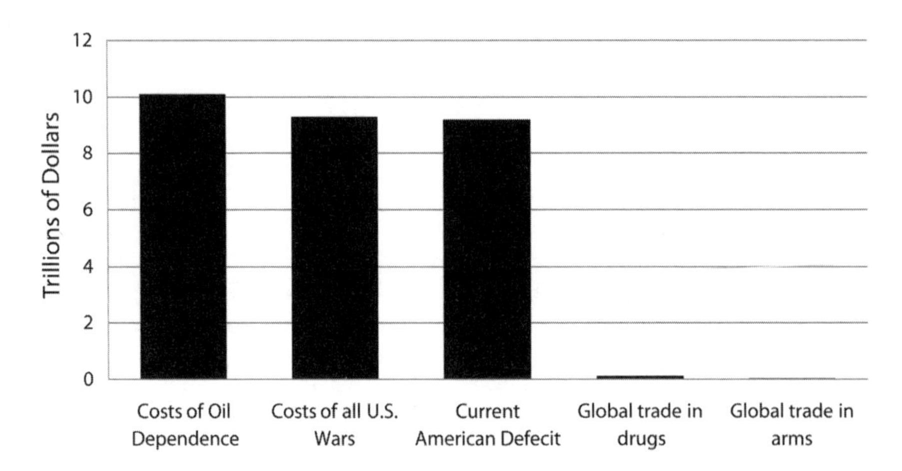

Figure 5
The Economic Costs of American Dependence on Foreign Oil

Considering just one "externality" associated with energy use, the cost of American dependence on foreign sources of oil is greater than the cost of all wars fought by the United States and the current national deficit.

Source: Numbers for the costs of oil dependence come from David Greene and Sanjana Ahmad, Costs of U.S. Oil Dependence: 2005 Update, Report to the U.S. DOE, ORNL/TM-2005/45 (January 2005). Costs for all U.S. wars ($9.3 trillion) come from the Institute for the Analysis of Global Security. Costs for the current American deficit ($9.2 trillion) come from the Congressional Budget Office. Estimates for the global trade in drugs ($122 billion) and arms ($36 billion) come from the United Nations. All figures have been adjusted to $2007.

flammable vapor clouds or pool fires—a fire burning so hot that it cannot be extinguished before all of the LNG has combusted.[109]

Gal Luft, the director of the Institute for the Analysis of Global Security, concluded that "if you locate LNG terminals close to residential areas, urban areas, then they become a major terrorist target. It's not just the terminals, but the whole LNG infrastructure from the tanker, to the terminal, to the truck."[110] Cindy Hurst, a lieutenant commander in the U.S. Navy Reserve, has cautioned that "despite the myriad security measures in place, it would be difficult to thwart people willing to die to carry out an attack [on a LNG facility]."[111] Similar to the attacks on the World Trade Center on September 11, 2001, no relevant industrial experience exists with accidents on the scale of a serious LNG explosion, so no one really knows for certain if public safety can ever be ensured.[112]

What is more, questions of LNG security are likely to emerge as imports begin to follow in the footsteps of petroleum markets—accounting for an increasing proportion of U.S. energy consumption. Researchers at ORNL estimated that from 1970 to 2004 American dependence on foreign supplies of oil has cost the country $5.6 to $14.6 trillion,[113] or more than the costs of all wars fought by the country going back to the Revolutionary War (including both invasions of Iraq). To those who may express dismay at such a high estimate, consider that researchers from the University of California–Davis and University of Alaska–Anchorage calculated that U.S. defense expenditures to exclusively protect oil in the Persian Gulf amount to about $27 billion to $73 billion per year (in 2004 dollars).[114]

Having the U.S. electricity sector, like the transportation sector, become more dependent on foreign sources of fuel could produce similar problems. Since the largest remaining reserves of LNG are located outside of the country, a growing dependence on natural gas could force a massive shift in U.S. military deployment to protect an entirely new collection of shipping lanes and distribution networks. Adam Serchuk, a senior program manager for the Energy Trust of Oregon, cautions that "the recent move toward natural gas–fired electricity generation, combined with recent financial interest in LNG facilities, could make the power sector dependent, for its marginal fuel on places like Venezuela, Algeria, Russia, and Indonesia."[115]

Nuclear Security

Issues relating to nuclear security involve both the risk of serious accident as well as theft and proliferation of fissile material. The safety record of nuclear plants is lackluster at best. One survey found that 63 nuclear accidents (defined as incidents that resulted in either death or more than $50,000 of property damage) have occurred worldwide from 1947 to 2007. The study documented that nuclear plants ranked first in economic cost among all energy accidents, accounting for 41 percent of energy accident related property damage from 1907 to 2007 (or $16.6 billion).[116] These numbers translate to more than one incident and $332 million in damages every year for the past three decades. The Chernobyl nuclear accident near Kiev, Ukraine, was responsible for at least 4,056 deaths, $7 billion in property damage, and 100,000 forced abortions.[117] The Chernobyl meltdown released more than 100 times the radiation than the atom bombs dropped on Nagasaki and Hiroshima, and most of the fallout concentrated near Belarus, Ukraine, and Russia, exposing more than 5 million people, including 1.6 million children, to unsafe levels of radiation. At least 350,000 people were forcibly resettled away from these areas, and cesium, ruthenium, radioactive iodine, cerium, plutonium, uranium, and strontium severely contaminated agricultural products, livestock, and soil as far away as Japan and Iceland (some milk in Poland is still undrinkable). After the accident, traces of radioactive deposits unique to Chernobyl were found in nearly every country in the northern hemisphere.[118] The international community has sponsored a multibillion dollar decontamination project, including the construction of a massive sarcophagus and 131 hydroelectric installations to prevent contaminated water from flowing downstream on the Pripiat and Dnieper rivers.[119]

The consequences of the accident at Chernobyl, moreover, are far from over. Fallout contaminated 6 million hectares of forest in the Gomel and Mogilev regions of Belarus, the Kiev region of Ukraine, and the Bryansk region of the Russian Federation. Three of the contaminants, cesium-137, strontium-90, and plutonium-239, are extremely long-lived and dangerous. Ninety-five percent of this contamination was accumulated in living trees, but 770 wildfires occurred in the contaminated zone from 1993 to 2001. Each one of these wildfires released radioactive emissions far into the atmosphere. A single, severe fire in 1992 burned 5 square kilometers of contaminated land (including 2.7 square kilometers of highly contaminated Red Forest next to the reactor), carrying highly toxic cesium dust particles into the upper atmosphere, distributing radioactive smoke particles thousands of kilometers, and dangerously exposing about 4.5 million people. Radiation levels were so high after this isolated fire that scientists throughout Europe initially thought there had been a second meltdown at Chernobyl Reactors One or Two, which remained in operation until 2000.[120]

Twenty-nine accidents have occurred since the Chernobyl disaster in 1986, and 71 percent of all nuclear accidents (45 out of 63) occurred in the United States, refuting the notion that severe accidents cannot happen within this country or that they have not happened since Chernobyl. Using extremely conservative estimates, nuclear power accidents have killed 4,100 people: more than died during September 11, 2001. Such accidents have involved meltdowns, explosions, fires, and loss of coolant, and have occurred during both normal operation and extreme, emergency conditions (such as droughts and earthquakes).

Using some of the most advanced probabilistic risk assessment tools available, an interdisciplinary team at MIT identified possible reactor failures in the United States and predicted that the best estimate of core damage frequency was around one every 10,000 reactor years. In terms of the expected growth scenario for nuclear power from 2005 to 2055, the MIT team estimated that at least 4 serious core damage accidents will occur and noted "both the historical and probabilistic risk assessment data show an unacceptable accident frequency. The potential impact on the public from safety or waste management failure . . . make it impossible today to make a credible case for the immediate expanded use of nuclear power."[121] The pressure to build new generators on existing sites to avoid complex issues associated with finding new locations only increases the risk of catastrophe, since there is a greater chance that one accident can affect multiple reactors.

Another assessment conducted by the Commissariat à l'Énergie Atomique in France concluded that no amount of technical innovation can eliminate the risk of human-induced errors associated with nuclear power plant operation. Two types of mistakes were deemed the most egregious: errors committed during field operations, such as maintenance and testing, that can cause an accident, and human errors made once small accidents begin that cascade to complete failure. When they looked at the safety performance of French Pressurized Water Reactors, the researchers concluded that human factors would contribute to about one-fourth (23 percent) of the likelihood of a major accident, implying that such a risk cannot be designed around.[122]

These risks are amplified with the next generation of nuclear systems. Former nuclear engineer David Lochbaum has noted that almost all serious nuclear accidents occurred with recent technology, making newer systems the riskiest. On July 26, 1959, the Sodium Reactor Experiment facility in California experienced a partial meltdown 14 months after opening. On January 3, 1961, Sl-1 Reactor in Idaho was slightly more than two years old before a fatal accident killed everyone at the site. The Fermi Unit 1 reactor began commercial operation in August 1966, but had a partial meltdown two months after opening. The St. Laurent des Eaux A1 Reactor in France started in June 1969, but an online refueling machine malfunctioned and melted 400 pounds of fuel four months later. The Browns Ferry Unit 1 reactor in Alabama began commercial operation in August 1974, and experienced a fire severely damaging control equipment six months later. Three Mile Island Unit 2 began commercial operation in December 1978 but had a partial meltdown three months after it started. Chernobyl Unit 4 started up in August 1984, and suffered the worst nuclear disaster in history on April 26, 1986, before its two-year birthday. Lochbaum attributes these accidents to previously unrecognized vulnerabilities, manufacturing defects, material imperfections, and shoddy construction.[123]

The risks may be especially acute for reactors in the United States since the domestic nuclear industry lacks qualified and experienced staff. The DOE has warned that the lack of growth in the domestic nuclear industry has gradually eroded important infrastructural elements such as experienced personnel in nuclear energy operations, engineering, radiation protection, and other professional disciplines; qualified suppliers of nuclear equipment and components, including fabrication capability; and contractor, architect, and engineer organizations with personnel, skills, and experience in nuclear design, engineering, and construction.[124] Since all commercial American reactors are light water reactors, system operators have little experience with gas cooled and other advanced reactor designs throughout the world.

Mistakes are not limited to the reactor site. Accidents at the Savannah River reprocessing plant have already released ten times as much radioiodine than the accident at Three Mile Island, and a fire at the Gulf United plutonium facility in New York scattered an undisclosed amount of plutonium into the vicinity, forcing it to permanently shut down. A similar fire at the Rocky Flats reprocessing plant in Colorado possibly released hundreds of particles of plutonium oxide dust into the environment.[125]

Furthermore, the August 2003 blackout revealed that operators of 15 nuclear reactors in the United States and Canada were not properly maintaining their backup diesel generators.[126] In Ontario during the blackout, reactors designed to automatically unlink from the grid and remain in standby mode instead went into full automatic shutdown, with only 2 of 12 reactors shutting down as planned. Since spent fuel ponds do not receive backup power from emergency diesel generators, when off-site power goes out, pool water cannot be recirculated to prevent boiling, evaporation, and exposure of fuel rods, increasing the risk of pool fires and explosions.

Stringent security regulations enacted after September 11 have reduced the risk of forcible entry, car or truck bombings, cyberterrorism, and aerial bombardment.[127] Yet the Nuclear Regulatory Commission found that 37 of 81 nuclear plants tested failed their 2003 Operational Safeguards Readiness Evaluation.[128] And while the industry purports that plant structures housing reactor fuel can withstand aircraft impact, multiple reports have cautioned that for too many plants the vital control building—the building that, if hit, could lead to a meltdown—is still located outside protective structures and vulnerable to attack.[129]

The Nobel Prize winning nuclear physicist Hannes Alven once said that "the military atom and the civil atom are Siamese twins." There is no shortage of terrorist groups eager to acquire the nuclear waste or fissile material needed to make a crude nuclear device, or a "dirty bomb." Commercial nuclear reactors already create an amount of plutonium equal to the global military stockpile every four years,[130] and the risks are not confined to the reactor site. All stages of the nuclear fuel cycle are vulnerable, including:

- Stealing or otherwise acquiring fissile material at uranium mines;
- Attacking a nuclear power reactor directly;
- Assaulting spent fuel storage facilities;
- Infiltrating plutonium stores or processing facilities;
- Intercepting nuclear materials in transit;
- Creating a dirty bomb from radioactive tailings.[131]

After three decades of searching, Pacific Gas and Electric is still unable to locate segments of one of their fuel rods missing from its Humboldt Bay nuclear power plant.[132] Since the collapse of the Soviet Union in 1991, authorities have documented 436 incidents of nuclear smuggling around the world, and those are only the incidents we know about.[133]

Human Rights

A final concern relates to human rights. Large, international, conventional energy companies—particularly oil and gas suppliers—have consistently employed private security firms to protect their operations and suppress dissent. In Indonesia, Myanmar, Nigeria, and Peru, some firms selling oil and gas to the U.S. market have denied free speech, employed torture, supported slavery and forced labor, sanctioned extra-judicial killings, and ordered executions. Shell gave guns to Nigerian security forces, and Chevron provided aid, helicopters, and pilots to an armed group that then gunned down nonviolent protestors on an oil drilling platform. Unocal admitted in court to knowing that the Burmese military committed acts of genocide to construct a pipeline for them, and British Petroleum, ExxonMobil, ConocoPhillips, and Shell continue to provide daily "security briefings" for mercenaries and supply vehicles, arms, food, and medicine to soldiers and police.[134]

The revenues acquired from oil and gas exports to the United States have also been used to propagate low-intensity warfare. Oil revenue has endowed the regime in Khartoum the means to expand its military and extend its military campaign in Darfur, Sudan, and allowed General Than Schwe to equip the Burmese military with light arms, helicopters, and armored vehicles in Myanmar. Proceeds from the Chad-Cameroon oil pipeline are allegedly helping fund conflicts in the Congo and Sudan, the oil and gas money from the Baku-Tbilisi-Ceyhan and South Caucasus Pipelines in Central Asia have enabled Azerbaijan to intensify its military campaign against Armenia. In this way, supporting the oil and gas fuel chain also furtively but directly endorses human rights abuses and international conflict.

No. 3: The Fuel Supply Challenge

Disruptions and interruptions in supply due to accidents, severe weather, and bottlenecks can all prevent fuels such as natural gas, coal, and uranium from being adequately distributed. Michael Pomorski, an energy trader with Cambridge Energy Research Associates, explained that such interruptions can drastically alter fuel prices, and that:

> Whatever fuel is chosen to generate electricity will always lead to a subsequent increase in demand, and thus price. The uncertainty of fuel prices—especially with the threat of a future carbon tax—makes choosing fuels for the generation of electricity exceptionally complicated and risky.[135]

Confirming Pomorski's explanation, an extensive study of electricity generation in Maryland found that utilities would likely have added 414 MW of coal capacity by 2025 without the existence of a carbon cap and trade system, but now that one exists they plan to build none.[136] Moreover, price escalation often occurs not because of true scarcity or shortage, but because of market speculation and intentional market manipulation. The rolling blackouts during the California energy crisis in 2001, for instance, had much less to do with the unsuitability or inaccessibility of natural gas and electricity and more to do with bad planning, ineptitude, and conniving.[137] During the Gulf War of 1991, fuel prices for oil and natural gas rose about 50 percent, not because of any physical exhaustion of supplies, but from oil and gas trader speculation.

While short-term future prices are always difficult to predict, the only certainty seems to be that as fossil fuels and uranium become scarcer, costs will rise in the long term. From 2002 to 2005, for example, operation and maintenance expenses for utilities rose by nearly $26 billion. Ninety-six percent of this increase was driven by rising fossil fuel prices, not because parts or labor had gotten more expensive. Aggregate fossil fuel costs nearly doubled in the four years between 2000 and 2004, from 2.3 ¢/kWh to 4.4 ¢/kWh.[138] But what accounted for these rapid price increases?

Natural Gas

Natural gas has by far been the fuel of choice for generators during the past decade. Many of the electricity generating units used for intermediate and "peaking" purposes (for example, to meet increased demand for air conditioning on hot, summer days) use natural gas for fuel. This is because natural gas generating units usually require a lower capital investment than nuclear or coal-fired plants, have shorter construction lead times, and tend to produce lower emissions than coal plants. Natural gas–fired units also can be turned on or off quickly, giving them operational flexibility to meet short-term peak electricity demands. Most U.S. coal plants today are 20 to 50 years old, and companies plan to replace them (in large part) with natural gas–fired generation.

The result of this trend has been that 90 percent of new power plants on order in 2003 and 2004 were gas-fueled.[139] The electricity sector's demand for natural gas has increased from 19 percent of total natural gas consumption in 2000 to 29 percent in 2005. This surge in demand for natural gas comes at a time when domestic natural gas production has begun to plateau. But consumption of natural gas is likely to increase even further for three reasons.

First, increased electricity demand in many areas has shrunk reserve margins to historically low levels. By 2005, reserve margins across the contiguous United States had dropped to 15 percent and, in some large states (like Texas and Florida), as low as 9 percent. Shrinking reserve margins coupled with increased electricity demands have forced many utilities to restart "mothballed" natural gas–fired generating units. And plans for new peaking units in large consumer states like Texas and Florida rely overwhelmingly on natural gas.

Second, because U.S. utilities have overinvested in gas-fired generating units, they hunger for new supplies of natural gas. Congress responded by authorizing greater drilling rights in the Gulf of Mexico in 2007 and hinted at granting greater access to federal lands where natural gas drilling is currently off-limits. Whether new drilling rights are granted or not, the tantalizing prospect of vast new sources of natural gas may lead utilities to believe that gas-fired units are safer investments than they really are.

Third, as pressure builds for the United States to adopt some form of binding greenhouse gas reduction targets, more generators will turn to natural gas because its carbon intensity is about half that of coal. Roger Garrett, director of Puget Sound

Energy's Resource Acquisition Group, for example, recently told industry executives that PSE had plans to invest in a significant number of new natural gas–fired combined cycle facilities partly because the company anticipates future binding carbon constraints.[140]

The overbuilding of gas-fired peaking plants has resulted in skyrocketing demand for natural gas, which, in turn causes prices to surge. Between 1995 and 2005, natural gas prices rose by an average of 15 percent *per year.*[141] Since 1997, the EIA has had to increase its projections for natural gas prices each year to conform to new data showing that the price was higher than expected.[142] As early as 2003, then Federal Reserve Chairman Alan Greenspan predicted continued strain in the long-term market for natural gas:

> Today's tight natural gas markets have been a long time in coming, and futures prices suggest that we are not apt to return to earlier periods of relative abundance and low prices anytime soon.[143]

There are significant costs associated with natural gas price volatility. In hearings before the House Committee on Natural Resources, the CEO of one large chemical company told Congress, "the recent history of natural gas prices is a study in

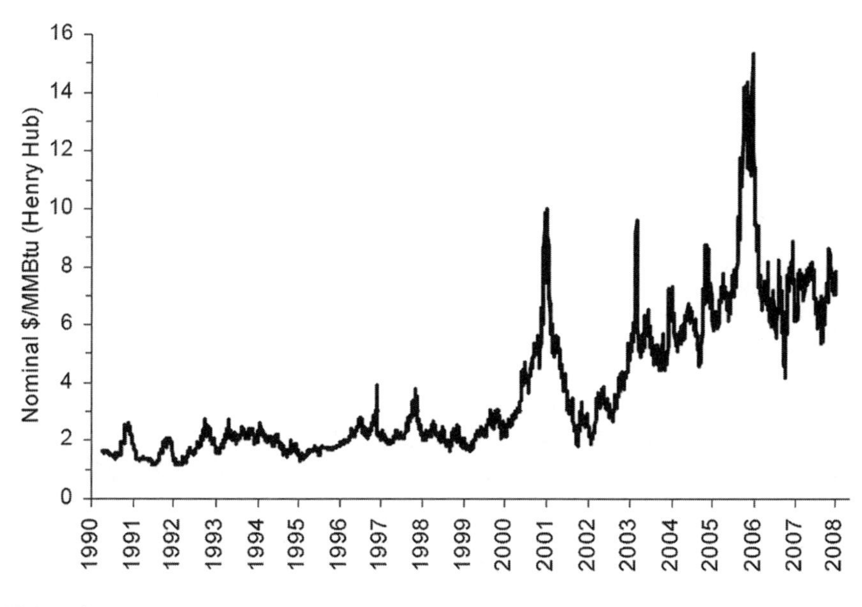

Figure 6
Natural Gas Price Fluctuations, 1990 to 2008

Natural gas futures prices at the Henry Hub (the pricing point for natural gas contracts traded on the stock exchange) shows extreme price fluctuations in natural gas from 2000 to 2008.

Source: Mark Bolinger and Ryan Wiser, Comparison of AEO 2008 Natural Gas Price Forecast to NYMEX Futures Prices (Lawrence Berkeley National Laboratory, January 7, 2008).

commodity price volatility."[144] For example, the price of natural gas jumped from $6.20 per million BTUs (British thermal units; MBtu) in 1998 to $14.50 per MBtu in 2001, then dropped precipitously for almost a year and then rebounded steadily from around $2.10 per MBtu in 2002 to more than $14.00 per MBtu near the end of 2005.[145] Hurricane Katrina caused similar price spikes when it disrupted natural gas refining and reprocessing infrastructure in the Southeast, especially since the Midwest and East Coast are heavily dependent on deliveries of natural gas and refined products via a few major pipelines emanating from the Gulf of Mexico.[146]

When natural gas prices swing wildly, utilities find it difficult to plan prudent investments or contract for bulk supplies. The enormous price spikes for natural gas seen over the past few years have made natural gas–fired plants uneconomic to operate, and have resulted in significant increases in electricity prices in several areas, much to the consternation of utility executives.[147] In October 2006, for example, Chesapeake Energy stunned the gas industry by announcing that it would shut off 100,000 cubic feet per day of unhedged gas production until natural gas prices rebounded. A week later, Questar Exploration and Production curtailed its output for the same reason.[148]

These unusual moves repudiated government (and industry) optimism about domestic natural gas output and reminded analysts that the gas market can be far more volatile, easily manipulated, and more expensive than forecasts predict. Indeed, in the fall of 2006 ratepayers in Illinois waged a modern-day version of the Boston Tea Party, sending tea bags to the state's utilities in protest of projected rate increases of 22 percent to 55 percent in 2007. In Boston, homeowners and small businesses have seen electricity prices rise by 78 percent since 2002, from 6.4 ¢/kWh to 11 ¢/kWh.[149] Across the United States, average retail electricity prices rose by 9.2 percent in 2006 alone, a trend likely to continue for the next several years.[150]

Natural gas–induced price spikes have been devastating to the U.S. economy. For some products, such as food processing, textiles, lumber, paper processing, chemical manufacturing, and cement mixing, natural gas can represent up to 15 percent of total costs.[151] Because natural gas accounts for nearly 90 percent of the cost of fertilizer, escalating natural gas prices create significant economic hardships for U.S. farmers. As well, some manufacturing and industrial consumers that relied heavily on natural gas moved their facilities overseas in the wake of price increases. The U.S. petrochemical industry, for example, relies on natural gas as a primary feedstock as well as for fuel. On February 17, 2004, the *Wall Street Journal* reported that the petrochemical sector had lost approximately 78,000 jobs to foreign plants where natural gas was much cheaper.[152] When the price of natural gas spiked in 2001, almost one-half of the country's methanol capacity and one-third of its ammonia capacity were shut down, and the Dow Chemical Company moved 1.4 billion pounds of production from the United States to Germany because of higher energy costs.[153] Even dairy producers in California had to suspend milk and cheese production until natural gas prices receded, and three of the state's sugar refineries went bankrupt.[154] Joseph Barton, former chair of the House Committee on Energy and Commerce, stated that he believed the country's higher natural gas prices have cost the economy $50 billion and more than 100,000 jobs in Texas, Ohio, New Jersey, and West Virginia.[155]

Coal

In 1866, William Stanley Jevons predicted the exhaustion of coal in the next few decades. "We have to choose," he wrote, "between a brief greatness and a long continued mediocrity." [156] Well, the "greatness" of coal has been more than brief, and it will not likely abate any time soon. Most analysts agree that between 500 and 550 billion tons of coal remains in the United States, enough to last between 200 and 250 years at current rates of consumption.[157]

Unlike natural gas and oil, which are drilled from the earth, coal must be beaten out of it. Prior to the 1990s, American coal production was fragmented and prone to flat or even declining coal prices and overcapacity, leading some investors to describe the commodity as "interesting as dirt." [158] However, techniques pioneered over the past 25 years, like surface and "strip" mining, significantly increased output and accompanied an industry consolidation, as coal companies were taken over in large numbers by conglomerates and integrated energy companies.[159]

Coal's future is not fundamentally a matter of resource availability, but one of environmental acceptability. The problem is not so much technical potential (known reserves) but achievable potential (known reserves that can be extracted at a competitive price and combusted within environmental guidelines). Harvard professor Mel Horwitch has noted that coal has transitioned from becoming "the great black hope" into "constrained abundance." [160] Unfortunately, the majority of the remaining coal reserves in the United States contain large quantities of pollution-creating sulfur.[161] Most remaining coal reserves are also "stranded." That is, miners have already exploited the most economical reserves, and the ones that remain, while they may be quite significant, are located in areas geographically distant from major consuming areas and thus the coal mined from them is more difficult to extract, process, and transport.[162]

Transportation bottlenecks and demand surges in developing countries such as India and China are also starting to provoke significant price fluctuations that affect both the market price and availability of coal. From 2001 to 2006, coal use around the world grew by 30 percent, and 88 percent of this increase came from Asia (with 72 percent of the increase from China alone).[163] In 2004 China shifted from a net exporter to an importer of coal, creating a severe shortage of ocean-worthy bulk carriers and causing the global price of coal (as well as other commodities such as iron, ore, and steel) to jump drastically as freight prices soared.[164]

In 2003, the cost of coal in Central Appalachia was $35 per ton. The price increased nearly 7 percent *each year* until, in 2006, a ton of coal in the same region cost close to $60 a ton (see Figure 7).[165] In some regions of the United States, coal prices actually doubled between 2002 and 2004. Such variability has been the result of a confluence of complicated factors including (a) dwindling reserves of coal in the eastern United States, (b) constricted rail service, (c) flooding and hurricanes hitting barge routes,

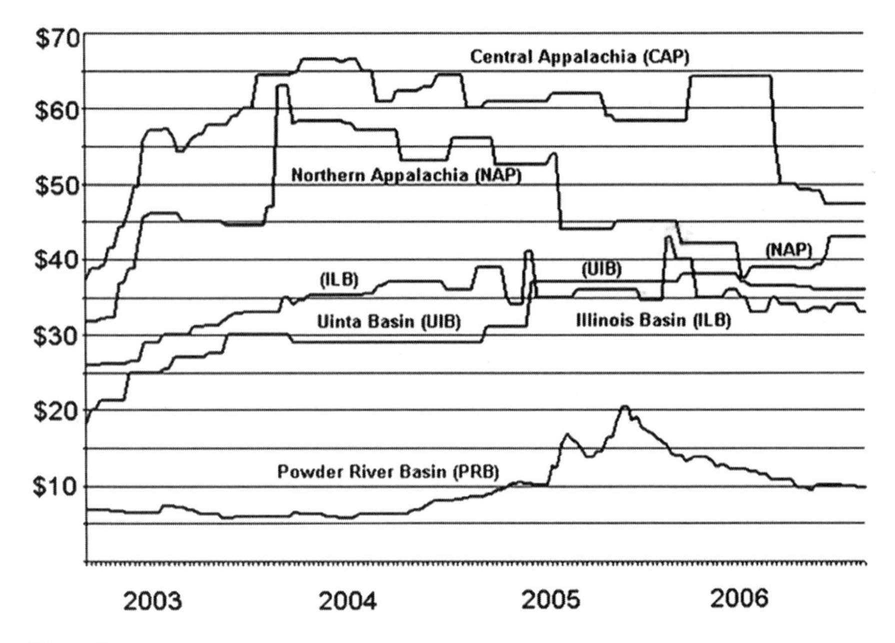

Figure 7
Coal Price Fluctuations 2003–2006 (in dollars per ton)

The average weekly price for a ton of coal, traded in five separate locations of the country, has shown considerable volatility.
Source: U.S. EIA Coal News and Markets 2006.

(d) bankruptcies and resulting consolidation and restructuring within the industry, (e) permitting, bonding, and insurance issues, (f) mine closures, (g) more stringent environmental regulations concerning mine planning and the posting of reclamation bonds, (h) restrictions on mountain top removal, and (i) rapidly rising global demand.[166]

A comprehensive 2005 IEA study revealed that more than half of the coal-powered plants reported significant price fluctuations in coal, varying by almost a factor of 20.[167] This volatility comes at a price to consumers in increased electricity rates and decreased coal production.

Uranium

Nuclear plants increase the country's dependence on imported uranium subject to large price spikes. The cost of uranium, for instance, jumped from $7.25 per pound in 2001 to $47.25 per pound in 2006, an increase of more than 600 percent.[168] The Nuclear Energy Agency reports that uranium fuel accounts for 15 percent of the lifetime costs of a nuclear plant, and 200 metric tons of uranium are required annually for every 1,000 MW reactor.

The DOE quietly acknowledged that domestic uranium production was currently at about 10 percent of its historical peak in 2000, and that most of the world's uranium reserves are becoming "stranded," and therefore much more difficult to extract.[169] The result is that investments in new nuclear plants would only make the United States more dependent on Africa, Russia, Canada, and Australia for sources of uranium. Admittedly, the chance that Canada and Australia will band together to become the new "OPEC of uranium" is as unlikely as it sounds, but Kazakhstan, Namibia, Niger, and Uzbekistan were responsible for more than 30 percent of the world's uranium production in 2006. And these countries have not had the most stable political regimes over the past several years.

The entire nuclear fuel cycle is therefore dependent on incredibly long lead times and geographically separated facilities. The time needed to bring major uranium mining and milling projects into operation average five or more years for exploration and discovery with an additional eight to ten years for production. Moreover, uranium conversion facilities operate only in Canada, France, the United Kingdom, and the United States, meaning it typically takes between five and seven years before uranium from the ground actually reaches a nuclear reactor.

The IAEA classifies uranium broadly into two categories: "primary supply" including all newly mined and processed uranium, and "secondary" supply encompassing uranium from reprocessing inventories (including highly enriched uranium, enriched uranium inventories, mixed oxide fuel, reprocessed uranium, and depleted uranium tails). The IAEA expects primary supply to cover 42 percent of demand for uranium in 2008, but acknowledge that the number will drop to between 4 and 6 percent of supply in 2025, as low-cost ores are expended.

But here lies a conundrum: the IAEA calculated that secondary supply can contribute only 8 to 11 percent of world demand. "As we look to the future, presently known resources fall short of demand," the IAEA stated, and "it will become necessary to rely on very high cost conventional or unconventional resources to meet demand as the lower cost known resources are exhausted."[170]

The obvious conclusion is that there simply will not be enough uranium to go around, even under current demand. Interestingly, however, the IAEA refused to state this obvious conclusion. While the agency recorded the total amount of uranium at around 3.6 gigagrams (Gg) in 2001, the number inexplicably jumped to 4.7 Gg in 2006. The increase was due not to new discoveries or improved technologies, but simply because of a clever redefinition of what the agency counts as uranium. The IAEA included in its new estimate the category of uranium that costs $80 to $130 per kilogram. This class comprises uranium ores of relatively low grades and of greater depth that have been so much harder to mine and would require such longer transport that the agency historically has not even counted them as usable stocks of uranium at all.

No. 4: The Environment Challenge

Despite three decades of "clean air" and "clean water" legislation in the United States, power plant pollution continues to be a serious threat to ecosystem health. Americans are experiencing a rise in respiratory illnesses (especially childhood asthma), and visibility continues to degrade at least in part as a result of power plant emissions. Coal-burning plants release an average of 68 percent of their waste into the environment, and more than one-third of the coal-fired power plants currently operating in the United States (approximately 123 GW out of 300 GW) do not have advanced pollution controls installed.[171] Operators have recognized that the industry is so damaging that roughly two-fifths of the capital cost for new power stations relates to compliance with existing environmental standards.[172] From 2002 to 2005, the electric power sector spent more than $21 billion seeking compliance with air and water pollution controls, and expects to spend another $47 billion from 2007 to 2025 to achieve compliance with the new environmental regulations.[173]

Fossil fuel and nuclear power plants are the nation's second largest users of water, produce millions of tons of solid waste, emit greenhouse gases, mercury, particulate matter, and other noxious pollutants into the atmosphere, and cause widespread social inequity. They concentrate environmental hazards among the poor and geographically disadvantaged, forcing Appalachia and Navajo Counties to pay the environmental costs of distributing "clean electrons" to Los Angeles and New York. The system makes the cleaner places more habitable and the displaced more resentful. These environmental problems—known long before the publication of this book —provoked *The Economist* to make the startling claim that "using energy in today's ways leads to more environmental damage than any other peaceful human activity."

Water

One of the most important, and least discussed, environmental issues facing the electricity industry is its water-intensive nature. The nation's oil, coal, natural gas, and nuclear facilities consume about 3.3 billion gallons of water each day.[174] In 2006, they accounted for almost 40 percent of all freshwater withdrawals (water diverted or withdrawn from a surface- or groundwater source), roughly equivalent to all the water withdrawals for irrigated agriculture in the entire United States.[175] If projected electricity demand is met using water-intensive fossil fuel and nuclear reactors, America will soon be withdrawing more water for electricity production than for farming.

A conventional 500 MW coal plant, for instance, consumes about 7,000 gallons of water per minute, or the equivalent of 17 Olympic-sized swimming pools every day.[176] Data from the Electric Power Research Institute (EPRI) confirm that every type of traditional power plant consumes and withdraws significant amounts of water. Conventional power plants use thousands of gallons of water for the condensing portion of their thermodynamic cycle. Coal plants also use water to clean and process fuel, and all traditional plants lose water through evaporative loss.

Newer technologies, while they withdraw less water, actually consume more. Advanced power plant systems that rely on recirculating, closed-loop cooling technology convert more water to steam that is vented to the atmosphere. Closed-loop systems also rely on greater amounts of water for cleaning and therefore return less water to the original source. Thus, while modern power plants using the best available technologies may reduce water withdrawals by up to 10 percent, they contribute even more to the nation's water scarcity by redistributing water from already-strained aquifers.[177]

Nuclear reactors, in particular, require massive supplies of water to cool reactor cores and spent nuclear fuel rods. Because much of the water is turned to steam, substantial amounts are lost to the local water table entirely. One nuclear plant in Georgia withdrew an average of 57 million gallons of water every day from the Altamaha River, but actually consumed 33 million gallons per day from the local supply (primarily as lost water vapor), enough to service more than 196,000 Georgia homes.[178]

With electricity demand expected to grow by approximately 50 percent in the next 25 years, continuing to rely on fossil fuel–fired and nuclear generators could create a water scarcity crisis. In 2006, the DOE warned that consumption of water for electricity production could more than double by 2030 to 7.3 billion gallons per day, if new power plants continue to be built with evaporative cooling. This staggering amount is equal to the entire country's water consumption in 1995.[179]

The electric utility industry's vast appetite for water has serious consequences, both for human consumption and the environment. Assuming the latest Census

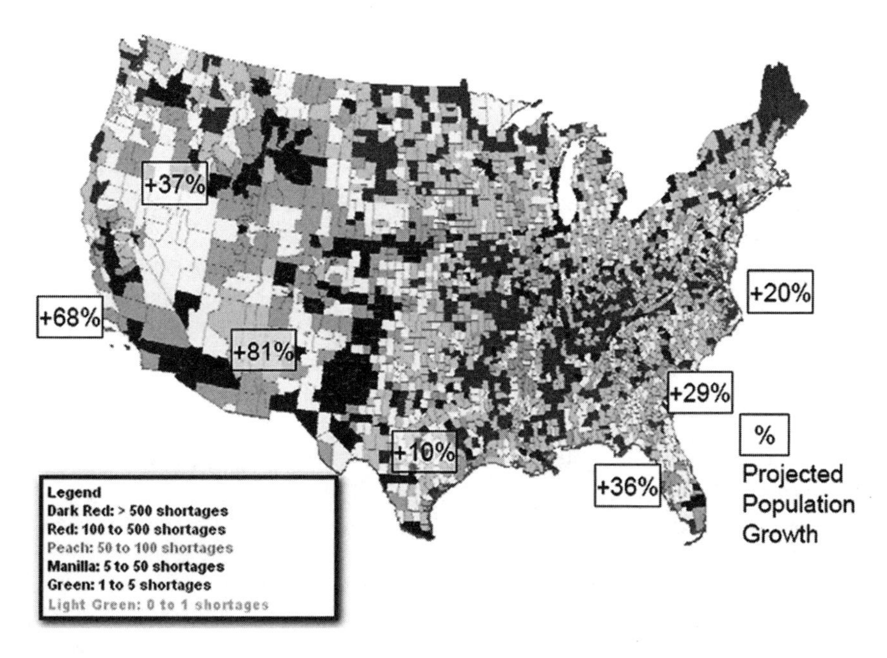

Map 2
Expected Water Shortages from Conventional Power Plants

Areas of the country expecting the most population growth are also the ones with the greatest needs for electricity. Dark areas on the map refer to counties that will be at risk of water shortages in 2025 as the result of conventional power plant operation.

Source: U.S. Department of Energy, *Energy Demands on Water Resources: Report to Congress on the Interdependency of Energy and Water* (Washington, DC: U.S. DOE, December 2006).

Bureau projections, the U.S. population is expected to grow by about 70 million people in the next 25 years. Such population growth is already threatening to overwhelm existing supplies of fresh and potable water. Few new reservoirs have been built since 1980, and some regions have seen groundwater levels drop as much as 300 to 900 feet over the past 50 years. Most state water managers expect either local or regional water shortages within the next 10 years, according to a recent survey, even under "normal" conditions. In fact, 47 states in the country reported drought conditions during the summer of 2002.[180]

Water shortages risk becoming more acute in the coming years as climate change alters precipitation patterns. In the Pacific Northwest, scientists expect global warming to induce a dramatic loss of snowpack as more precipitation falls as rain. Numerous studies have suggested that the hydrology of the region will be fundamentally altered with increased flood risks in the spring and reductions of snow in the winter. Consequently, power retailers in the region have expressed concern that large hydroelectric and nuclear facilities will have to be shut down due to lack of adequate water for electricity generation and cooling.[181]

Unplanned shutdowns as a result of water scarcity have already occurred. During the steamy August of 2006, the record heat sparked unplanned reactor shutdowns in Michigan and Minnesota as nuclear plant operators scrambled to find enough water to cool radioactive fuel cores.[182] Xcel Energy, one of the power sector's largest players, even had to cancel the construction of a $1.2 billion coal facility in Pueblo, Colorado, because of concerns that there was not enough water available to service the facility.[183]

The water-related impacts from conventional energy sources are not limited to water scarcity. The Argonne National Laboratory has documented how power plants have withdrawn hundreds of millions of gallons of water each day for cooling purposes and then discharged the heated water back to the same or a nearby water body. This process of "once-through" cooling presents potential environmental impacts by impinging aquatic organisms in intake screens and by affecting aquatic ecosystems by discharge effluent that is far hotter than the surrounding surface waters.[184] Drawing water into a plant often kills fish and other aquatic organisms, and the extensive array of cooling towers, ponds, and underwater vents used by most plants severely damage riparian habitats.

In some cases, the thermal pollution from centralized power plants can induce eutrophication—a process where the warmer temperature alters the chemical composition of the water, resulting in a rapid increase of nutrients such as nitrogen and phosphorous. Rather than improving the ecosystem, such alterations usually promote excessive plant growth and decay, favoring weedy species over others and reducing water quality. In riparian environments, the enhanced growth of choking vegetation can collapse entire ecosystems. This form of thermal pollution has been known to decrease the aesthetic and recreational value of rivers, lakes, and estuaries and complicate drinking water treatment.[185]

As well, America's fleet of coal- and oil-fired power plants produce more than 100 million tons of sludge waste every year. Seventy-six million tons of these wastes are primarily disposed on-site at each power plant in unlined and often unmonitored wastewater lagoons and landfills. These wastes are highly toxic, containing concentrated levels of poisons such as arsenic, mercury, and cadmium that can severely damage human nervous and endocrine systems.[186] In mining the fuel needed to power America's existing conventional plants, the coal industry discharges between 70 million and 2.5 billion tons of fine coal into the nation's streams, creeks, and rivers every year.[187]

Nuclear facilities can be even worse. To produce fuel for nuclear reactors, uranium is often "leached" out of the ground by pumping a water solution through wells to dissolve the uranium in the ore. The uranium is then pumped to the surface in a liquid solution. About 20 such in-situ leach facilities operate in the United States.[188] At the reactor site, electricity generation using nuclear technology creates wastewater contaminated with radioactive tritium and other toxic substances that can leak into nearby groundwater sources. In December 2005, Exelon Corporation reported to authorities that its Braidwood reactor in Illinois had since 1996 released millions of gallons of tritium-contaminated wastewater into the local watershed, prompting the company to distribute bottled water to surrounding communities while local drinking water wells were tested. The incident led to a lawsuit by the Illinois Attorney General and the State Attorney for Will County who claimed that "Exelon was well aware that tritium increases the risk of cancer, miscarriages and birth defects and yet they made a conscious decision not to notify the public of their risk of exposure." [189]

Similarly, in New York, Entergy's Indian Point Nuclear Plant (on the Hudson River) emptied thousands of gallons of radioactive waste into underground lakes from 1974 to 2005. The Nuclear Regulatory Commission (NRC) accused Entergy of improperly maintaining two spent fuel pools that leaked tritium and strontium-90, cancer causing radioactive isotopes, into underground watersheds, with as much as 50 gallons of radioactive waste seeping into water sources per day.[190] As one local resident put it, "With great sadness . . . commercial and recreational fishing, boating and swimming, among other activities we enjoy, has been degraded by the failure of Entergy to perform their due diligence in keeping pollutants of any kind from entering the Hudson River." [191]

Air

The connection between power production and air pollution was vividly documented by the August 2003 Northeast blackout, which not only shut off electricity for 50 million people in the United States and Canada but also stopped the pollution coming from fossil fuel-fired power plants across the Ohio Valley and the Northeast. In effect, the power outage established an inadvertent experiment for gauging the atmosphere's response to the grid's collapse. Twenty-four hours after the blackout, sulfur dioxide (SO_2) concentrations in the Northeast declined 90 percent, particulate matter dropped by 70 percent, and ozone concentrations fell by half.[192] The blackout reveals that conventional electricity generation is by far the largest source of air pollutants that harm human health and injure the natural environment. The five most serious pollutants from conventional power plants are sulfur dioxides (SO_2) and nitrogen oxides (NO_x), which form ozone, mercury (Hg), particulate matter (PM), and greenhouse gases (primarily CO_2).

Indeed, each year the American electricity industry emits almost 8 tons of air pollution for each person in the country, more than 100 times their weight.[193] To get an idea of just how much gas this represents, consider that it would take every American more than a century of farting continuously before we removed the equivalent amount of gas from our bodies.[194] Using just the marginal cost of avoiding this pollution (and not accounting for its actual damage), the power industry induces $201.9 billion in damages every year (see Tables 2 and 3).

In all likelihood, the true costs are much, much greater. In a national survey of air quality, the American Lung Association warned that 81 million people live in areas of the United States with unhealthy short-term levels of air pollution. Sixty-six million of these Americans live in areas with unhealthy year-round levels of particle pollution, 136 million in areas with unhealthy levels of ozone, and 46 million in counties with all three forms of pollution.[195]

Lung diseases now affect more than 10 percent of the population and are the third leading cause of death for Americans. The total annual cost for lung disease is estimated to exceed $80 billion in health care expenditures.[196] Nationwide, the EPA has designated 474 counties "non-attainment areas" for unsafe levels of ozone and 224 counties as unsafe areas for fine particulate matter.[197] That is, these counties are so degraded that no additional pollution is permitted under federal law. These areas include all the eastern states from Maine south to Georgia as well as Alabama, Arizona, California, Colorado, Illinois, Indiana, Kentucky, Louisiana, Michigan, Missouri, Nevada, Ohio, Tennessee, Texas, West Virginia, and Wisconsin. It may come as no surprise, then, that at the high end of the range, the health costs related to power plant pollution could approach $700 billion each year.[198]

Of course, not everything can be quantified monetarily. Researchers at the Harvard School of Public Health estimated that the air pollution from conventional energy sources kills between 50,000 and 70,000 Americans every year.[199] In Texas, nearly two-thirds of the pollution from all of the state's facilities is emitted by 100 different units located within two miles of a single school.[200] Such proximity matters, as the laws of physics dictate that the maximum health impacts from emission sources will be observed closest to the sources.

Children are particularly vulnerable to the pollution from fossil fuels. Young children spend more time outdoors, and they breathe 50 percent more air per pound of body weight than adults. Since they breathe polluted air when their respiratory systems are still developing, they become especially susceptible to chronic and life-lasting damage. Exposure to fine particles is associated with many childhood illnesses, and children are less likely to recognize symptoms, leading to delays in treatment and worsening possible damage. Study after study has shown that high periods of air pollution are correlated with a 26 percent increase in sudden infant death syndrome; that infants in high pollution areas are 40 percent more likely to die from

Table 2

Major Air Pollutants from Conventional Power Plants

Every year, including last year, power plants belched billions of tons of pollutants into the atmosphere. If converted to a solid, these pollutants would amount to a greater volume than all of the trash Americans put into landfills annually.

Pollutant	Electricity Related	Quantity/ yr	Impact	Number of Major Emitting Facilities	Amt/kWh	Amt/capita
Sulfur Dioxides (SO_2)	69%	9.2 million tons	Causes asthma, heart disease, respiratory problems, and acid rain	836	3.79 g/kWh	32 kg/year
Nitrogen Oxides (NO_x)	22%	4.0 million tons	Induces respiratory illnesses, haze, acid rain, and deterioration of water and soil quality	897	1.66 g/kWh	14 kg/year
Mercury (Hg)	40%	44.2 tons	Ingestion can cause neurological and developmental damage	376	0.0023 g/kWh	0.151 g/year
Particulate Matter (PM)	51%	125,000 tons	Increases the risk of direct morbidity and mortality when inhaled	424	0.16 g/kWh	1.6 kg/year
CO_2	39%	2,178 million tons	Contributes to global greenhouse gas inventory and climate change	899	893 g/kWh	7,878 kg/year

Source: Figures taken mostly from Paul J. Miller and Chris Van Atten, *North American Power Plant Air Emissions* (Montreal: Commission for Environmental Cooperation of North America, 2004), with adjustments made based on Granger Morgan, Jay Apt, and Lester Lave, *The U.S. Electric Power Sector and Climate Change Mitigation* (Washington, DC: Pew Center on Global Climate Change, June 2005); P. L. Spath, M. K. Mann, and D. K. Kerr, *Life Cycle Assessment of Coal-Fired Power Production* (Golden, CO: National Renewable Energy Laboratory, 1999, NREL/TP-570-25113); S. Pacca and A. Horvath, "Greenhouse Gas Emissions from Building and Operating Electric Power Plants in the Upper Colorado River Basin," *Environmental Science & Technology* 36, no. 14 (2002): 3194–3200.

Table 3

Estimated Cost of Power Plant Air Pollution, 2007

Using mostly the control costs for the pollution from power plants—that is, the costs of avoiding it through abatement technologies or mitigating it by planting trees—the amount of damage done by power plants last year was close to the amount of revenue created by the electricity industry.

Pollutant	Cost/ton	Justification	Tons Emitted/yr	Total Cost/yr
Sulfur Dioxides (SO_2)	$2,600	Avoided cost of installing flue gas scrubbing systems, baghouse filters, electrostatic precipitators, or flue gas desulfurization	9.2 million tons	$23.9 billion
Nitrogen Oxides (NO_x)	$11,255	Avoided cost of installing selective catalytic reduction equipment	4.0 million tons	$44.9 billion
Mercury (Hg)	$960 million	Damage cost of removing mercury from water through amalgam separators	44.2 tons	$42.4 billion
Particulate Matter (PM)	$63,000	Avoided cost of installing best available control technology	125,000 tons	$7.9 billion
Carbon Dioxide (CO_2)	$38	Marginal cost of planting trees to sequester CO_2	2,178 million tons	$82.8 billion

Source: Cost estimates for all pollution sources but PM come from U.S. Energy Information Administration, *Electricity Generation and Environmental Externalities: Case Studies* (Washington, DC: U.S. Department of Energy, September 1995). Prices for PM come from Stefan Schaltegger, Martin Bennet, and Roger Burritt, *Sustainability Accounting and Reporting* (New York: Springer, 2006), at 66. All prices adjusted for $2007.

respiratory causes; and that, though children make up roughly 25 percent of the population, they comprise almost half of all asthma cases.[201] The EPA has calculated that the number of deaths related to asthma in children has tripled between 1979 and 1996, and that asthma is the third ranking cause of hospitalization among children under the age of 15. Asthma also accounts for one-third of all pediatric emergency rooms visits and is the fourth most common cause of pediatric visits to the doctor's office.[202]

Sulfur Dioxide (SO_2) and Nitrogen Oxides (NO_x)

Sulfur dioxide (SO_2) is a pollutant responsible for lake- and forest-damaging acid precipitation and is a precursor to the development of small particles that damage human health. In 2003, the EPA estimated that roughly 40 million Americans lived in areas with unhealthy levels of SO_2.[203] The electric utility sector was responsible for 69 percent of sulfur dioxide emissions in 2006. Of these SO_2 emissions, 96 percent were from coal facilities, 3 percent from oil generators, and 1 percent from natural gas units. Older plants, many built before the Dodgers left Brooklyn in 1957 and Neil Armstrong walked on the moon in 1969, emit ten times as much SO_2 as newer facilities.

The wide variation in emissions rates reveals two things. First, emissions depend greatly on the quality of fuel and level of pollution control. A power plant using flue gas desulfurization controls and burning subbituminous coal mined in the western part of the country can reduce emissions by more than 90 percent. However, only 31 percent of coal-fired capacity in the United States is equipped with this or an equivalent scrubber technology. Second, a relatively small number of facilities are responsible for the majority of emissions. In fact, only 240 facilities, or just 27 percent of all the coal-fired units in the United States, produce 90 percent of the industry's emissions.

Nitrogen oxide (NO_x) emissions react with volatile organic compounds in the atmosphere (gasoline vapors or solvents, for example) and produce compounds that can result in severe lung damage, asthma, and emphysema.[204] Power plants were responsible for 22 percent of national NO_x emissions in 2006, and of these, 93 percent were from coal plants, 2 percent from oil units, and 5 percent from natural gas stations. NO_x emissions vary between power plants in ways similar to SO_2 emissions. Newer facilities release as little as 190 tons per year, while older plants release up to 45,300 tons, or more than 235 times the amount from cleaner plants. Such variation is due to the amount of nitrogen present in the fuel, quantity of excess air used in combustion, air temperature, and level of postcombustion pollution controls.

NO_x is also a major source of ground-level ozone (smog). It contributes to acid rain and pollutes surface water. In 2003, the EPA estimated that more than 70 million Americans lived in areas with unhealthy deposits of NO_x.

Emissions of SO_2 and NO_x create further problems when they react together in the atmosphere to form compounds that are transported long distances and induce acidification of lakes, streams, rivers, and soils.[205] Many parts of the country (especially the Ohio Valley and mid-Atlantic states) have hazardous concentrations of sulfur and nitrogen deposits.[206] Acid rain from SO_2 and NO_x compounds can render many bodies of water unfit for certain fish and wildlife species. Acidic deposition can also mobilize toxic amounts of aluminum, increasing its availability for uptake by plants and fish, which are then ingested by humans.

Concern over the significant environmental impacts of SO_2 and NO_x led the EPA to implement a "cap-and-trade" system that limits the aggregate emissions of SO_2 and NO_x and distributes allowances to regulated entities. Companies that reduce their emissions beyond the caps can sell allowances to other companies that have not reduced their emissions enough under the existing caps. The EPA periodically reduces the emissions caps to ensure that the total amount of pollution decreases over time.

One of the most tragic aspects of acid rain, however, is its irreversibility. Once polluted, streams, rivers, and habitats seldom recover. When locations become

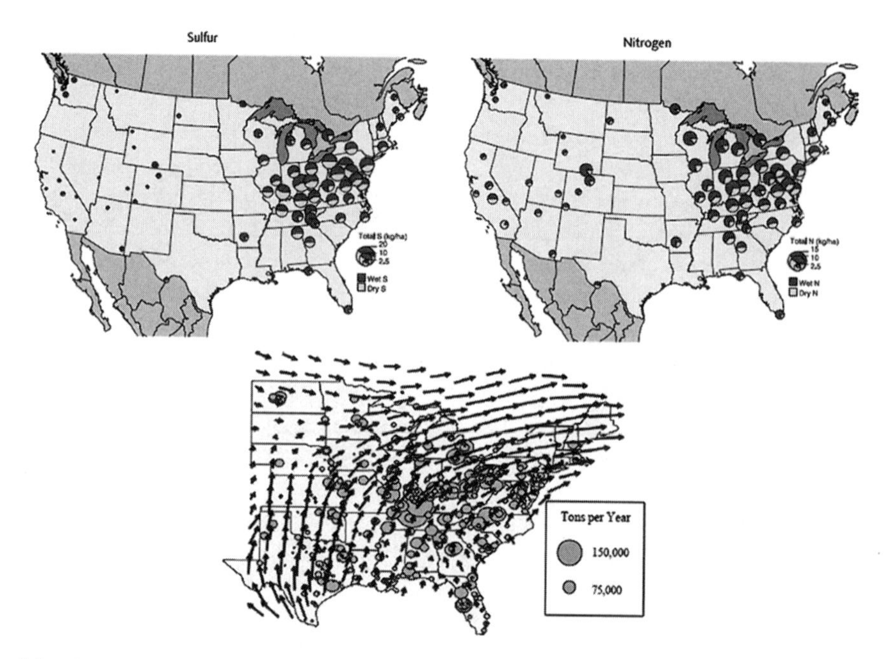

Map 3
Ozone, Sulfur Dioxide, and Nitrogen Oxide Pollution in the United States

Arrows show air flow when high ozone levels are present in the Northeast. Circles show locations and magnitude of NO_x and SO_x emissions from power plants.

Source: Cost estimates for all pollution sources but PM and mercury come from U.S. Energy Information Administration, *Electricity Generation and Environmental Externalities: Case Studies* (Washington, DC: U.S. Department of Energy, September 1995). Prices for PM come from Stefan Schaltegger, Martin Bennet, and Roger Burritt, *Sustainability Accounting and Reporting* (New York: Springer, 2006), at 66. All prices adjusted for $2007.

acidified, the diversity of invertebrates sharply declines, habitats fall apart, densities of salmon and trout fall, and bird breeding is severely impaired. Scientists have found a complicated, nonlinear relationship between continued emissions and polluted areas. In Europe, industry reduced emissions of SO_2 by roughly 80 percent between 1977 and 2007, and sulfur deposition declined in many places by more than 50 percent. However, ecologists found that the mean pH in acidified streams still increased by 0.3 to 0.4 units every 10 years. The basic lesson is that, once polluted, streams take longer to recover, and even light acidification can completely impair recovery of these fragile ecosystems.[207] Moreover, many of the damaged forests, ecosystems, and habitats are hundreds of miles away from the nearest serious polluter, demonstrating how far-reaching and widely dispersed acid rain pollution can be.

Despite the immense progress made under the Clean Air Act Amendments of 1990 in the United States, the EPA warns that surface water sulfate concentrations have actually increased in the Blue Ridge provinces of Virginia and that some parts of the Northern Appalachian Plateau region continue to experience dangerously high levels of stream acidification.[208] Acidification has been so severe in central Ontario and Quebec that streams and rivers have not responded to reductions in sulfate deposition. About 95,000 lakes and streams remain damaged and Atlantic Salmon in Nova Scotia have become extinct in 14 rivers and severely threatened in another 20. New York State, the Adirondacks, and the Catskill mountains are the most sensitive regions to acid inputs in the United States. There, recent reductions in SO_2 emissions have neither improved water quality nor increased the capacity of rivers and streams to neutralize acids. Nearly 25 percent of surveyed lakes in the Adirondacks no longer support any fish. In Western Pennsylvania, acid deposition has been linked with the deterioration of tree health and excessive mortality of mature sugar maples and red oaks. In Virginia, a 13-year trout stream study estimated that even reductions of 40–50 percent beyond existing Clean Air Act levels will not support recovery of chronically acidic streams, and will not halt the destruction of streams that are currently suffering worsening levels of acidification. In the Great Smoky Mountains National Park in Tennessee and North Carolina, chronic loading of sulfate and nitrate has made forest sites in the region so vulnerable that scientists estimate that within 80 to 150 years soil calcium reserves will be inadequate to support the growth of healthy trees and merchantable timber.[209]

Acid rain and ozone are not just bad for the environment. During years with hot summers, rates of ozone generation increase and areas with more than 135 million people violate ozone nonattainment levels designed to ensure air quality. During these events, studies have linked short term ozone exposure to hospital admissions and doctor visits for asthma and other respiratory problems. A 1996 Harvard School of Public Health study found that exposure to ozone was linked to 10,000 to 15,000 hospital admissions and between 30,000 to 50,000 emergency room visits in 13 U.S. cities.[210] Studies have also found the inverse to be true. Health data from 5,000 individuals living in Los Angeles, California, confirmed that a 1 percent reduction in ozone engendered substantial benefits for human health with fewer

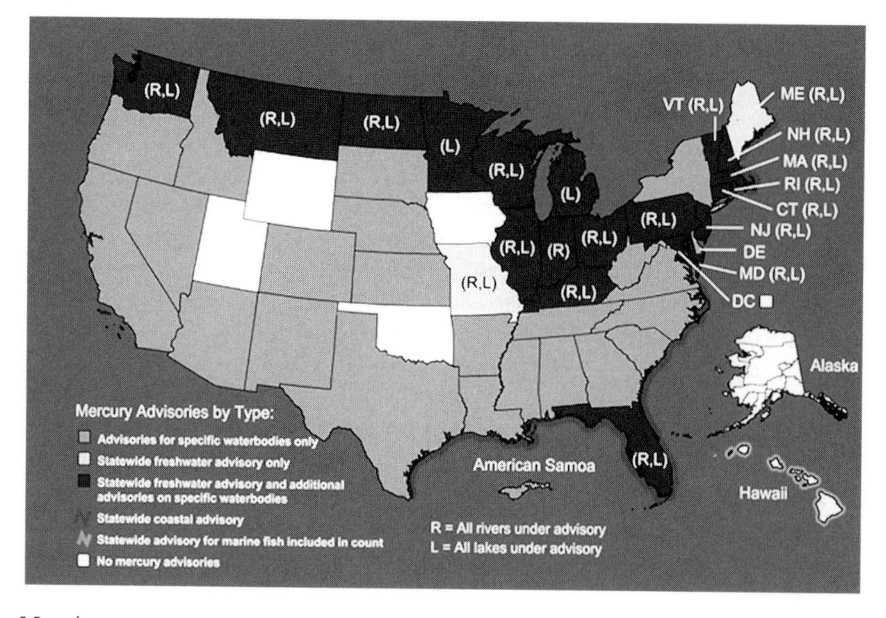

Map 4
Advisories for Mercury Contamination in the United States

All but eight states in the United States reported health threatening mercury contamination in their waters in 2003.

Source: U.S. Environmental Protection Agency.

deaths, fewer emergency room visits, and roughly 23,000 avoided instances of respiratory-related diseases.[211]

Mercury (Hg)

Power plants are responsible for 40 percent of national mercury emissions, with some facilities emitting close to one ton of mercury every year. American coal-fired power plants release about 50 tons of mercury into the nation's air. The greatest concentrations are found in the southern Great Lakes and Ohio River valley, the Northeast, and scattered areas in the South, with the most elevated concentrations in the Miami and Tampa Bay areas.[212]

A comprehensive EPA study on mercury noted that epidemics of mercury poisoning following high doses in Japan and Iraq have demonstrated that neurotoxicity is of greatest concern when developing fetuses are exposed it. Dietary mercury is almost completely absorbed into the blood and distributed to all tissues including the brain; it also readily passes through the placenta to the fetus and fetal brain.[213]

Most Americans do not ingest mercury directly, but accumulate small amounts of the poisonous metal through the consumption of fish. In 2003, 43 states had to issue mercury advisories to warn the public to avoid consuming contaminated fish from in-state water sources.[214] The EPA estimates that as many as 3 percent of women of childbearing age eat sufficient amounts of fish to be at risk from mercury exposure. One study projected that 630,000 infants are born with dangerous levels of mercury every year throughout the country.[215]

Particulate Matter

Particulate matter is not a specific pollutant itself, but instead a mixture of fine particles of harmful pollutants such as soot, acid droplets, and metals. Particulate matter (PM) is the generic term for the mixture of these microscopic solid particles and liquid droplets in the air. Because its makeup is often complex, PM is by far the most difficult pollutant to detect and monitor. The most widely studied are those particles with an aerodynamic diameter of 10 microns (PM_{10}) or less and those with a diameter of 2.5 microns ($PM_{2.5}$) or less.

The science linking PM with increased risks of cancer, illness, and death is as incontrovertible as it is disturbing. PM is arguably the most studied of all atmospheric pollutants. The medieval English evidently believed that burning coal made the air unhealthy by dispersing particulates, and that charcoal and wood should be used to heat the home. King Edward I even issued a proclamation in 1306 that the use of coal was punishable by death, and at least one man was executed under this law.[216] Modern studies of particulate pollution began with "bad air days" in Donora, Pennsylvania, in 1948, where 19 people died from increased SO_2 concentrations, and in 1952 in London, England, where 4,000 people died from inhaling smoke particles. These incidents prompted long-ranging epidemiological, toxicological, and laboratory dose-response studies. Their findings were confirmed, yet again, when the burning of 650 oil wells in Kuwait during the 1991 Persian Gulf War affected U.S. soldiers with respiratory illness, asthma, and emphysema.

Inhalation of PM is strongly associated with heart disease and chronic lung disease.[217] Since microscopic solids or liquid droplets are so small, they can get deep into the lungs and cause serious health problems. Numerous scientific studies have linked PM exposure to the following:

- Irritation of the airways, coughing, or difficulty breathing;
- Decreased lung function;
- Aggravated asthma;
- Development of chronic bronchitis;
- Irregular heartbeat;
- Nonfatal heart attacks;
- Premature death in people with heart or lung disease.[218]

Roughly half of the nation's 250,000 tons of PM emissions come indirectly from the NO_x and SO_2 emitted from power plants, which react in the atmosphere to form dangerous PM particles. Some power plants may be releasing up to 400 tons of PM every year, when both these primary and secondary conditions are included in estimates.[219]

In 1993, Harvard researchers published the results of a 16 year, six-city study that tracked the health of more than 8,000 individuals for a period of 14 to 16 years. They observed a nearly linear relationship between particle concentrations in the air and increased mortality rates, indicating that even relatively low levels of PM contributed to adverse health effects. The risk of an early death in high-level areas was 26 percent higher than in areas with the lowest levels of pollution, even after controlling for other risk factors such as occupation and smoking. The risk of cardiopulmonary disease in high-level areas was an astounding 37 percent higher than in low level areas.[220]

A follow-up study in 1995 by the American Cancer Society and Harvard Medical School involved 151 metropolitan areas and 552,138 adults. After controlling for differences in age, sex, and cigarette smoking, and tracking individual exposure to sulfate and fine particulate air pollution, the researchers found that PM was directly associated with cardiopulmonary and lung cancer mortality. The researchers calculated a 17 percent increase in mortality risk in areas with higher concentrations of fine particles relative to areas with lower concentrations, a 15 percent increase in mortality risks in areas with higher concentrations of sulfate aerosols, and a 31 percent higher risk of premature death from cardiopulmonary disease. For

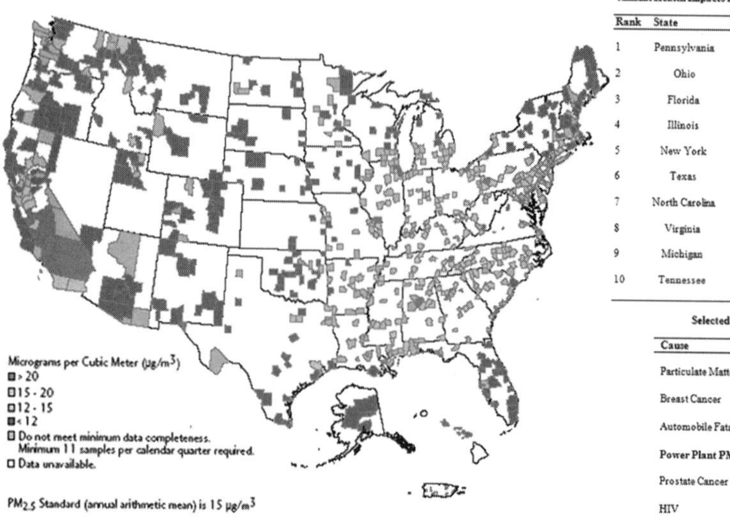

Annual Health Impacts from Power Plant PM Pollution (10 Worst States)

Rank	State	Mortality	Hospital Admissions	Heart Attacks
1	Pennsylvania	1,825	1,664	3,329
2	Ohio	1,743	1,638	2,873
3	Florida	1,416	1,367	2,145
4	Illinois	1,356	1,333	2,361
5	New York	1,212	1,191	2,455
6	Texas	1,160	1,105	1,791
7	North Carolina	1,133	1,013	1,603
8	Virginia	989	895	1,421
9	Michigan	981	968	1,728
10	Tennessee	952	804	1,276

Selected Causes of Death in U.S., 2007

Cause	Estimated Annual Deaths
Particulate Matter	65,638
Breast Cancer	40,910
Automobile Fatalities	36,710
Power Plant PM Pollution	32,684
Prostate Cancer	30,142
HIV	18,017
Drunk Driving	16,694

Map 5

Particulate Matter Pollution in the United States and Associated Deaths

Particulate matter pollution from power plants poses a serious threat to human health in every state in the United States, with the unhealthiest levels of pollution occurring in Ohio and Pennsylvania.

Source: Data compiled from the Centers for Disease Control, Mothers Against Drunk Driving, National Transportation Safety Board, Environmental Protection Agency, and Clean Air Task Force.

subjects who had never smoked, the increased risk of premature death from this disease was 43 percent. For women who resided in the more polluted cities and had never smoked, their risk of cardiopulmonary disease increased 57 percent.[221]

In 1996, Natural Resource Defense Council (NRDC) researchers extrapolated the results of earlier epidemiological studies to estimate the extent of premature death due to air pollution from PM in 239 cities. The researchers concluded that 64,000 people die prematurely from heart and lung disease attributable to particle air pollution, with lives being shortened an average of one to two years in the most polluted cities.[222]

New evidence from the EPA suggests that these studies all *underestimate* the severity of PM inhalation.[223] In a survey of almost 1,000 epidemiological and toxicological studies assessing the health impacts of $PM_{2.5}$ or $PM_{2.5-10}$ in the United States and Canada, the EPA found a clear, causal, undisputable relationship that long-term exposure to fine particles is associated with mortality and morbidity. They confirmed the link between PM pollution and increased bronchitis in children and decreased lung function. They discovered new evidence linking long-term exposure to PM with atherosclerosis development and cystic fibrosis. They found a significant association between acute fine PM exposure and hospitalization for cardiovascular disease and respiratory disease in 204 counties. As a result of this new evidence,

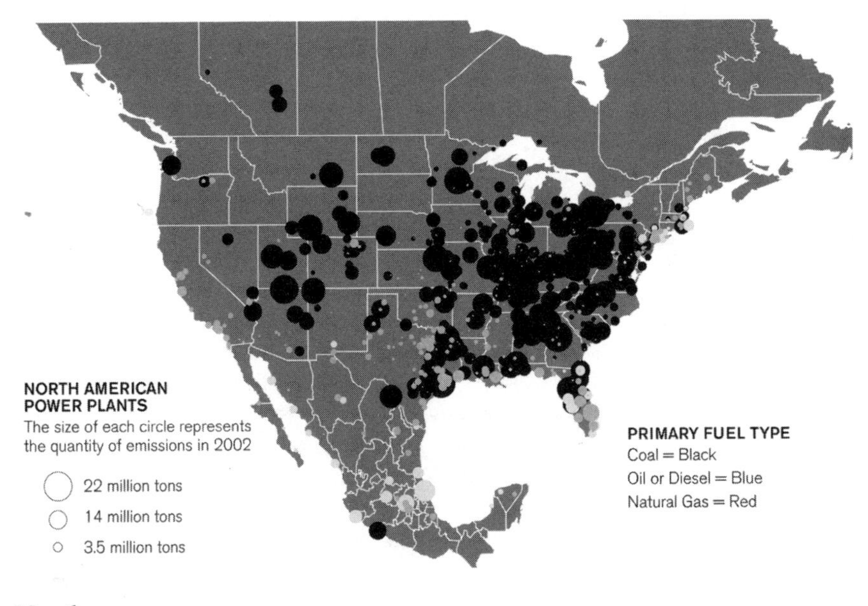

Map 6
National Distribution of Power Plant CO_2 Emissions

Coal, oil, diesel, and natural gas power plants continue to emit massive amounts of carbon dioxide every year. The largest sources of these emissions are located in the South and Northeast.

Source: Commission for Environmental Cooperation.

EPA revised their PM mortality and morbidity estimates upward to add 1,368 deaths to the original projections.

About 80 million Americans now live in areas where PM emissions are considered dangerous.[224] One meta-analysis of 80 epidemiologic studies on PM found that even minimum "healthy" rates were still bad for you. In other words, PM levels well below current U.S. National Ambient Air Quality Standards could be harmful to human health as "there is no evidence of a safe threshold level."[225]

Using the EPA's own math, PM pollution from power plants is responsible for 32,684 premature deaths each year, as well as nearly 22,000 hospital admissions, more than a half million asthma attacks (resulting in 26,000 hospital emergency room visits), 38,000 heart attacks, and 16,000 cases of chronic bronchitis.[226] These figures document that power plant pollution killed more Americans last year than prostate cancer, HIV/AIDS, or drunk driving. The deaths attributed to PM are not those people who are on their deathbeds. One survey of 800 scientific studies on PM published between 1997 and 2001 confirmed that premature death did not imply the advancing of life of those who were about to die anyway. In more than 90 percent of cases, premature deaths were advanced by months and years.[227]

Greenhouse Gas Emissions

Despite all of the political and media attention surrounding climate change and global warming, carbon-intensive fuels continue to dominate electricity generation in the United States. Fossil fuel power plants in the United States emitted 2.25 billion metric tons of CO_2 in 2003, more than 10 times the amount of CO_2 compared to the next-largest emitter, iron and steel production.[228] From 1990 to 2005, CO_2 emissions from power plants increased 21.7 percent. Coal and natural gas accounted together for 45 percent of total U.S. greenhouse gas emissions in 2005.[229] The electric utility sector emitted 40 percent of the country's CO_2, with coal responsible for 87 percent of emissions, natural gas 11 percent, and oil 2 percent. Of all U.S. industries, electricity generation is, by substantial margins, the single largest contributor of the pollutants responsible for global warming. Almost every state in the country was home to at least one power plant with significant CO_2 emissions (see Map 6), and even the most advanced ultrasupercritical pulverized coal plants still emit prodigious amounts of carbon (about 738 grams of CO_2/kWh).[230]

Strikingly, these figures likely underestimate the country's carbon emissions since U.S. territories are not included in national energy statistics. The numbers do not include coking coal, industrial coal, petroleum coke, natural gas, residual fuel oil, and distillate used for industrial processes. They also exclude emissions associated with the international transport of conventional fuels. David Hawkins and his colleagues from NRDC estimate that over their 60 year life spans, new coal generating facilities could collectively introduce about as much CO_2 as was released by all of the coal burned since the dawn of the Industrial Revolution.[231]

Natural gas facilities also contribute indirectly to global warming by emitting significant amounts of methane during the production process. Natural gas, when not separated from oil deposits, is often burned off at the well site or flared, releasing CO_2, carbon monoxide, NO_X, and SO_2. When not flared, operators usually vent unprocessed gas directly into the atmosphere. A staggering 5 percent of world natural gas production is lost to flaring and venting, making the gas production industry responsible for roughly 10 percent of global methane emissions.[232]

Moreover, the trend toward LNG poses similar environmental concerns. John Hritcko, the vice president for Strategy and Development for Shell U.S. Gas & Power Company, recently estimated that by 2025, the United States is expected to import the equivalent of today's global LNG trade just to satisfy its own domestic demand.[233] LNG is more energy intensive than normal natural gas since it requires the conversion of natural gas to a liquid, transportation across the ocean, and regasification. Greenpeace recently concluded that the energy-intensive fuel cycle for LNG exacts an energy penalty of 18 to 22 percent—contributing an additional 11 to 18 percent in CO_2 emissions, compared to the same amount of nonliquefied natural gas.[234]

Nuclear energy is not much of an improvement, despite recent claims by the Nuclear Energy Institute that nuclear power is "the Clean Air Energy." Reprocessing and enriching uranium requires a substantial amount of electricity, often generated from fossil fuel–fired power plants, and uranium milling, mining, leeching, plant construction, and decommissioning are all greenhouse gas intensive. In order to enrich natural uranium, it is converted to uranium hexafluoride, UF_6, and then diffused through permeable barriers. In 2002, the Paducah uranium enrichment plant in Kentucky emitted 193.7 metric tons of Freon, a potent greenhouse gas, through leaking pipes and other equipment.[235] Data collected from one uranium enrichment company revealed that it takes a 100 MW power plant running for 550 hours to produce the amount of enriched uranium needed to fuel a 1,000 MW reactor (of the most efficient design currently available) for just one year.[236] According to the *Washington Post,* two of the nation's most polluting coal plants in Ohio and Indiana produce electricity exclusively for the enrichment of uranium.[237]

Thus, when one takes into account the carbon-equivalent emissions associated with the entire nuclear life cycle, nuclear plants do contribute significantly to climate change. An assessment of 103 life cycle studies of greenhouse gas equivalent emissions for nuclear power plants found that the average CO_2 emissions over the typical lifetime of a plant are about 66 grams for every kWh, or the equivalent of some 183 million metric tons of CO_2 in 2005.[238] If the global nuclear industry were taxed at a rate of $24 per ton for the carbon equivalent emissions associated with its life cycle, the cost of nuclear power would increase by about $4.4 billion per year.[239]

Furthermore, the carbon equivalent emissions of the nuclear life cycle will only get worse, not better, since over time reprocessed fuel is depleted, necessitating a shift to fresh ore, and reactors must utilize lower quality ores as higher quality ones are depleted. The Oxford Research Group projects that because of this inevitable eventual shift to lower quality uranium ore, if the percentage of world nuclear capacity remains what it is today, by 2050 nuclear power would generate as much carbon dioxide per kWh as comparable natural gas–fired power stations.[240]

The impacts of fossil fuels and nuclear power on climate change will likely be severe. In its most recent report, the Intergovernmental Panel on Climate Change (IPCC)—a forum made up of thousands of the world's top climate scientists—concluded that continued emissions of greenhouse gases will contribute directly to the following:

- Changes in the distribution, availability, and precipitation of water, resulting in severe water shortages for millions of people;
- Destruction of ecosystems, especially the bleaching of coral reefs and widespread deaths of all types of migratory species;
- A significant loss of agricultural and fishery productivity;
- Damage from floods and severe storms, especially among coastal areas;
- Deaths arising from changes in disease vectors and an increase in the number of heat waves, floods, and droughts.[241]

The IPCC's analysis was supported by U.S. scientists Klaus S. Lackner and Jeffrey D. Sachs[242] and the United Nations,[243] which also found that greenhouse gas emissions would induce the following:

- Major changes in wind patterns, rainfall, and ocean currents;
- Alteration of ocean chemistry as the ocean surface becomes more acidic, stunting coral growth;
- Habitat destruction and species extinction, especially for species with limited range or mobility;
- Enhanced disease transmission, especially among diseases regulated by temperature and precipitation;
- A shift in the growing seasons for crops, specifically among marginal soils that have low buffering potential;
- Increased frequency of natural hazards, hurricanes, flooding, and droughts;
- Rising ocean levels due to the thermal expansion and melting of land ice;
- The ever-present risk of abrupt and catastrophic changes such as the collapse of the North Atlantic Ocean's thermohaline circulation or collapse of Greenland's ice sheet.

Plausible scenarios include droughts and massive crop failures, millions of people being displaced from coastlines due to rising sea levels, substantial depletion of groundwater supplies, and disastrous wildfires and storms.

Worryingly, a research team headed by the NASA/Goddard Institute for Space Studies and the Columbia Center for Climate Systems Research recently confirmed the likelihood of these types of impacts. After analyzing 30,000 sets of data relating to biological and physical changes affecting the planet over a span of three decades, the researchers identified that many of the adverse impacts from climate change—

earlier flowering of plants, premature breeding of birds, cannibalism and declining populations of polar bears in North America, changing precipitation patterns in South America, unstable glacial melting in the Alps, long-term changes in the productivity of fisheries and forests in Europe and Africa, and changing species migration patterns in Asia, Australia, and Antarctica—are already occurring.[244]

Policymakers should not underestimate the impacts of global warming on the U.S. economy. The Pew Center on Global Climate Change estimates that, in the Southeast and southern Great Plains, the financial costs of climate change could reach as high as $138 billion by 2100. Pew researchers warn that "waiting until the future" to address global climate change might bankrupt the U.S. economy.[245] Alan Carlin, a senior economist at the EPA, warns that the cost of removing CO_2 after it has been emitted may be as high as $3,500 per ton.[246] If his estimate is correct, then the cost of removing all of the CO_2 emitted from power plants *just in 2007* would exceed $7.7 trillion.

Take a breath, and consider just how much money that is. For $7 trillion, the U.S. government could mail everyone in the country a check for $21,000. And if the average American spent $1,000 of that $7 trillion every second, it would take her almost three decades to spend it all. If she could shop only during business hours, it would take more her more than 120 years before the money ran out.[247]

Indeed, one study published in *Energy Policy* already concluded that *any* use of fossil fuels will be prohibitively expensive by 2050. The authors assessed the global rate of fossil fuel depletion and multiplied the amount by a corresponding carbon factor. Looking at oil, gas, and coal for 62 countries, they found that a fossil fuel–based strategy became completely economically unsustainable, even taking into consideration only greenhouse gas emissions.[248]

Land

In addition to the environmental damage caused by fossil fuel combustion, the production of fossil fuels and uranium—the drilling, mining, processing, and transportation—produces a substantial amount of pollution and toxic waste. In the United States, there are more than 150 refineries, 1,298 coal mines, 4,000 offshore platforms, 410 underground gas storage fields, 125 nuclear waste storage facilities, 160,000 miles of oil pipelines, and 1.4 million miles of natural gas pipelines. Each can degrade its surrounding natural environment and negatively influence the health and safety of Americans.

Each stage of the oil and gas fuel chain—exploration, onshore and offshore drilling, refining, and transportation—poses serious and unavoidable risks to ecosystems and human health. Exploration necessitates heavy equipment and can be quite invasive, as it involves "discovering" oil and gas deposits found in sedimentary rock through various seismic techniques such as controlled underground explosions, special air guns, and exploratory drilling. Less intrusive techniques, such as remote sensing relying on airplanes and satellites, have a limited success rate and are not widely used.[249]

Construction of access roads, drilling platforms, and their associated infrastructure frequently induces environmental impacts beyond the immediate effects of land clearing: they open up remote regions to loggers and wildlife poachers. About 1,000 to 6,000 acres of land are deforested for every one kilometer of new oil and gas roads built through forested areas around the world, and in Ecuador, an estimated 2.5 million acres of tropical forest were colonized due to the construction of just 500 kilometers of roads for oil production.[250]

The production and extraction of oil and gas, which are themselves toxic (both contain significant quantities of hydrogen sulfide, a substance that is potentially fatal and extremely corrosive to equipment such as drills and pipelines), is even more hazardous. Drilling for oil and gas involves bringing large quantities of rock fragments, called "cuttings," to the surface, and these cuttings are coated with drilling fluids, called "drilling muds," which operators use to lubricate drill bits and stabilize pressure within oil and gas wells. The quantity of toxic cuttings and mud released for each facility is gargantuan, ranging between 60,000 to 300,000 gallons per day. In addition to cuttings and drilling muds, vast quantities of water contaminated with suspended and dissolved solids are also brought to the surface, creating what geologists refer to as "produced water." The average offshore oil and gas platform in the United States releases about 400,000 gallons of produced water back into the ocean or sea every day. Produced water contains lead, zinc, mercury, benzene, and toluene, making it highly toxic and requiring operators to often treat it with chemicals, increasing its salinity and making it fatal to many types of plants, before releasing it into the environment. The ratio of waste to extracted oil is staggering: every one gallon of oil brought to the surface yields eight gallons of contaminated water, cuttings, and drilling muds.[251] The next stage, refining, involves boiling, vaporizing, and treating extracted crude oil and gas with solvents to improve their quality. The average U.S. refinery processes 3.8 million gallons of oil per day, and about 11,000 gallons of its product (0.3 percent of production) escapes directly into the local environment every day of operation, where it can contaminate land and pollute water.[252] Moreover, refineries are also the second largest source of noxious air emissions in the country, after electricity generation.

The U.S. Geological Survey (USGS) estimated that there are more than 2 million oil and natural gas wells in the continental United States. But the most intense areas of oil and gas production are off the shores of the Gulf of Mexico and along the northern coast of Alaska. Offshore oil and natural gas exploration and production in the Gulf of Mexico exposes aquatic and marine wildlife to chronic, low-level releases of many toxic chemicals through the discharge and seafloor accumulation of drilling muds and cuttings, as well as the continual release of hydrocarbons around production platforms.[253] Drilling operations there generate massive amounts of polluted water (an average of 180,000 gallons per well every year), releasing toxic metals including mercury, lead, and cadmium into the local environment.[254] One study found that mercury levels in the mud and sediments beneath the oil platforms in the Gulf of Mexico were 12 times greater than acceptable levels under federal EPA standards.[255] This exposure to these chronic environmental perturbations continues to threaten marine biodiversity, and human health, over wide areas of the Gulf.

The NRDC also noted that the onshore infrastructure required to sustain oil and natural gas processing in the United States has destroyed more coastal wetlands and salt marsh than can be found in the total area stretching from New Jersey through Maine. Similarly, the U.S. Minerals Management Service has documented 70 oil and natural gas spills between 1980 and 1999 from just the *production* of oil and natural gas. Oil and natural gas spills also accounted for more than 3 million gallons of fuel released into the Gulf of Mexico alone.[256]

In Alaska, independent studies undertaken by three groups of ecologists and the National Academies of Science have concluded that oil and gas production on the North Slope in Alaska disrupts tundra surfaces and alters the hydrological processes of wetland ecosystems responsible for the spawning and development of wildlife. North Slope oil production undermines nutrient availability for tundra plants that is essential for food and habitat for wildlife, and prematurely thaws the ice and permafrost, releasing large amounts of methane that further accelerates global warming.[257] For especially remote areas, such as the Arctic National Wildlife Refuge, the operation of oil and gas refineries would release discharged solids, drilling waste, and dirty diesel fuel into the ecosystem's food chain. The fugitive emissions and flares from these facilities tend to create acidification downwind and induce localized climate change. Furthermore, the roads, raised pipelines, and other transportation corridors needed to support refinery operations disrupt the migration routes of large animals, accelerate thermokarst dissolution, and fundamentally alter predation patterns, water purity, and soil chemistry. The arctic environment is especially sensitive to such changes because the ecosystem operates on a simplified food chain with slower rates of photosynthesis and decomposition.

During the summer months, domestic natural gas production and imported natural gas far exceed demand, so excess supply is placed in large underground storage facilities.[258] Around 400 natural gas storage facilities have been constructed in the United States, storing an estimated 8.25 trillion cubic feet of natural gas. The three principal types of underground storage sites used are depleted oil and gas reservoirs, aquifers, and salt cavern formations.[259] The DOE and energy companies have spent an estimated $4 billion to create artificial caverns and salt domes below the surface to store oil.

Oil and natural gas storage facilities, in addition to significantly adding to the cost of natural gas and oil infrastructure, are also inefficient and susceptible to serious accidents that can pollute the air and water of local communities. Since natural gas is typically a combination of methane, ethane, propane, and a mix of other heavy hydrocarbons, its storage and transportation necessitates infrastructure-intense facilities that are sometimes prone to failure and leakage. The EIA quietly admitted in 2004 that LNG storage operators must "boil-off" significant quantities of natural gas every day to maintain adequate pressure—meaning that approximately 0.25 to 0.50 percent of their inventory is lost *every day* due to vaporization.[260] One report from the LBNL noted that leaks in natural gas storage facilities can occur due to improper well design, construction, maintenance, and operation. The report cautioned that leakage from natural gas storage structures can be especially hazardous when they cause natural gas to migrate into drinking water aquifers or escape to the surface, creating a "significant safety risk." Leaked natural gas can significantly endanger life and property, water resources, vegetation, and crops.[261]

Indeed, In January 2001, hundreds of explosions rocked the Yaggy field—a natural gas salt formation storage site in Hutchinson, Kansas—when natural gas escaped from one of the storage wells and erupted into a 7 mile wall of fire (burning an estimated 143 million cubic feet of natural gas). Cleanup for the disaster necessitated the construction of 57 new vent wells extending a distance of more than 9 miles.[262]

Overpressurization (needed to enlarge gas bubbles and obtain higher delivery rates) is another main cause of leakage, as many underground natural gas storage projects tend to be operated at pressures exceeding their original designs. Such leaks can become excessively costly: The Gulf South Pipeline Company's Magnolia facility, a $234 million salt-cavern facility, opened in 2003 only to permanently close a few months later after a well collapsed.[263]

Pipelines are prone to catastrophic failure as well. Faulty joints connecting pipeline components, malfunctioning valves, operator error, and corrosion induce frequent leaks and ruptures. Looking back from 1907 to 2007, natural gas pipelines are the type of energy infrastructure most frequent to fail, accounting for 33 percent of all major energy accidents worldwide.[264] Internationally, between 20 million and 430 million gallons of oil were spilled in reported incidents *each year* between 1978 and 1997, with the number of significant spills ranging from 136 to 382 annually over this period.[265] The U.S. Department of Transportation has noted that oil and gas pipelines fail so often here that they expect 2,241 major accidents and an additional 16,000 spills every ten years.[266] Company records for the single Trans-Alaska Pipeline in the Arctic show 642 spills totaling 1.2 million gallons of spilled petroleum since operation began in 1977.

The cumulative impacts of oil and gas development in Azerbaijan, where oil and gas exploration, production, transportation, and refining have been going on since the 1860s (longer than the United States), may explain how the country earned the prestigious 2007 title of "worst polluted place in the world." The suburbs of Baku, Azerbaijan, are known for their extremely high levels of birth defects, cancer, and premature mortality resulting directly from the oil and gas industry.[267]

Coal extraction, processing, and transportation have an even more direct affect on water and land resources. Of the more than one billion tons of coal mined in the United States annually, roughly 70 percent comes from surface mines. Mountaintop removal (a newer technique for mining coal that uses heavy explosives to blast away the tops of mountains) has destroyed streams, blighted landscapes, and diminished the water quality of rural communities. Failing coal slurry impoundments, acid mine drainage, aquifer disruption, saline pollution from coal-bed methane recovery, and occupational safety and health hazards (including mine-related deaths) are among the other impacts of continued reliance on coal-fired electricity production.[268]

Photograph 1
Mountaintop Removal Coal Mining near Kayford Mountain, West Virginia

A massive dragline, dwarfed by the huge scale of the operation, at work in West Virginia. *Source:* Ohio Valley Environmental Coalition.

Coal is transported an average distance of 500 miles before it is combusted in power plants.[269] Accidents can thus occur all along coal's long haul. Roadway fatalities are twice as high on coal-hauling roads as on normal roads. Between 2000 and 2004, Kentucky documented 704 accidents involving coal trucks, resulting in the deaths of 53 people and the injury of more than 530.[270]

Coal hauling wreaks infrastructural damage as well. Dallas Burtraw and his colleagues at Resources for the Future found that the cost of transporting coal on roads was almost as costly to the pavement as air pollution is to human health. They estimated $715,113 in extra maintenance needed for every mile of road frequently traversed by fully loaded coal-hauling trucks.[271]

Nuclear facilities do not avoid substantial land degradation either. Because nothing is burned or oxidized during the fission process, nuclear plants convert almost all their fuel to waste with little reduction in mass during chain reactions. Nuclear power plants have at least five waste streams. They create spent nuclear fuel at the reactor site; produce tailings and uranium mines and mills; routinely release small amounts of radioactive isotopes during operation; catastrophically release large quantities of pollution during accidents; and create plutonium waste.[272]

The most radioactive of these wastes, spent fuel, must be stored at individual reactor sites in large pools of water for at least ten years, after which they are relocated to large concrete casks that provide air-cooling, shielding, and physical protection. While there are many different cask types, those in the United States typically hold 20 to 24 pressurized water reactor fuel assemblies, sealed in a helium atmosphere inside the cask to prevent corrosion. Decay heat is transferred by helium from the fuel to cooling fins on the outside of the storage cask.

Why not reprocess or reuse nuclear waste, readers may inquire? When they initially designed American reactors, nuclear engineers expected that the plutonium from spent fuel would be recycled at reprocessing centers or removed and reused in fast-neutron reactors. The federal government did begin commercial reprocessing of nuclear waste in 1966 at a facility in West Valley, New York, but the operation ended in disaster. The plant was repeatedly criticized for lax security measures and for exposing its employees to the highest doses of radiation ever reported to the Occupational Safety and Health Administration (established in 1970). The cost of reprocessing was originally estimated to be $15 million but later reported to be $600 million, the probability of a major earthquake in the area was deemed too great a risk to justify continued operation, and the plant was very inefficient.[273] After reprocessing only 640 tons of spent fuel but accumulating more than 600,000 gallons of high-level waste, the facility was closed in 1972. It was not until 2002 that the West Valley facility was stabilized to the point that it could be safely decommissioned, and remaining cleanup was estimated to cost an additional $5 billion and take another 40 years.[274] The experience with West Valley, coupled with an assortment of accidents at other enrichment facilities mentioned above in the section on "nuclear security," have served as powerful impediments to further reprocessing in the United States.

The second option, reuse at fast-neutron reactors, also called "fast breeders" because they were designed to produce more plutonium than they consumed, was rejected on national security grounds. Because of the link between plutonium and nuclear weapons, the potential application of fast breeders leads to concern that nuclear power expansion would usher in an era of uncontrolled weapons proliferation. The United States signed the Nuclear Non-Proliferation Treaty in 1968 to partially address the issue, and President Jimmy Carter banned civilian reprocessing of nuclear fuel in the United States in 1977 over the same concern.[275]

The responsibility for permanently storing nuclear waste falls exclusively to the federal government, but it is clearly failing in its role. The Nuclear Waste Policy Act (NWPA) of 1982 obligated utilities to pay a fixed annual fee of a tenth of a cent for every kWh from nuclear generation into the Nuclear Waste Fund to cover the costs of waste disposal. In return, the federal government and DOE were required to take and dispose of spent nuclear fuel in a permanent geologic repository beginning in 1998 (the fund has since collected about $26 billion, or $750 million per year). Pursuant to the NWPA, nine states were initially identified as potential sites for long-term repositories, with the idea that one in the east and one in the west would be chosen. Regulators quickly abandoned all but one of these sites for obvious political reasons, leaving Yucca Mountain in Nevada the only option.

Unfortunately, Yucca was also deemed the least optimal of the nine sites because the National Academies of Science reported it had the greatest risk of releasing radiation.[276] Still, because it was the only alternative, the federal government began funding a permanent storage facility at Yucca Mountain in 1985. In 2008, the project had already cost $8 billion and was still 20 years behind schedule, underfunded, and, according to Nevada Senator Harry Reid, who opposes it, "a dying beast."[277] Even if miraculously completed, Yucca would have only enough initial space for 63,000 tons of spent fuel, yet at least 105,000 tons will need storage by 2039.[278]

Worried that the government would not meet its responsibility to build a permanent storage facility, several electric utilities operating commercial nuclear reactors went before the D.C. Circuit court in 1996 to seek a ruling on the extent of the government's obligations under the NWPA. Under that case, *Indiana Michigan Power Co. v. Department of Energy*, the court ruled that the government had an unconditional obligation to accept waste by January 1, 1998. Without seeking a rehearing, the government nonetheless informed utilities that it would not accept the deadline.[279] Facing growing quantities of nuclear waste and limited storage space, utilities responded on January 31, 1997, and petitioned the U.S. Court of Appeals for a writ of mandamus to require the federal government to begin accepting highly radioactive spent fuel from the utilities by the following January. The government refused, and by 2004, about 20 utilities had suits pending against the DOE in Federal Claims Court for damages totaling $8.5 billion, with total liability projected to reach as much as $40 to $80 billion.[280] By February 2008, the number of lawsuits pending against the DOE relating to nuclear storage had jumped to 60.[281]

The DOE has relied upon on-site storage as a stopgap remedy until Yucca Mountain is finalized or the United States finds a long-term solution to nuclear waste. As a result, about 56,000 tons of spent nuclear fuel are scattered in dry casks and storage pools in more than 120 different locations in the United States. Twenty-six reactors were projected to be out of pool storage space in 1998 and 80 will reach maximum pool capacity by 2010. One ton of highly radioactive waste is generated for every 4 pounds of usable uranium, and each reactor consumes an average 32,000 fuel rods over the course of its lifetime.[282] The costs of expanding on-site storage are therefore enormous, with each dry cask costing about $35,000 to $65,000 per ton. When Congress requested in 2007 that the DOE study the potential for making temporary storage for high-level nuclear waste more permanent, ostensibly to demonstrate that the nation was capable of "moving forward" with some element of a nuclear waste policy, the DOE uncharacteristically balked. Stating that interim storage was "clearly not the solution," the DOE argued that the NWPA legally prevented them from taking spent fuel until after Yucca Mountain was completed, and refused to accept nuclear waste.[283]

Regardless of whether the nuclear waste problem is resolved in favor of on-site or centralized storage, keep in mind that the costs of nuclear waste are not borne solely

Map 7
Nuclear Spent Fuel Storage Installations in the United States

In 2006, the Nuclear Regulatory Commission reported that 42 unclassified storage sites existed for nuclear waste spread across 30 states.

Source: U.S. Nuclear Regulatory Commission.

by our generation, or even by generations over the next millennium. Typically, a single nuclear plant will produce 30 tons of high-level waste each year, and this waste can be radioactive for as long as 250,000 years. Taking just one-tenth of that, 25,000 years, and assuming the cost of storing 1 ton of nuclear waste was just $35,000 per year (the lowest end of the estimates), each plant in the United States has an additional price tag of $875 million.

The degradation of land connected with the nuclear fuel cycle is not limited to waste production and storage. Both nuclear reactors and uranium enrichment facilities must also be tediously decommissioned, a process that is freakishly expensive, time-consuming, dangerous for workers, and hazardous to the natural environment.

After a cooling off period that may last as long as 50 to 100 years, reactors must be meticulously dismantled and cut into small pieces to be packed in containers for final disposal. Nuclear plants often have an operating lifetime of 40 years, but the industry reports that decommissioning takes an average of 60 years. While it will vary along with technique and reactor type, the total energy required for decommissioning can be as much as 50 percent *more* than the energy needed for original construction. To give an idea for how capital intensive it is and how long it takes, there are currently 13 nuclear power plant units that have permanently shut down and are in some phase of the decommissioning process, but not one of them is complete.[284]

Taking just a few examples, the relatively small Humboldt Bay nuclear facility was shut down in July 1976 but will not be completely decommissioned until 2012 or 2013 (making it the first ever decommissioned reactor in the country). Zion Units 1 and 2 were permanently shut down in 1998, but the plant will not even be able to begin decommissioning until 2013. Peach Bottom Unit 1 was shut down in October 1974 but will not begin decommissioning until 2034. Decommissioning at nuclear accident sites is even more expensive and time consuming. The decommissioning of Three Mile Island Unit 2, which shut down permanently after an accident in 1979, will not even start until 2014. Fuel rods at Chernobyl, the site of the famous nuclear accident in 1986, are still being removed and operators caution it could take until 2138 before the power plant is completely decommissioned.

Decommissioning of uranium enrichment facilities—large complexes of buildings with thousands of pieces of equipment, enrichment cascades, piping, and electrical wiring—requires a precarious six stage and very labor-intensive process: careful characterization of every square centimeter of each building, disassembly, removal of uranium deposits from process equipment, decontamination, melt refining and recycling of metals, and treatment of wastes; and all in an extremely radioactive environment! The process generates its own low-level radioactive and hazardous wastes and can further contaminate soil and groundwater. The Capenhurst gaseous diffusion plant in the United Kingdom, decommissioned by the British Nuclear Fuels Corporation in 1994, required the entire facility to be treated with gaseous chlorine trifluoride (ClF_3) to remove deposits of uranium on equipment before every piece of the plant was removed, cut up into pieces, and decontaminated using a series of aqueous, chemical baths.

Decommissioning of the three enrichment facilities in the United States—all of the gaseous diffusion type, with one retired facility located near Oak Ridge, Tennessee; one operating facility near Paducah, Kentucky; and another retired one near Portsmouth, Ohio—will require the same ClF_3 treatment, since deposits of highly enriched uranium have become littered throughout the process buildings. This radioactive debris is accompanied by significant amounts of asbestos and polychlorinated biphenyls, which is probably why the National Research Council estimated that decommissioning the three facilities will cost $18.7 billion to $62 billion, with an additional $2 billion to $6 billion to cover the disposal of a large inventory of depleted uranium hexafluoride (depleted UF_6), which must be converted to uranium oxide (U_3O_8).[285]

The U.S. GAO recently surveyed how well the decommissioning process was going at these enrichment facilities, and found that the earliest it will be completed for all three plants is 2040. By then, the GAO warned that the cost of decommissioning, funded by taxpayers, will have *exceeded* the plants' revenues by at least $4 billion to $6.4 billion. As of 2004, these plants, heavily contaminated with radioactive particles and large caches of spent hexafluoride fuel, still required extensive cleanup of 30 million square feet of space, miles of interconnecting pipes, and thousands of acres of land.[286]

This could be why physicist Alvin M. Weinberg compared nuclear power to a "Faustian bargain," since it creates an unbreakable commitment where society receives a perpetually risky supply of energy in exchange for yielding political power to a small cadre of technocrats and national security agencies.[287] Unlike Faust, who was ultimately able to renege on the bargain, society has bound itself in perpetuity to the remarkable belief that it can devise social institutions that are stable for periods equivalent to geologic ages, exhibiting supreme confidence in the ability of human organizations to outlast radioactive waste.[288] Nuclear waste will remain dangerously radioactive for hundreds of thousands of years—longer than our civilization has practiced Catholicism or cultivated agriculture. The half-life of uranium-238, one of the largest components of spent fuel, is about the same as the age of the Earth at 4.5 billion years.

As the next chapter shows, however, we need not commit ourselves to a power system characterized by thousands of premature deaths, millions of tons of highly toxic and radioactive waste, billions of dollars of environmental damage, and the copious but inescapable amount of human suffering associated with it.

CHAPTER 2

The Big Four Clean Solutions

Supreme Court Justice Oliver Wendell Holmes once remarked, "a page of history is worth a volume of logic." The history of the electricity sector tells us two things: the challenges facing the current system are vast, unappealing, and growing; and society already has all of the technologies it needs to tackle them.

This chapter shows how four clean power technologies can do everything fossil-fueled and nuclear plants currently cannot. It argues that energy efficiency and demand-side management (DSM), renewables, distributed generation (DG), and combined heat and power (CHP) can best respond to the "demand," "infrastructure," "fuel supply," and "environment" challenges outlined in Chapter 1.

The best, cleanest power source is energy efficiency, a resource historically proven, through decades of experience and thousands of programs, to be the most cost-effective way of responding to increases in demand. Energy-efficiency measures have shorter lead times, need lower capital requirements, and offer consumers and utilities quick returns on investment compared with building conventional power plants, new oil and gas fields, and coal mines. Energy-efficiency technologies can reduce utility planning and risk, improve load factors, increase cash flows, reduce capital expenditure, and minimize financial risk associated with excess generating and transmission capacity.[1]

Wind, hydroelectric, solar, biomass, and geothermal plants can displace the need for pollution-belching base load and peaking power facilities, reducing dependence on foreign supplies of fuel and insulating society from price volatility. They offer improved environmental performance and reduce the burdens placed on air, water, and land from electricity generation, enhancing human health and our quality of life.

DG and CHP systems strengthen the reliability of the power grid and decrease transmission congestion. They reduce the capital needed for energy production and displace power plant and T&D construction. They help conserve the amount of land needed for power generation and transmission facilities, and improve overall system efficiency.

A Primer on Clean Power Solutions

While all four clean power sources can be considered a class of the same species, each of them has very different stripes.

Generally, the term "energy efficiency" refers to the reduction of electricity consumption as the result of improved performance or increased deployment of more efficient equipment. It is "technologically providing more desired service per unit of delivered energy consumed." [2] Energy efficiency enables the most economically efficient use of energy to perform a certain task (such as light, torque, or heat) by minimizing unit of resources per unit of output. Energy efficiency can include substituting resource inputs or fuels (CHP for conventional coal or insulation instead of nothing), changing habits and preferences (lowering thermostats or better maintaining industrial boilers), or altering the mix of goods and services to demand less energy (opening the windows at night rather than using air conditioning or walking to work).[3] As one influential report on efficiency put it, "energy-efficiency opportunities are typically physical, long-lasting changes to buildings and equipment that result in decreased energy use while maintaining constant levels of energy service." [4] Thus, energy efficiency is "doing more with less through smarter technologies" and not what people sometimes negatively consider "doing less, worse, or without." [5]

When implemented by utilities, energy efficiency is usually called "demand-side management," or DSM. DSM technically refers to programs that allow utilities to better match their demand with their generating capacity. By shifting electricity loads for utilities, system reliability can be enhanced and new power plant construction avoided or delayed. Current programs aim at limiting peak electricity loads, shifting peak loads to off-peak hours, or encouraging consumers to change demand in response to the cost of providing power. Mechanisms include price-based programs giving customers varying rates that reflect the value and cost of electricity during different time periods, and incentive-based programs that pay participating customers to reduce their loads at times requested by program sponsors.[6]

A close and sometimes synonymous variant is "demand response" or "load management." These techniques refer to curtailment or other immediate steps that are aimed at reducing only peak electricity load. Significant consumer benefits and less opportunity for market manipulation by electricity providers accrue from lower peak electricity prices. Load management programs sometimes involve direct control over residential air conditioners and water heaters, with customers permitting the utility to use a timer or radio controlled switch to shut off equipment during peak periods in exchange for incentives.[7]

Renewable generators utilize sunlight, wind, falling water, biomass, waste, and geothermal heat to produce electricity. Renewable energy resources use "fuels" greatly abundant in the United States, and thus are less susceptible to supply chain interruptions and shortages. Vikram Budhraja, former president of Edison Technology Solutions, notes that "the beauty of renewable energy" is that "it has no fuel

Table 4
The Four Types of Clean Power Solutions

Clean power solutions can be classified according to four categories: energy efficiency and demand-side management technologies and programs, renewable power generators, distributed generation technologies, and combined heat and power systems.

Energy Efficiency

	Definition	Approach	Technologies
Energy Efficiency	Substitute fuels and technologies, change habits and preferences, and alter services to improve the efficient use of energy	Information distribution, loans, leasing, rebates, and performance contracting	Energy controls, higher performance windows, more efficient lighting, improved heating and air conditioning, energy-efficient design and certification schemes
Demand-Side Management (DSM)	More accurately match electricity demand with generation	Minimize unneeded capacity	Time-of-use rates, net metering, real time pricing, improved forecasting
Load Management	Reduce electricity consumption during peak times use	Enhance utility control of nonutility energy-using devices	Timers and radio controlled switches on appliances

Renewable Power Generators

Source	Description	Fuel	Size of Individual Units
Onshore Wind	Wind turbines capture the kinetic energy of the air and convert it into electricity via a turbine and generator	Wind	1.5 kW to 2.5MW
Offshore Wind	Offshore wind turbines operate in the same manner as onshore systems but are moored or stabilized to the ocean floor	Wind	750 kW to 5 MW

Solar photovoltaic (PV) systems	Solar photovoltaic cells convert sunlight into electrical energy through the use of semiconductor wafers	Sunlight	1 W to 100 MW
Solar Thermal	Solar thermal systems use mirrors and other reflective surfaces to concentrate solar radiation, utilizing the resulting high temperatures to produce steam that directly powers a turbine. The three most common generation technologies are parabolic troughs, power towers, and dish-engine systems.	Sunlight	5 kW to 320 MW
Geothermal (conventional)	An electrical-grade geothermal system is one that can generate electricity by means of driving a turbine with geothermal fluids heated by the Earth's crust	Hydrothermal fluids heated by the Earth's crust	25 MW to 1,400 MW
Geothermal (advanced)	Deep geothermal generators utilize engineered reservoirs that have been created to extract heat from water while it comes into contact with hot rock and returns to the surface through production wells	Hydrothermal fluids heated by the Earth's crust	10 MW to 1,500 MW

Biomass (combustion)	Biomass generators combust biological material to produce electricity, sometimes gasifying it prior to combustion to increase efficiency	Agricultural residues, wood chips, forest wastes, energy crops	20 to 50 MW
Biomass (landfill gas)	These biomass plants generate electricity from landfill gas and anaerobic digestion	Municipal and industrial wastes and trash	30 kW to 10.5 MW
Hydroelectric	Hydroelectric dams impede the flow of water and regulate its flow to generate electricity	Water	200 kW to 6,809 MW

Distributed Generation and Combined Heat and Power Systems

	Steam Turbines	Reciprocating Engines	Stirling Engines	Natural Gas Turbines	Microturbines	Fuel Cells
Primary Fuel Options	Variety of Fuels	Gasoline, Natural Gas, Kerosene, Propane, Alcohol, Fuel Oil, Digester Gas	Any type of Combustible Fuel	Natural Gas, Distillate Oil, Coal-Derived Gas, Naphtha, Crude Oils	Natural Gas, Digester Gas	Natural Gas, Hydrogen
Typical Cycle	Simple Cycle	Simple Cycle	Cheng Cycle	Combined Cycle	Closed Cycle	N/A
Thermal Efficiency	27.6% to 33.4%	30.1% to 37.2%	40.3% to 98.1%	30% to 43%	25% to 30%	17% to 22%
Capacity	10 kW to 800 MW	3 kW to 10 MW	50 W to 25 kW	2.5 kW to 350 MW	20 kW to 500 kW	1 kW to 100 MW

Source: Benjamin K. Sovacool, "Distributed Generation (DG) and the American Electric Utility System: What Is Stopping It?" *Journal of Energy Resources Technology* 130, no. 1 (March 2008): 16–19.

input," and can instead rely on widely available resources mostly free for the taking.[8]

Wind turbines convert the flow of air into electricity and are most competitive in areas with stronger and more constant winds, such as locations offshore or in regions of high altitude.

Solar photovoltaic (PV) panels, also called "flat plate collectors," convert sunlight into electrical energy through the use of semiconductor wafers, and are often used in arrays and integrated into buildings.

Solar thermal systems, sometimes called "concentrated" or "concentrating" solar power, use mirrors and other reflective surfaces to concentrate solar radiation, utilizing the resulting high temperatures to produce steam to then power a turbine.

An electrical-grade geothermal system is one that can generate electricity by means of driving a turbine with geothermal fluids heated by the Earth's crust.

Biomass generators combust agricultural residues, wood chips, forest wastes, energy crops, municipal and industrial wastes, and trash to produce electricity. Biomass generation also includes advanced combustion techniques such as biomass gasification (in which the biomaterial is gasified prior to its combustion to increase efficiency) and co-firing (in which biomass burns with another fuel, such as coal or natural gas, to increase its density), as well as the electrical generation from landfill gas and anaerobic digestion.

Two types of hydroelectric facilities exist: large-scale facilities that consist of a dam or reservoir impeding water and regulating its flow, and run-of-the-river plants that create a small impoundment to store a day's supply of water.[9] Smaller hydroelectric systems, also referred to as "run-of-the-mill," "micro-hydro," and "run-of-the-river" hydropower, consist of a water conveyance channel or pressured pipeline to deliver water to a turbine or waterwheel that powers a generator, which in turn transforms the energy of flowing water into electricity. Because they operate on a much smaller scale, use smaller turbines, and require much less water, run-of-the-mill hydroplants escape many of the challenges raised by their larger counterparts.

The category of electricity known as "ocean power" includes shoreline, near-shore, and offshore "wave extraction" technologies and ocean thermal energy conversion (OTEC) systems. Because they are a much newer technology than other renewables, comprehensive cost analyses and product reviews are limited. Such plants do not currently exist in the commercial sector and are not discussed much in this chapter.

The third and fourth clean power sources, DG and CHP systems, utilize fossil fuels, but still have manifold benefits compared to conventional technologies. DG entails producing power on-site, in small increments, close to the end user. The technique emphasizes the deployment of small-scale generating facilities, often having installed capacities of a few to a hundred kW. DG technologies tend to be owned not just by utilities or traditional power providers but also by residential owners, commercial enterprises, and industrial firms. DG units can provide power for almost any objective and from any location. They can utilize many modes of operation, running as base-load generators, peaking units, or emergency backup facilities. They can be placed almost anywhere on the transmission system, from the low-voltage end of the distribution system to placement at or near load centers.[10]

The six most prevalent forms of DG technologies are steam turbines, reciprocating engines, stirling engines, natural gas turbines, microturbines, and fuel cells. Steam turbines, the oldest and most widely used technology for electricity generation, combust fuel to heat water that produces steam, which turns turbine blades on a shaft to work a generator. Internal combustion engines (also called endothermic engines and reciprocating engines), adapted from automobiles, utilize reciprocating or piston-driven combustion to drive an electric generator while the heat from its exhaust, cooling water, and oil generates steam in a boiler. Stirling engines shuttle nitrogen or helium back and forth between hot and cold parts of the engine, using magnets within a linear alternator to produce power. Natural gas turbines have been adapted from aircraft designs emphasizing thrust and low weight to power designs emphasizing efficiency and improved torque to generate electricity. Much like natural gas turbines, microturbines use small high-speed turbo alternators based on truck, miniature helicopter, and airplane technology. Fuel cells produce electricity electrochemically by converting energy from a chemical reaction between a fuel, such as liquid hydrogen, and an oxidant, such as liquid oxygen.

Sometimes considered a form of DG, but often given their own classification, CHP systems produce thermal energy and electricity from a single fuel source, thereby recycling normally wasted heat through cogeneration (where heat and electricity are both used) and trigeneration (where heat, electricity, and cooling are produced). Recycled thermal energy is then used directly for air preheating, industrial processes that require large amounts of heat, space cooling, and refrigeration. As a result, CHP technologies consistently have double the efficiency of traditional utility power plants that vent heat into the environment as a waste product. Most CHP systems in the United States possess a generating capacity of about 50 MW and are deployed in energy-intense industries such as paper processing, chemical manufacturing, and petroleum refining.[11] Utilities and local municipalities use large CHP units for district heating and cooling in the downtown areas of cities and large sports complexes. Some commercial operators use CHP to maximize the energy efficiency of complexes such as hospitals, supermarkets, and universities.[12]

No. 1: The Demand Challenge

Chapter 1 noted that demand for power is expected to double by 2040, and that the country will need to build hundreds to thousands of expensive coal, natural gas, and nuclear plants to meet it. The historical record suggests that energy efficiency, DSM, and load management practices represent the most feasible way of responding to increases in electricity demand. As the Pulitzer Prize winning author and energy analyst Daniel Yergin put it, these measures are the "cheapest, safest, most productive energy alternative readily available in large amounts." [13] Increasing energy efficiency, another study concluded, "is generally the largest, least expensive, most benign, most quickly deployable, least visible, least understood, and most neglected way to provide energy services." [14]

History Speaks Loudly

The DOE recently calculated the benefits of DSM and found that it lowers wholesale electricity prices as costly power plants are displaced and total demand on the system decreases. Generating peak electricity is extremely expensive, often exceeding $5,000 to $10,000 per installed kW (meaning a 100 MW plant can cost $750 million to build and require $75 million per year to operate), implying that DSM should be profitable for *all* utilities.[15] To prove it, ORNL researcher Eric Hirst estimated the costs and benefits of DSM programs for three types of utilities: a "surplus" utility, an "average" utility, and a "deficit" utility. Surplus utilities are those with excess capacity and few planned retirements, as well as slow projected growth in fossil fuel prices and incomes.[16] Deficit utilities have little excess capacity, many planned retirements, and rapid growth in prices and incomes. Average utilities fall in the middle. Hirst found in each case that DSM programs raise electricity prices but reduce electricity costs, and the overall percentage reduction in cost far exceeds the increase in price. He found 2 to 1 cost benefits for the surplus utility (i.e., every dollar spent on DSM yielded two dollars of savings), 5 to 1 cost benefits for the average utility, and 8 to 1 cost benefits for the deficit utility. In other words, energy efficiency may cause electricity prices (per kWh) to go up slightly, but because people and companies use fewer kWhs, their bills actually go down. For some utilities, cost savings can be as high as $8 for every $1 invested in DSM.

For these reasons, a few states have experimented with large-scale DSM programs. In the state of New York, cost-effective DSM and load management policies have saved more than 1,000 gigawatt hours (GWh) of electricity and displaced 880 MW of peak demand in the past five years.[17] Montana, Idaho, and Oregon conserved energy totaling more than 50,000 GWh per year in the mid-1990s with a retail value of $2 billion to consumers.[18] California's DSM programs operating between 1990 and 1992 delivered 112 percent of planned energy savings, meaning they cost-effectively saved *more* electricity than intended.[19] The Massachusetts Department of Energy Resources discovered their load management program lowered participants' electricity costs by $20 million in 1999, with benefits to Massachusetts customers exceeding $6 million in just 13 hours on one high-cost day.[20]

Looking at energy as a whole (and not just electricity) to highlight the importance of energy efficiency techniques, total primary energy use per capita in the United States in 2000 was almost identical to energy use per capita in 1973. Over the same 27-year period, economic output (measured in terms of GDP per capita) increased 74 percent. National energy intensity, or energy use per unit of GDP, fell 42 percent.

Economists at ACEEE and ASE have argued that three-fifths of this decline can be attributed to energy efficiency improvements. If the United States had not

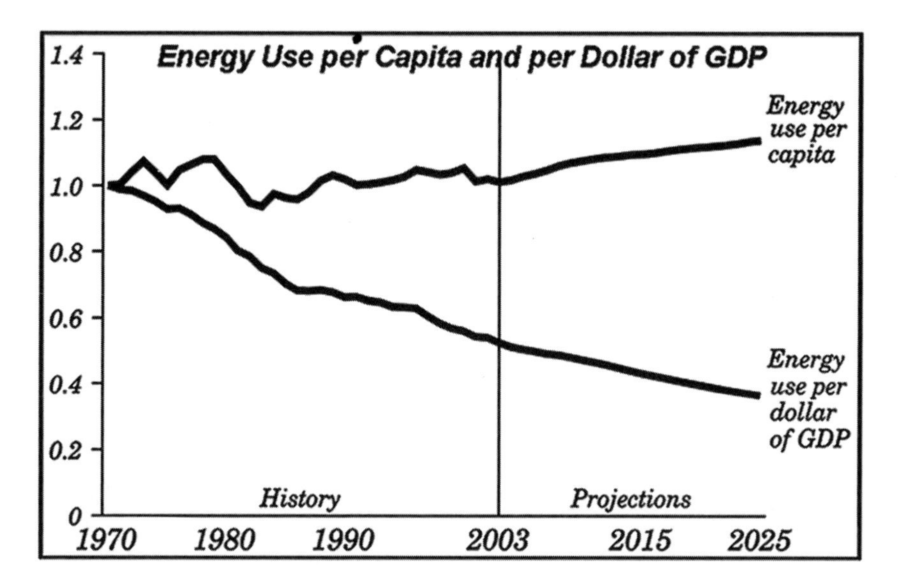

Figure 8
Energy Intensity of the U.S. Economy, 1970 to 2025

The amount of energy needed to produce every unit of GDP has declined significantly from the 1970s to today, thanks in large part to energy efficiency improvements. *Source:* Energy Information Administration.

dramatically reduced its energy intensity over the years, consumers would have spent at least $530 billion more on energy purchases in 2007 (or more than $1.4 billion *every day*).[21] To put the numbers in even greater perspective, the U.S. economy in the 1950s consumed more than 20,000 BTU for every inflation-adjusted dollar of GDP. By 2000, however, it consumed only about 12,000 BTU per inflation-adjusted dollar. Although the U.S. economy has grown by 126 percent since 1973, its energy consumption has grown by only 30 percent. In other words, when applied to the electricity industry the gains made by energy efficiency outdo *every single source of electricity generation* today, including coal, natural gas, and nuclear power (see Figure 9).

What accounted for the dramatic change? Energy Star and other equipment standards, established by the federal government in the early 1970s, required the DOE to test and standardize the energy usage of appliances. These standards required manufacturers to improve the energy efficiency of gas furnaces by 25 percent between 1972 and 2001, central air conditioners by 40 percent, and refrigerators by 75 percent. From 2000 to 2015, adherence to appliance standards will save consumers around $85 billion on a net present value basis.[22] Some states, such as California, implemented their own more stringent standards on top of federal regulations. State regulators estimate that California's standards displaced 8.6 percent of peak load within the state per year, and cost about 1 percent as much as utilities would have paid to build and run new power stations.[23]

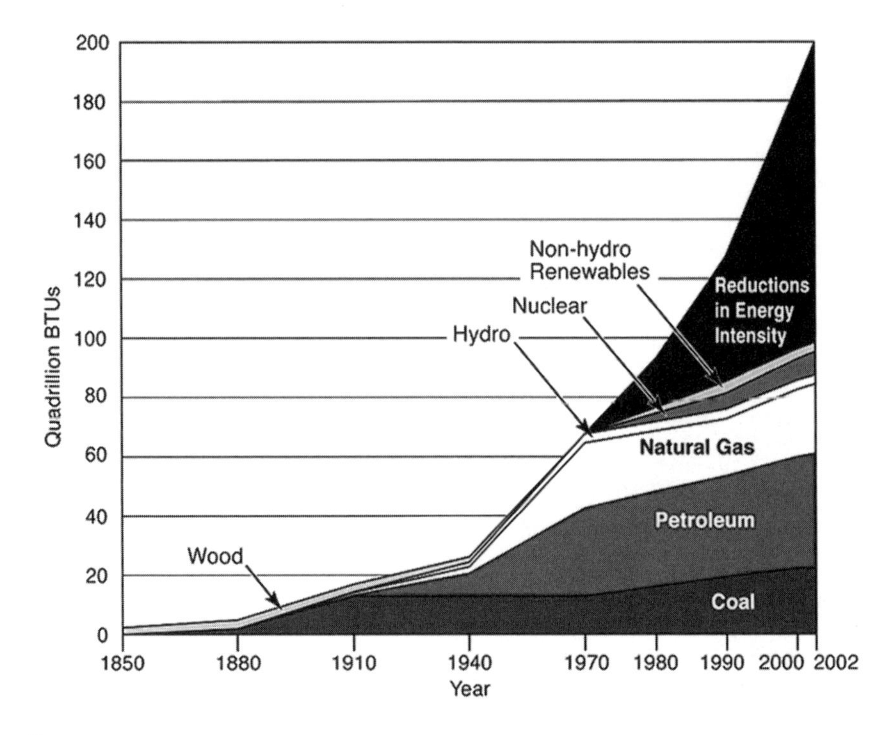

Figure 9
Amount of Energy Saved from Energy Efficiency Improvements, 1970 to 2002

Energy efficiency measures and programs have saved more energy than the production of electricity from any single source, including coal and nuclear.

Source: Marilyn A. Brown, "Energy Myth One: Today's Energy Crisis is 'Hype,'" in *Energy and American Society—Thirteen Myths,* ed. Benjamin K. Sovacool and Marilyn A. Brown (New York: Springer, 2007), 38.

Individuals did their part, too, purchasing more fuel-efficient appliances; they insulated and weatherproofed their homes; and they adjusted thermostats to reduce energy consumption. These measures led to a decrease in per capita residential energy use of 27 percent (and 37 percent per household) despite a 50 percent increase in new home size since 1970 and the growing use of air conditioning, electronic equipment, and a multitude of "plug loads." Businesses retrofitted their buildings with more efficient heating and cooling equipment and installed energy-management and control systems, accounting for a decline of 25 percent of energy use per square foot of commercial building space. Factories adopted more efficient manufacturing processes and employed more efficient motors for conveyors, pumps, fans, and compressors.

Some utilities discovered that they could save electricity cheaper than the cost of operating existing plants, meaning efficiency can improve cash flow by displacing operating costs, appeasing investors, and saving consumers money at the same time.[24] Southern California Edison ran a huge energy efficiency campaign that made money even when credited solely with the value of unburned fuel.[25]

Worldwide, Lee Schipper, a director of Research at the World Resources Institute, found that Sweden and the United States had almost the same standard of living, but the Swedes accomplished theirs by consuming *half* as much energy, even though Sweden had more severe winters and produced more energy-intensive products per capita.[26] Japan achieved economic growth of 39 percent with almost no increase in energy usage between 1978 and 1982, and Germany achieved growth of 16 percent GDP with a reduction of 30 percent in energy usage over the same time.[27]

The above examples show that energy efficiency measures take many forms, at almost any scale, and can be successfully implemented by almost any group of people or institution. Collectively, these gains in energy efficiency, prompted by high fuel costs and government policies, represent one of the great economic success stories of the past century.[28] As two researchers put it, efficiency measures are "a free lunch you're paid to eat." [29] Most amazingly, such measures captured their benefits with no reduction in quality of life or energy services. Sociologists Allan Mazur and Eugene Rosa assessed levels of per capita energy consumption and factors such as life expectancy and gross national product for 55 countries. They found that life-style indicators did not have any high association with high energy consumption, and that reductions in energy use occurred without any deterioration of indicators of health and health care, education, culture, and general satisfaction.[30]

(But We Need to Listen)

Despite these impressive gains, there is much more potential in energy efficiency than some ever imagined. A recent survey conducted by the National Association of Regulatory Utility Commissioners (NARUC) found cost-effective energy efficiency potential in *all* regions of the country, with the most untapped potential in the Northeast and South, where electricity costs are highest (meaning energy efficiency efforts are more economical than areas where energy is cheaper).[31] Another study projected that cost-effective energy efficiency programs could reduce consumption by around one trillion kWh by 2020, offsetting almost all projected growth in electricity use—and the needed capacity additions to achieve it.[32] The ASE found that aggressive investments in energy efficiency could free up enough electricity to eliminate the need to construct more than 1,300 power plants in the next 20 years.[33] One study published in *Electricity Journal* projected that a national DSM program aimed at reducing peak demand by just 5 percent would yield $3 billion in net generation, transmission, and distribution savings per year and displace some 625 infrequently used peaking plants and associated delivery infrastructure.[34]

Focusing narrowly on just one technology, air conditioners, and assessing four key installation issues (equipment sizing, refrigerant charging, adequate airflow, and sealing ducts), ACEEE estimated electricity savings as high as 24 to 35 percent per household. If implemented in conjunction with utilities as part of a DSM program, ACEEE estimated savings could displace 41 GW of peak demand capacity nationwide, with an equivalent cost to utilities of $75 per installed kW per year, or about one-hundredth the cost of building and operating conventional peaking plants.[35]

Other broader studies confirm DSM's cost-effectiveness. The IEA reviewed 40 large-scale commercial DSM programs implemented in the 1990s and found that they saved electricity at an average cost of 3.2 ¢/kWh, well below the cost of supplying electricity (regardless of the fuel source). Researchers at Lawrence Berkeley National Laboratory assessed 40 large commercial DSM programs run by utilities and calculated that energy efficiency measures avoided electricity generation at an effective rate of 3.2 ¢/kWh.[36] A more recent survey of DSM programs in the past five years found the average cost of saving energy at between 2.1 and 3.0 ¢/kWh in the United States and Europe.[37] Economists from Resources for the Future and Stanford University reviewed the performance of energy efficiency measures such as appliance standards, financial incentives, and information programs in the United States and found that they saved electricity at an average rate of between 2.89 and 3.28 ¢/kWh.[38] Similarly, the EIA found an average cost of 2.6 ¢/kWh for demand-side management, load management, and energy efficiency programs in 2006.[39]

Three important points need to be made about the current potential of energy efficiency. First, one kWh saved is more valuable than one kWh generated. One thousand MW worth of energy efficiency and 1,000 MW of large-scale generation are not equivalent. Richard Cowart, director of the Regulatory Assistance Project, found that every dollar invested in energy efficiency

- Mitigated against uncertainty and reduced load, wear, and maintenance needs on the entire fossil fuel chain, even in hours when reliability problems were not anticipated by system managers;
- Depressed the costs of locally used fuels such as oil, coal, and natural gas;
- Reduced demand across peak hours, the most expensive times to produce power;
- Lessened costly pollutants and emissions from generators;
- Improved the reliability of existing generators;
- Moderated transmission congestion problems;
- Operated automatically through customers coincident with the use of underlying equipment or load, meaning they are always "on" without delay or the needed intervention by system operators to schedule or purchase the resource.[40]

Discounting all of these benefits and looking purely at reductions in peak demand, the New York Independent Systems Operator (ISO) determines a reserve criterion of 18 percent during times of peak demand to ensure overall system reliability. Accordingly, each 1.0 MWh (megawatt-hour) of peak demand that customers avoid through energy efficiency means that utilities can subtract 1.18 MWh of total capacity needed. Quite literally, every single kWh avoided through energy efficiency equates to 1.18 kWh of avoided supply.[41]

Second, a majority of energy efficiency savings can be accomplished with small and targeted programs. In the ISO–New England service area, for example, about 9 percent of the system's total generating capacity is tapped 1 percent of the time. The price of power during this most expensive 1 percent accounts for 16 percent of the total annual dollars spent on the spot market.[42] These numbers reveal that a very small fraction of supply accounts for a large fraction of total cost. Nationwide, NARUC calculated that just 0.4 percent of industrial customers account for 30 percent of total demand. Consequently, relatively small DSM programs directed at a miniscule proportion of the nation's electricity customers or generators can produce mammoth benefits in terms of total demand reductions.

Third, the "whole" of energy efficiency is greater than its "parts." The potential for energy efficiency savings becomes even greater when one departs from looking at individual technologies and instead focuses on optimizing and integrating entire energy systems, reaping cumulative benefits instead of isolated savings. Amory and Hunter Lovins' house in the cold Rocky Mountains of Colorado has no conventional heating system because its owners designed it with the most energy-efficient insulation, windows, and ventilation products available. These recover heat losses to within about 1 percent of the free heat gains from natural light, people, lights, and appliances. The house is so efficient, when one of the owners wants to turn up the heat, he or she jokingly remarks that it can be harnessed "from a 50-watt dog, adjustable to 100 W by throwing a ball." Promoting energy efficiency measures bundled together, in this instance, displaced the need to install, maintain, and fuel a conventional heating system and its consequent furnace, ducts, fans, pipes, pumps, wires, control systems, and fuel-supply equipment.[43]

Similarly, a California prison integrated about 3 acres of solar panels with energy efficiency and demand response mechanisms. Engineers configured the prison to consume less power during costly peak periods and more power during cheaper non-peak periods. During times of peak demand, they utilize maximum output from the solar panels primarily to sell power back to the grid. Integrating energy efficiency measures together in this manner yielded benefits that exceeded costs by 70 percent within the first year, even when all state subsidies were excluded.[44] Both examples demonstrate that energy-efficient systems can accomplish more than individual technologies. "All these potentials don't add; they multiply," comments energy guru Amory Lovins, and "even the most advanced industries are barely scratching the surface of how much energy efficiency is available and worth buying."[45]

No. 2: The Infrastructure Challenge

Chapter 1 noted that the current electric utility system is aging, congested, unreliable, and expensive. DG technologies displace the need to invest in costly T&D infrastructure and improve system reliability. Energy efficiency, renewables, and DG technologies have quicker construction lead times that avoid the risk of cost overruns. Renewables benefit from more rapid learning curves, meaning that every increase in capacity improves the overall efficiency of the electric utility system and their prices go down, especially since the cost of their "fuel" is virtually free.

Improved T&D Reliability

Deploying DG systems in targeted areas provides an effective alternative to constructing new transmission and distribution lines, transformers, local taps, feeders, and switchgears, especially in congested areas or regions where the permitting of new transmission networks is difficult. One study found that up to 10 percent of total distribution capacity in high growth scenarios could be cost-effectively deferred using DG technologies within ten years.[46] The IEA even concluded that "by moving portable power generators to distribution substations, utilities have been able to cope with rapid load growth more quickly than by upgrading distribution facilities."[47]

Pacific Gas and Electric Company (PG&E), the largest investor-owned utility in California, built an entire power plant in 1993 to test the grid benefits of a 500 kW DG plant. PG&E found that the generator improved voltage support, minimized power losses, lowered operating temperatures for transformers on the grid, and improved transmission capacity. The benefits were so large that the small-scale generator was twice as valuable as estimated, with projected benefits of 14 to 20 ¢/kWh.[48] The experience convinced PG&E to consider the use of DG to displace T&D infrastructure. Using conventional approaches, planners proposed an upgrade of 230-kV and 60-kV lines serving seven substations in the San Francisco area, estimated to cost PG&E $355 million (in 1990 dollars). However, PG&E ultimately discovered that a cheaper alternative was to strategically deploy distributed 500-kW DG plants connected to distribution feeders. By investing in such locally sited DG projects, PG&E found that it could defer a significant number of its transmission upgrades and ultimately saved $193 million (or more than half the present cost of the expansion plan) by installing DG technologies.[49]

Since modern DG technology enables utilities to remotely dispatch hundreds of scattered units, they also improve the ability of utilities to handle peak load and grid congestion problems. Another PG&E analysis, comparing fifty 1-MW distributed plants to one 50-MW central plant in Carissa Plains, California, found that the grid advantages (in forms of load savings and congestion) more than offset the disadvantages (in terms of high capital cost and interconnection) of installing the new generation.[50]

The Congressional Budget Office concluded that DG systems can provide utilities with a variety of important ancillary services as well, including system control (how operators control generators on the grid), reactive supply (the injection of power needed to maintain required voltages), and spinning reserves (alternating existing generation capacity to offset load imbalances). Because of their smaller size, DG technologies can be started up more quickly and deployed more easily than centralized systems. Smaller units have lower outage rates, decreasing the need for reserve margins and spinning reserves. Since DG technologies can be constructed more quickly, they enable utility managers to respond better to supply and demand fluctuations, especially when used with advanced functions like real time pricing and net metering.[51]

Indeed, researchers at the University of Albany and NREL have determined that dispersed solar PV resources are so valuable they could have prevented the $6 billion blackout affecting 40 million people spread across the Northeast, Midwest, and Canada in 2003. After running thousands of simulations, they found that had distributed solar PV facilities been operating on August 14, the blackout most likely would not have occurred. The researchers noted that the indirect cause of peak demand on that day—hot temperatures and greater air conditioning loads—are also the best source for solar PV generation. If just a few hundred MW of solar PV had been operating, the researchers concluded that power transfers and losses would have been reduced, voltage support enhanced, and uncontrolled events would not have cascaded into a complete blackout.[52]

Modularity

As Chapter 1 highlighted, classic grid systems are "lumpy systems" in the sense that additions to capacity are made in primarily large lumps (gargantuan power plants, new transmission networks). These plants have long lead times and uncertainties, making planning and construction difficult, especially when the balance of supply and demand can change rapidly within a short period of time. They can also be extremely capital intensive: a typical 1,100 MW light water reactor can cost as much as $3 billion when licensing and construction expenses are included. Yet we have a hard enough time predicting the weather or the outcome of political elections; imagine the difficulty of projecting how an entire industry will look 5, 10, or even 20 years from now.

In the 1970s and 1980s, excessively high forecasts of growth in demand for electricity led to overbuilding of generating plants and massive electric system cost overruns in many states. One infamous example was in Washington State, where the Washington Public Power Supply System (WPPSS) began a construction program for as many as seven new nuclear power plants in the early 1970s. WPPS believed that regional electricity requirements would grow by 5.2 percent each year well into the 1990s and started building nuclear power plants to meet their projections.

At the same time, the massive backlog of nuclear power plant orders after the 1973 oil crisis caused an equally massive shortage of skilled nuclear engineers and architects (69 plants were ordered in 1973 and 1974). Problems of plant design, poor craftsmanship, and strikes caused even longer delays. Five-year construction estimates lengthened to 10- and 12-year periods. One WPPS project started in 1970 was not finished until 1984, and the WPPS annual report in 1981 projected that $23.7 billion was needed to complete one of its plants after $5 billion had already been expended, all the while electricity growth dropped 65.4 percent below original projections.[53] By the mid-1980s, WPPS faced financial disaster and all but one of the plants was cancelled, leading to the country's largest municipal bond default at the time. The entire experience came to be called the "WHOOPS" fiasco (as a play on the WPPS acronym) and is an enduring lesson of the risk associated with investing in large power plants. Consumers across the Northwest are still paying for WHOOPS in their monthly electricity bills.

While WHOOPS is perhaps the most spectacular example, similar "boom and bust" cycles in power plant construction and cost overruns occurred in many states during the 1980s, and directly produced the high electricity rates that spurred the "electric restructuring" movement of the mid-1990s.[54] Between 1972 and 1984, more than $20 billion in construction payments flowed into 115 nuclear power plants that were subsequently abandoned by their sponsors.[55] The Shoreham Nuclear Power Plant adjacent to the Wading River in East Shoreham, New York, cost ratepayers $6 billion but was closed by protests in 1989 before the plant could generate a single kWh of electricity.

Cost overruns are not limited to nuclear reactors or the 1980s. In November 2006, Duke Energy announced that the price tag for the company's proposed coal-fired power plants near Charlotte, North Carolina, had soared to $3 billion. Just two months prior, the company had reported to state utility regulators that the two plants would cost only $2 billion. Charlotte's daily newspaper speculated that such a substantial cost discrepancy (a 50 percent increase) raised the possibility that the total expense for the plants could continue ballooning during the five years that the utility estimated it would take to build the facilities.[56]

In contrast, clean power technologies tend to have quicker lead times—taking between a few months to five years to implement. In the early 1990s, the New England Electric System enlisted 90 percent of small businesses in a retrofit program in two months. On the other coast, PG&E marketers signed up one-fourth of new commercial buildings for design improvements, and then one year later raised the bar and captured it all in the first nine days of the new program.[57] Homeowners in many towns and cities in Vermont retrofitted 40 percent of all houses with woodburning iron stoves in less than three years, and the Chinese installed 9 million biogas plants between 1972 and 1978.[58] These examples suggest that people can do many small things for themselves very quickly if they have the incentive and opportunity.

On the supply side, the quicker lead times for renewables and DG enable a more accurate response to load growth, and minimize the financial risk associated with borrowing hundreds of millions of dollars to finance plants for 10 or more years before they start producing a single kWh of electricity. FPL says it can take as little as 3–6 months from groundbreaking to commercial operation for new wind farms.[59] In 2005, Puget Sound Energy (PSE) proved that FPL's boast was achievable in practice when it brought eighty-three 1.8 MW wind turbines to its Hopkins Ridge Wind Project from groundbreaking to commercial operation in exactly 6 months and 9 days.[60] And in Nevada, Ormat Nevada Incorporated commissioned a 20 MW geothermal power plant only 8 months after groundbreaking.

Solar panels can be built in various sizes, placed in arrays ranging from watts to megawatts, and used in a wide variety of applications, including centralized plants, distributed substation plants, grid-connected systems for home and business use, and off-grid systems for remote power use. PV systems have long been used to power remote data relaying stations critical to the operation of supervisory control and data acquisition systems used by electric and gas utilities and government agencies. Solar installations may require even less construction time than wind or geothermal facilities since the materials are prefabricated and modular. John Ravis, a project finance manager for TD BankNorth, recently told industry analysts that utility-level PV systems can come online in as little as two months if the panels are available.[61]

Utilities and investors can cancel modular plants easier, so abandoning a project is not a complete loss (and the portability of most DG systems means recoverable value exists should the technologies need to be resold as commodities in a secondary market). Smaller units with shorter lead times reduce the risk of purchasing a technology that becomes obsolete before it is installed, and quick installations can better exploit rapid learning, as many generations of product development can be compressed into the time it would take to build one giant power plant.[62] In addition, outage durations tend to be shorter than those from larger plants and repairs for reciprocating gas and diesel engines take less money, time, and skill. As one study concluded, "technologies that deploy like cell phones and personal computers are faster than those that build like cathedrals. Options that can be mass produced and adopted by millions of customers will save more carbon and money sooner than those that need specialized institutions, arcane skills, and suppression of dissent."[63]

Technological Learning

Economist Alfred Marshall theorized in 1890 that the larger manufacturing firms become, the greater they can create economies of skill, machinery, and materials (a concept later known as "economies of scale"). The isolated worker, Marshall noted, often throws a number of small things away that could have been collected and used in a factory. When a hundred sets of furniture or clothing have to be cut on exactly the same pattern, greater care will be spent so that only a few small pieces are wasted. Marshall surmised that increased manufacturing output would allow firms to purchase items in greater quantities, save on carriage and freight charges, sell in larger quantities, and minimize time and trouble. In cotton spinning and calico weaving, Marshall concluded, manufacturing in volume was the way to go.[64]

Sociologist Donald MacKenzie found something similar in the adoption of new technologies. In his survey of military technology, he found that the more the military adopted a particular technology, the more momentum it gained, the more constituents pushed for it, the more policymakers awarded it with further R&D contracts, and the more it improved.[65]

Marshall and Mackenzie could have found the same advantages in the adoption of clean power technologies, where the potential for cost savings from "learning" is relatively high. The Institute of Electrical and Electronics Engineers (IEEE) estimated that achieving 20 percent market share for renewables would bring large scale development of renewable energy and nationwide standards that would, in turn, lower costs. Such a "learning by doing" approach was estimated to reduce the expense of producing, installing, and maintaining renewable energy technologies.[66]

This "learning effect" was confirmed by the DOE's projection of significant continued improvements in the competitiveness of wind technology over the next decade. DOE forecasted cost reductions due to discounts for large-volume purchases of materials, parts, and components as well as from the learning effects that flow from deploying wind technology to meet greater cumulative electricity volumes.[67] In fact, researchers from Resources for the Future estimate that 15 percent market penetration of renewables in 2020 could further lower the construction costs for wind turbines by more than 20 percent and decrease the cost of biomass generators by nearly 60 percent.[68] Janet Sawin from the Worldwatch Institute found that every doubling of manufacturing volume for wind technologies corresponded with an 8 to 10 percent reduction in cost. During the precommercial stage of wind power in California, for example, costs fell from 7 to 20 percent for every doubling of capacity, and in Germany and Denmark, every doubling of capacity resulted in a 15 to 20 percent decline in cost.[69]

Older windmill designs relying on lattice structures have given way to modern turbine designs utilizing easy-to-manufacture steel and aluminum foundations that house nacelles that can yaw into the wind and blades that can alter their pitch to maximize production. These learning improvements lowered the cost of wind-produced electricity for most areas of the country. According to the California ISO and CEC, in 1993 the average cost of wind energy throughout the United States was 7.5 ¢/kWh, less than one-fifth the 1980s average price of 39 ¢/kWh. In 2005, the average price of wind energy in California was 3.5 ¢/kWh. These costs compare to around 6 ¢/kWh for popular gas-turbine plants built from 2000 to 2004. The DOE recently concluded the same when they looked at the performance of wind turbines across country as a whole. Utilizing a database of wind power performance encompassing 128 projects installed between 1998 and 2007 (and constituting 8,303 MW of capacity, or about 55 percent of all capacity built in that decade), DOE researchers found that wind prices have dropped significantly from 6.3 ¢/kWh in 1999 to just 4.0 ¢/kWh in 2007, with some of the best turbines producing power at 2.4 ¢/kWh.[70]

The benefits of "learning by doing" become even greater when you consider a concept known as "capacity factor." A capacity factor is the ratio of a generating facility's actual output over time compared to its theoretical output if it were operating at maximum efficiency. The EIA estimated that the average capacity factor for all power plants in the United States was approximately 55 percent.[71] That is, over a long period of time, an average power plant actually contributes to the electricity grid only 55 percent of its theoretical maximum output. Nuclear and hydroelectric generators have boasted the highest capacity factors, occasionally exceeding 90 percent. Coal ranks near the middle, with a capacity factor of about 60 percent. Less reliable natural gas generators have much lower capacity factors of 29 percent.

This low percentage is, in part, because gas-fired plants are generally used as "peaking" units.

Historically, all forms of electricity generation have followed the same general trend: the more the technologies get deployed, the higher their capacity factor and the lower their costs. When coal and steam boilers were generating just a few GWh of electricity in the early 1930s, they had capacity factors in the low twenties. But by 1997, when the deployment of coal-fired units reached thousands of GWs of capacity, their capacity factor had jumped to 61 percent.[72]

Nuclear reactors also prove the concept. The World Nuclear Association notes that nuclear generators had a capacity factor of around 10 percent when just 22 GW were deployed. Yet their capacity factor rose to 30 percent with the deployment of 53 GW and close to 90 percent once installed capacity reached 97 GW.[73] Similarly, the capacity factor for hydroelectric generators and geothermal plants rose in direct correlation with the amount of total installed capacity.

The interrelationship between rising capacity factors and installed capacity suggests that deploying more clean power technologies will significantly improve their capacity factors. Recent experience with wind energy confirmed this rule. In 2000, wind turbines reported capacity factors in the low teens. But by 2006, when installed wind energy had more than tripled in the United States, wind turbines registered capacity factors in the mid-thirties (see Figure 10). Entire wind farms in Oahu, Hawaii, and San Gorgonio, California, have even achieved capacity factors of 36 and 38 percent.[74] In a 2006 analysis, the EIA observed that wind turbine capacity factors appeared to be improving over time and concluded that "capacity factor grows as a function of capacity growth."[75] Indeed, a comprehensive assessment of wind turbine performance across a sample of 170 wind projects built between 1983 and 2006 (totaling 91 percent of nationwide installed capacity in 2006, or 10,564 MW), found that average capacity factors for wind hovered around 22 percent in 1998 but jumped to 31 percent in 2003 and 35 percent in 2006. Using the most updated data available, actual operating performance shows that modern wind turbines commonly exceed capacity factors in excess of 40 percent. Out of 58 projects installed between 2004 and 2006, more than one-quarter achieved capacity factors in the low to mid-forties, with average capacity factors for Hawaii reaching an astounding 45 percent, those in the Heartland averaging 40.8 percent, and those in California averaging 36.9 percent.[76]

Solar energy appears to follow this same pattern. In the early 1980s, with 10 MW of solar panels installed globally, the average capacity factor was around 9 percent. By 1995, however, after more than 70 MW had been installed, the average capacity factor of panels jumped to almost 15 percent, and in the past five years have surpassed 21 percent.[77] Researchers from the Institute for Energy Policy and Economics found that "over the last 10 years 'learning by doing' has led to a simplification of industrial manufacturing processes . . . As a result, costs have fallen considerably."[78] Berkeley researcher Gregory F. Nemet concluded that the cost of solar panels has declined by a factor of nearly 100 since the 1950s, more than any other technology, precisely because of this learning. PV electrical systems cost more than $60 per watt

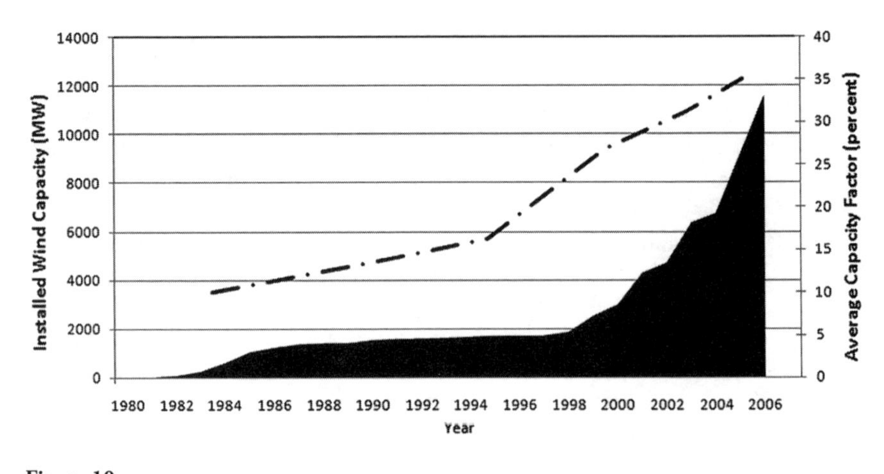

Figure 10
Wind Energy Capacity Factors and Installed Capacity, 1980 to 2006

The overall efficiency of wind turbines grows the more the technology gets used, implying that further expansion of wind energy will improve performance. The dashed line shows improvements in capacity factor, while the shaded area shows installed capacity.

Source: Christopher Cooper and Benjamin K. Sovacool, *Renewing America: The Case for A National Renewable Portfolio Standard* (New York: Network for New Energy Choices, 2007).

in 1976, but only around $3 per watt in 2004. As manufacturing plants got larger, engineers became more comfortable with the technology, module efficiency improved, and logistics and silicon mining became more efficient.[79]

No. 3: The Fuel Supply Challenge

All clean power sources utilize widely abundant and nondepletable forms of fuel and help decentralize the security risks of the grid, lowering the cost of fossil fuels, insulating the system from price shocks, and improving system resilience when outages do occur. Investments in energy efficiency and renewables also enhance economic growth, providing more jobs and local revenue than equal investments in conventional systems.

M. King Hubbert, the famous geophysicist who predicted that American oil production would peak about 1970, often remarked that it would be incredibly difficult for people living now, accustomed to exponential growth in energy consumption, to assess the transitory nature of fossil fuels. Hubbert argued that proper reflection could happen only if one looked at a time scale of 3,000 years. On such a scale, Hubbert thought that the complete cycle of the world's exploitation of fossil fuels would encompass perhaps 1,100 years, with the principal segment of this cycle covering about 300 years (see Figure 11).[80] Indeed, some are already projecting that, at current rates of consumption, the world has less than 200 years of fossil fuel supply and 65 years of natural gas, 70 years of uranium, and 164 years of coal left.[81] For

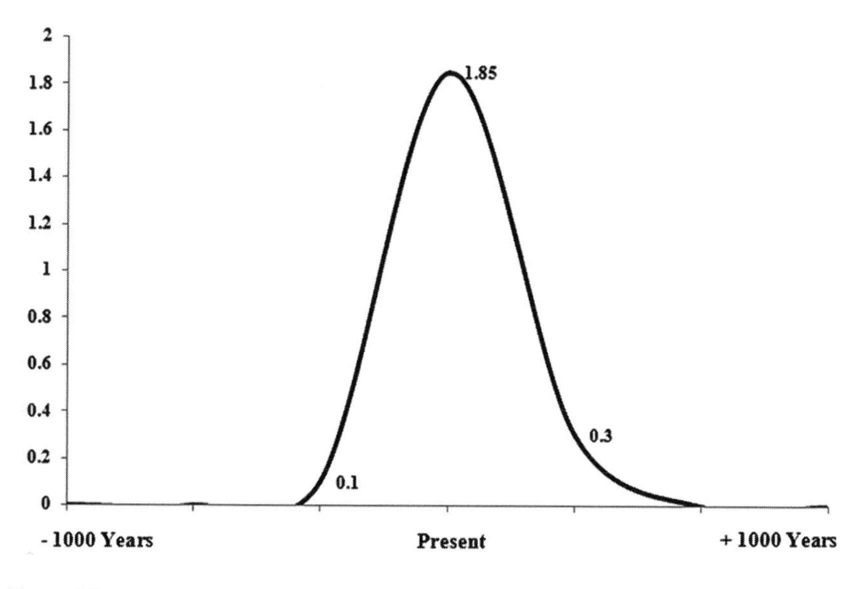

Figure 11
Rate of Global Fossil Fuel Depletion, 1008 to 3008

Hubbert's "Candle" Curve Depicting the Rate of Fossil Fuel Depletion for the next thousand years (in 100,000 TWh/year). Note that fossil fuels will almost completely run out at current rates of consumption in about 400 years.

Source: Adapted from M. King Hubbert, "Energy Resources of the Earth," *Scientific American* (September 1971): 61.

this reason, Tim Jackson, from the Centre for Environmental Strategy at the University of Surrey, has referred to fossil fuel and uranium reserves as "thermodynamic time-bombs." [82]

Thankfully, the Earth receives radiation from the Sun in a quantity far exceeding humanity's needs. By heating the planet, the Sun generates wind and creates waves. The Sun powers the evapotranspiration cycle, allowing for the generation of power from hydroelectric sources. Plants photosynthesize, creating a wide range of "biomass" products. These resources, in contrast to fossil fuels, have no candle-like curve. As German Parliamentarian Hermann Scheer put it, "Our dependence on fossil fuels amounts to global pyromania. And the only fire extinguisher we have at our disposal is renewable energy." [83]

Fortuitously, the United States has an enormous cache of renewable energy resources. A comprehensive study undertaken by the DOE calculated that 93.2 percent of all domestically available energy was in the form of just wind, geothermal, solar, and biomass resources. The amount of renewable resources found within the country, in other words, amounted to a total resource base the equivalent of 657,000 billion barrels of oil, more than 46,800 times the annual rate of national energy consumption (see Figure 12).[84] Perhaps an even more amazing feature of this estimate is that it was validated by researchers at USGS, ORNL, Pacific Northwest

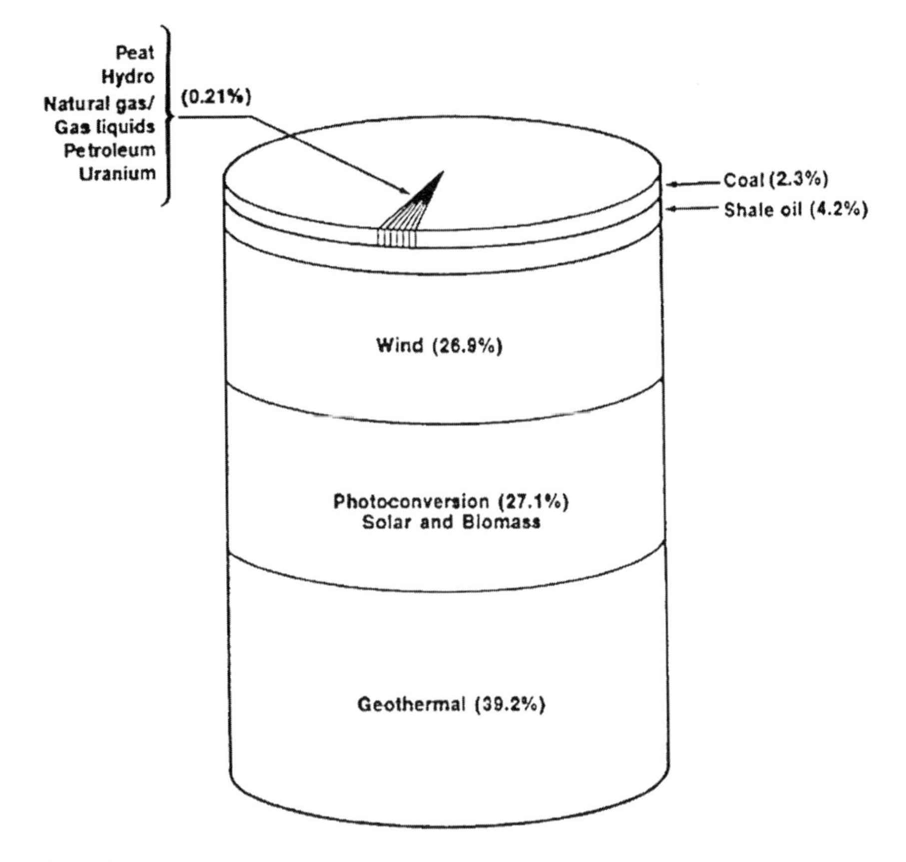

Figure 12
Domestic U.S. Energy Resources and Reserves

Looking at technically available energy reserves, solar, geothermal, biomass, and wind account for more than 90 percent of the U.S. resource base.

Source: U.S. Department of Energy, *Characterization of U.S. Energy Resources and Reserves* (Washington, DC: DOE/CE-0279, 1989).

National Laboratory, Sandia National Laboratory, National Renewable Energy Laboratory (NREL), the Colorado School of Mines, and Pennsylvania State University.

Using published, nonpartisan, and peer-reviewed estimates from the DOE, EPA, NREL, ORNL, and the Energy Foundation (and not estimates from manufacturers and trade associations), the country has 3,730,721 MW of *achievable* renewable energy potential by 2010.[85] That is, the estimate assumes the utilization of existing, commercially available technologies that would only need to be built to realize their potential (see Table 5). Two things pop out when looking at Table 5. First, the table shows that renewable resources have the capability to provide 3.7 times the total amount of installed electricity capacity operating in 2008. Second, the country has so far harnessed only a whopping 2.9 percent of this potential generation.

Table 5
Achievable Potential for Commercially Available Renewable Energy Generators
The United States possesses so much renewable power potential that existing technologies could meet almost 4 times the country's power needs.

	Electricity Generation (thousand kWh) in 2006	Grid-Connected Installed Capacity (MW) in 2007	Achievable Installed Capacity (MW) by 2010
Onshore Wind	25,781,754	12,600	1,497,000
Offshore Wind	0	0	791,000
Solar PV	505,415	624	710,000
Solar Thermal	N/A	354	98,000
Geothermal	14,842,067	3,100	2,800
Biomass (combustion)	50,064,892	9,733	465,000
Biomass (landfill gas)	5,509,189	539	1,370
Hydroelectric	288,306,061	80,000	165,551
Total	**385,009,378**	**106,950**	**3,730,721**

Source: U.S. Department of Energy, National Renewable Energy Laboratory, Oak Ridge National Laboratory, Environmental Protection Agency, and the Energy Foundation.

What is more, unlike the distribution of coal, natural gas, and uranium, which are heavily concentrated in some states, plentiful renewable resources exist in almost every single county and city in the United States (see Maps 8 through 13).

Wind Power

The nation possesses an exceptional abundance of onshore wind resources. The fuel potential for wind energy, particularly in areas with frequent and strong winds, remains largely untapped. The Midwest and the Great Plains have been called the "Saudi Arabia of wind" and theoretically hold enough technical potential to fulfill the entire country's energy needs.[86]

The energy potential for offshore wind is even larger. Offshore wind turbines can harness stronger, more consistent winds than those that course through mountain passes or across open plains.[87] Moreover, offshore wind farms may be able to utilize more efficient and larger blades, exploiting the basic physical laws that make long blades more efficient and economical. Wind power resources on the eastern continental shelf are estimated to be more than 400 GW, or several times the electricity used by eastern coastal states.[88] The DOE estimates that more than 900,000 MW of wind generation capacity, roughly equivalent to the current amount of total installed electricity capacity for the *entire* country, exists within 50 miles of the country's coasts.[89] The largest areas of Class 7 wind power, the strongest class, are located in Alaska, with regions in the Aleutian Islands containing winds strong

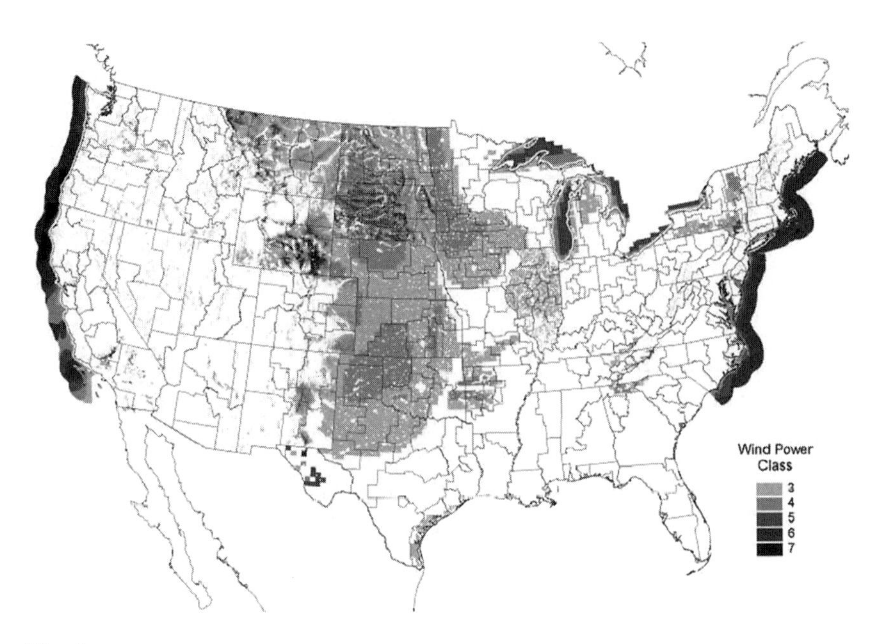

Map 8
Wind Energy Potential for the United States

The country has an exceptional amount of wind energy potential in the Midwest, around the Great Lakes, and around the nation's shoreline. The darker areas show the highest wind classes, or areas with the most wind.
Source: National Renewable Energy Laboratory.

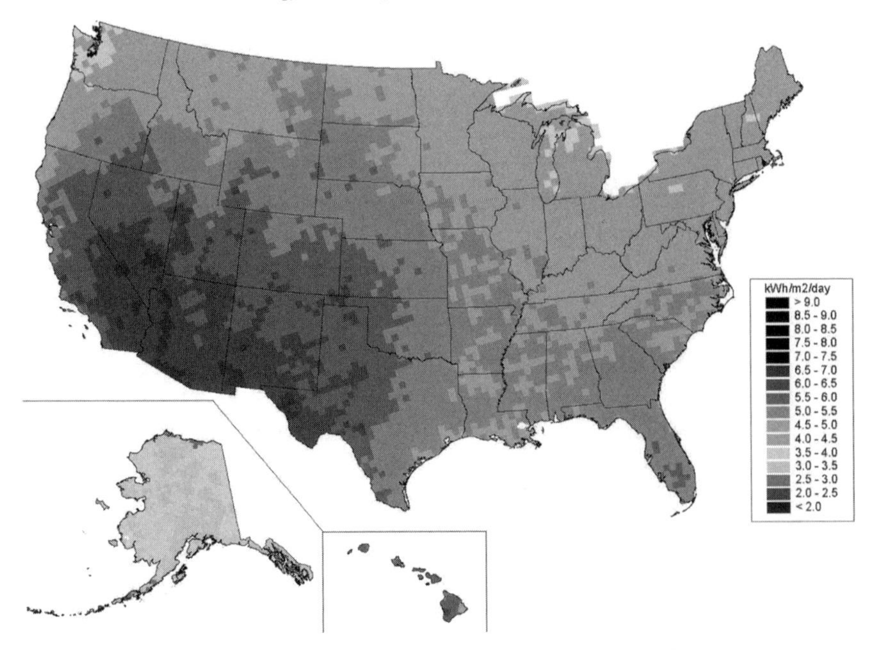

Map 9
Solar PV Potential for the United States

Solar PV potential for flat plate collectors, facing south at a latitude tilt, is especially high in the West and Southwest. The darker areas show the areas with the most solar insolation.
Source: National Renewable Energy Laboratory.

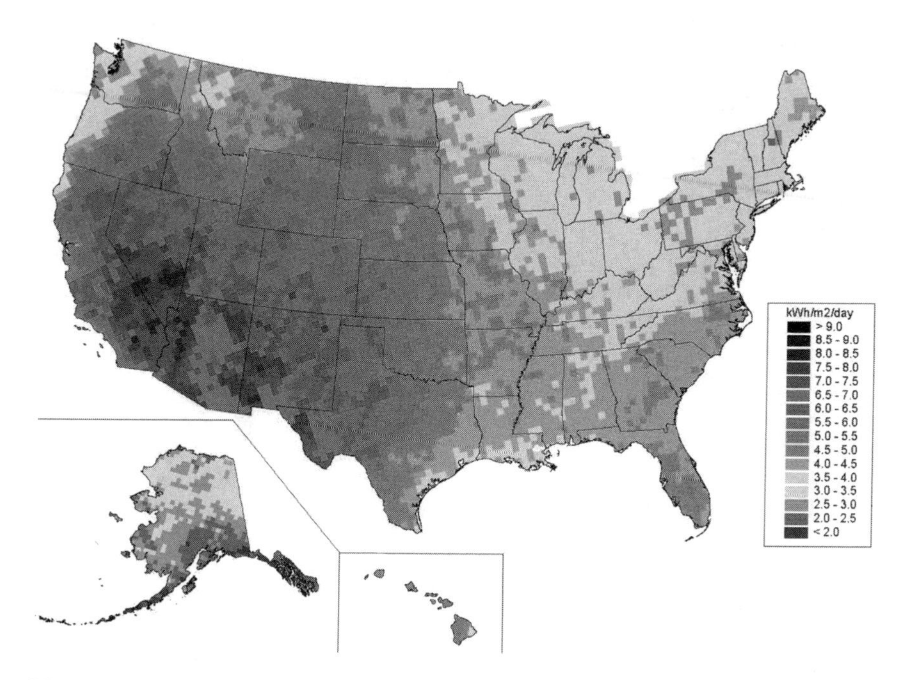

Map 10
Solar Thermal Potential for the United States

The potential for concentrated solar thermal systems is greatest in the Southwest, with considerable potential around California, Nevada, and Arizona. The darkest areas show the counties with the most solar radiation.

Source: National Renewable Energy Laboratory.

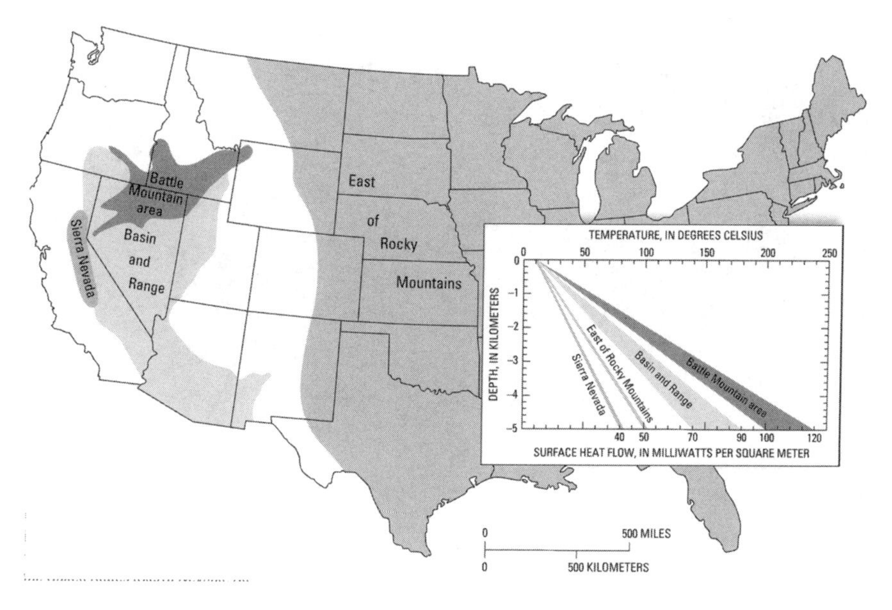

Map 11
Geothermal Electric Potential for the United States

Some of the best geothermal electric resources in the world exist in the Sierra Nevada and Battle Mountain ranges east of the Rocky Mountains.

Source: U.S. Geological Survey.

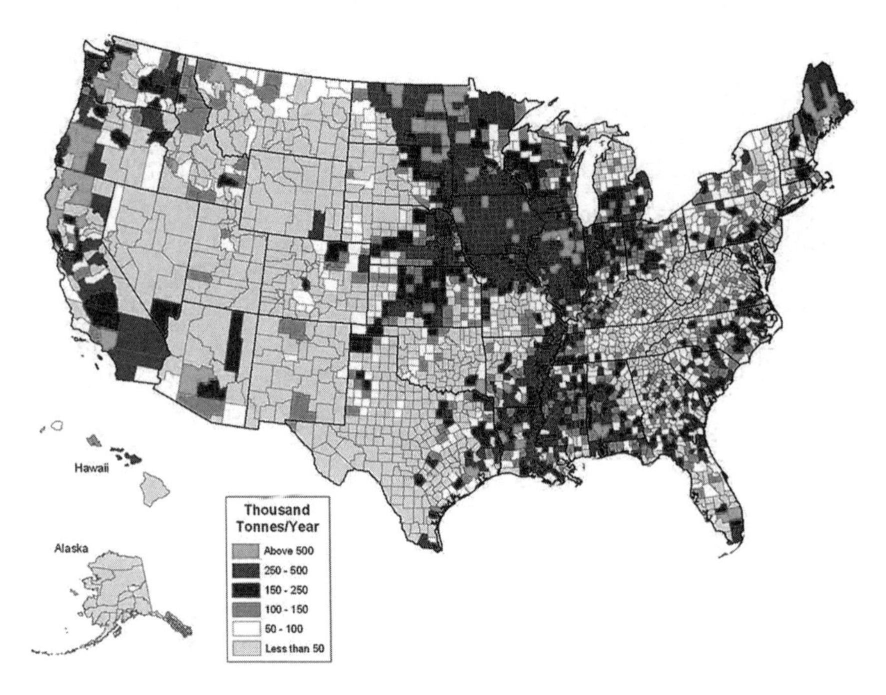

Map 12
Bioelectric Potential for the United States

Bioelectric resources in the form of agricultural wastes and energy crops are widely available in almost every county and near almost every city in the United States. The darker areas show the counties with the most sustainable biomass potential.

Source: National Renewable Energy Laboratory.

Map 13
Hydroelectric Potential for the United States

Often neglected as a renewable resource, the country could easily double its hydroelectric capacity by drawing on the motion of streams, rivers, and lakes with resources available in every state. The darker areas show the streams and rivers with the most potential.

Source: U.S. Department of Energy, *Water Resources of the U.S. with Emphasis on Low Head/Low Power Resources* (Washington, DC: DOE/ID-11111, April 2004), 48.

enough to produce power around the clock.[90] Of course, much of this wind resource would have to be transmitted from distant locations to load centers.

Solar Power

Solar energy is by far the largest single source of energy. The Earth intercepts 170,000 TWh (terrawatt-hour) of power from the sun each day, more than 4 times the primary energy consumption of all humans on the planet. Biological systems (such as plants) capture less than 0.1 percent of this energy via photosynthesis.[91]

Compared to other electricity technologies, solar panels are highly durable and reliable, operate quietly, lack any moving parts, and require minimum maintenance. Solar panels and solar thermal systems have immense technical potential in the American southwest. For example, the DOE estimates that concentrating solar systems would need to occupy only 15 percent of federal lands in Nevada to supply all of the electricity needs of the United States.[92] A rather inefficient and poorly configured solar PV array covering half a sunny area 100 miles by 100 miles could also meet all of the nation's electricity demand.[93]

Naturally, a solar array powering the nation would not have to be constructed that way. Researchers at the DOE estimate that the industry could supply every kilowatt-hour of our nation's current electricity requirements by building PV modules on only 7 percent of the country's available roofs, parking lots, highway walls, and buildings (without substantially altering appearances or requiring new sources of land).[94] Another study estimated the potential of grid-connected solar PV on rooftops alone at 500 GW—roughly half the country's current installed capacity.[95]

The raw amount of potential solar resources in a region provides only a crude idea of the value of solar energy. Not all electricity is created equal. A better metric for determining the availability of renewable resources in any given region is the "effective load carrying capability," or ELCC. The ELCC refers to the difference between the amount of energy a generating unit produces and the amount of energy that can actually be used by consumers at any given time. For example, nuclear and hydropower units have relatively low ELCCs because they are producing about the same amount of electricity 24 hours a day. In times of low demand, these units continue to produce energy, even if no one is using it. The excess energy must be stored, fed into the grid as reserve capacity, or wasted. By one estimate, U.S. power stations waste one-fifth more energy than Japan uses every year.[96]

Because solar generators tend to produce the greatest amount of energy during the same times consumer demand is highest, solar has an amazingly high ELCC relative to other technologies.[97] In many parts of the country, solar PV has an ELCC above 70 percent. In many parts of the Southeast, solar's ELCC exceeds 60 percent (see Map 14).[98] Researchers in Sacramento, California, estimated that the ELCC for solar PV within the city was so high that the actual value of solar energy was more than $6,000 per kW.[99] That is, because solar PV generated electricity at periods of high demand, its value was greater than electricity generated by other units throughout the day.

NREL researchers compared the recorded ELCC of solar PV deployed by utilities in nearly every region of the country to earlier theoretical estimates of ELCC. Not only did NREL find that actual ELCC closely matched expectations, they discovered that *valuable* amounts of solar PV are available in every region of the United States (see Figure 13).[100] A report analyzing affordable energy options for Alabama, Florida, Georgia, North and South Carolina, and Tennessee calculated that solar power systems covering just 0.1 percent of the region's land area could generate as much energy as 35 nuclear plants.[101]

Geothermal Power

The Earth's interior reaches temperatures greater than 4,000 degrees Celsius, with geothermal energy flowing continually to the surface. At present, only high- and moderate-temperature systems have been used to generate power, and they are located primarily near plate-boundary zones, as geothermal electric potential is highly dependent on rock porosity, permeability, subterranean reservoir temperature, and geomorphic pressure.[102]

For fossil fuels and uranium, conventional exploitation generally involves digging, crushing, and processing huge amounts of rock to recover relatively small amounts of a particular element. In contrast, geothermal energy is tapped by means of a liquid carrier, generally the water in the pores and fractures of rocks, that naturally reaches the surface at hot springs or can be brought to the surface by drilled wells. It is thus accomplished without large-scale movement of rock. Geothermal sources are also usable over a wide spectrum of temperature and volume, whereas most other natural resources can be reaped only if deposits exceed minimum size or grade for profitable exploitation.[103] Modern geothermal systems continue to improve their performance through advances in deeper drilling, vapor-dominated technology, permeability of fractures, and reservoir design.

While more than 40 operational geothermal plants provide about 3,000 MW of current power nationwide, the energy content of domestic geothermal resources to a depth of 3 kilometers is estimated to be equivalent to a 30,000-year supply of energy using current rates of consumption.[104] The USGS estimates that the nine western states can provide more than 20 percent of national electricity needs (up to 150,000 MWh) using existing geothermal technology.[105] Researchers at NREL similarly found that 2,800 MW of hydrothermal capacity could be developed using today's technology and 30,000 MW by 2050, as well as more than 100,000 MW of geopressured potential and 130,000 MW of deep geothermal potential.[106]

Bioelectric Power

Since biomass plants can combust a variety of fuels, the likelihood of fuel shortages is rare. Many industrial and agricultural processes produce significant amounts

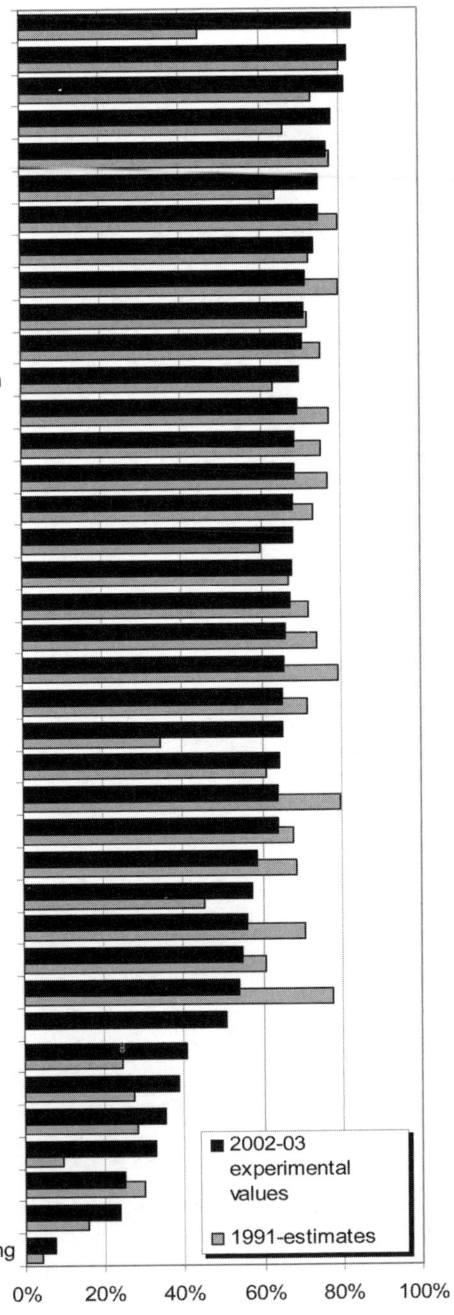

Figure 13
Effective Load Carrying Capability of Solar PV for Major Electric Utilities

Researchers had projected the theoretical ELCC for solar PV systems in 1991. When they followed through with more careful analysis in 2003, they found that the actual amount of ELCC potential closely matched their earlier estimates.

Source: National Renewable Energy Laboratory.

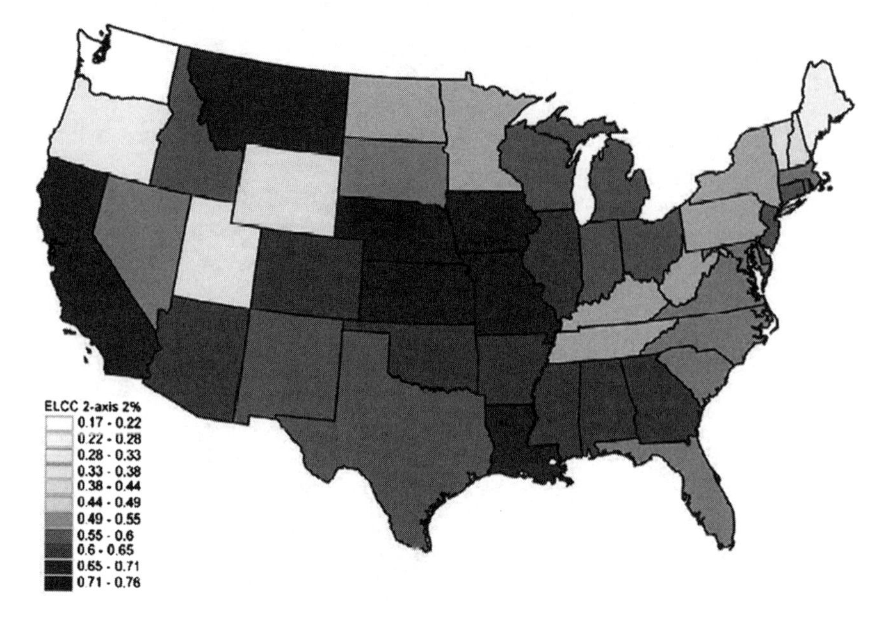

Map 14
Effective Load Carrying Capability for Solar PV by State

At least 33 states have solar PV resources with extremely high ELCC potential. The darker areas demonstrate better ELCC performance.
Source: National Renewable Energy Laboratory.

of combustible by-products, including tobacco residue, chicken carcasses, coffee grounds, peach pits, sawdust, scarp wood, and rice hulls. The Midwest possesses exceptionally large reserves of biomass fuel in the form of crop residues and energy crops, and the United States as a whole already has a surplus of available land that can be used for dedicated bioenergy crops. As one example, federal programs have idled 60 million acres over the past 10 years, and the Conservation Reserve Program removes 36 million acres from production each with the primary goal of erosion control. The vast majority of this land is planted in perennial grasses that could be harvested as energy crops while still meeting the goal of erosion prevention.

The DOE recently calculated that the nation's forestlands could produce 368 million dry tons of fuel annually, including 52 million dry tons of fuelwood harvested from forests, 145 million dry tons of residues from wood processing mills and pulp and paper mills, 47 million dry tons of urban wood residues including construction and demolition debris, 64 million dry tons from logging and site clearing operations, and 60 million dry tons of biomass from fuel treatment operations to reduce fire hazards. These numbers likely underestimate the potential of biomass as an energy source since the DOE excluded all forestland areas not currently accessible by roads as well as all environmentally sensitive areas.

The DOE also found that agricultural lands could produce nearly 1 billion tons of biomass and still meet food, feed, and export demands. Their projection included 428 million dry tons of annual crop residues, 377 million tons of perennial crops, 87 million tons of grains for biofuel, and 106 million tons of animal manure, process residue, and other currently wasted feedstocks. The DOE assumed that all cropland was managed with no-till methods and that only 55 million acres of cropland, idle cropland, and cropland pasture were dedicated to the production of perennial bioenergy crops.[107]

Furthermore, a study undertaken by the University of Tennessee found that the Tennessee Valley Authority (TVA) region featured extensive biomass and waste gas potential. The study suggested, for instance, that forest residues, agricultural residues, and energy crops alone could provide 22.2 billion kWh of electricity. It also concluded that TVA possessed a significant number of landfills and wastewater treatment plants where methane could be easily captured to produce electricity.[108]

Hydroelectric Power

The hydrologic cycle produces vast amounts of water that fall to the earth in the form of ice, rain, and snow that run into lakes, rivers, streams, ponds, marshes, and creeks. An estimated 3 million miles of rivers and streams traverse the nation, and every inch of rain releases 17.4 million gallons of water into them each square mile. By far the most used and mature of all renewable energy technologies, hydroelectric facilities harness the power of this moving water, operate in 48 of the 50 states, and offer a source of inexpensive power. Even though 80,000 MW of installed hydroelectric capacity exists, David Pimentel and his colleagues from Cornell University projected that the development and rehabilitation of existing dams in the United States could produce an additional 60 billion kWh per year.[109] An analogous DOE study found a staggering 165,551 MW of hydroelectric capacity could still be developed after excluding national battlefields, parks, parkways, monuments, preserves, wildlife refuges, management areas, and wilderness reserves.[110]

Lower Prices

As Chapter 1 noted, the conventional energy system exposes the American economy to significant variations in the cost of fossil fuels and uranium. The IEA estimated that even a modest 10 percent increase in fossil fuel prices could reduce GDP growth in the United States by more than $200 billion a year.[111] Clean power sources, especially energy efficiency and renewables, lock in the cost of electricity since they need not rely on volatile supplies of fuel. In the current era of restructuring, natural disasters, and price spikes, many manufacturers and utilities regard certainty as the most important factor in determining whether to invest in certain energy technologies. The more uncertainty there is about future fuel costs, the higher risk premiums placed on investment returns.

As a particularly insidious example of how utilities are not used to anticipating fossil fuel variability, consider the situation faced by MidAmerican—a large utility in the Midwest—in the early 1990s. MidAmerican fought against a state statute that required them to sign 30-year contracts with renewable energy providers guaranteeing a fixed rate of 6 ¢/kWh over the lifetime of the contract. At the time, MidAmerican's ratepayers were charged around 3 ¢/kWh. Rather than embracing the long-term contract, MidAmerican filed five lawsuits and litigated for close to 15 years to have the statute overturned. By the time MidAmerican reached a compromise with regulators, the utility's rates had already risen to around 9 ¢/kWh (33 percent more than the contract they had fought in court) and are expected to rise further.[112]

The experience with MidAmerican shows how a greater use of clean power technologies would insulate the U.S. economy from rising prices. The use of renewables and energy efficiency diversifies the "fuels" used to generate electricity, thereby minimizing the risk of fuel interruptions, shortages, and accidents. Together, clean power technologies can increase security by lessening the number of large and vulnerable targets on the grid, providing insulation for the grid in the event of an attack, and minimizing foreign dependence on oil and natural gas. It is far more certain that the sun will shine and the wind blow tomorrow than that saboteurs will not blow up a power station or snip a few transmission lines. Renewables are thus far more resilient and less attractive targets to possible attackers.

Clean power systems are safer in another way as well. One study assessing major energy accidents from 1907 to 2007 found that nonhydroelectric renewable resources are the *safest* of all energy technologies. Not one single major energy accident in the past century involved small-scale renewable energy systems or energy efficiency, whereas fossil-fueled, nuclear, and larger hydroelectric facilities caused 279 accidents totaling $41 billion in damages and 182,156 deaths.[113] An investigation of energy-related accidents in the European Union found that nuclear power was 41 times more dangerous than equivalently sized coal, oil, natural gas, and hydroelectric projects. Nuclear plants were at risk of killing about 46 people for every GWy of power produced (mostly from the Chernobyl accident).[114] A database of major industrial accidents from 1969 to 1996 compiled by the Paul Scherrer Institute found that 31 percent, or 4,290 out of 13,914, were related to the fossil fuel sector.[115] Another study concluded that about 25 percent of the fatalities caused by severe accidents worldwide in the period 1970 to 1985 occurred in the fossil fuel energy sector.[116]

Less obviously, clean power technologies bring economic and political benefits beyond insulating the economy from fuel interruptions and terrorist attack. In 2003, the EU implemented its own mandatory greenhouse gas control program, a move that only shortly preceded the entry into force of the Kyoto Protocol. Emissions reduction strategies undertaken by countries in Europe (as well as Japan and other industrialized countries that have ratified Kyoto) will increase electricity prices. Countries such as the United States and Canada, who refuse to abide by the protocol, will enjoy a competitive advantage in the form of lower energy costs and costs of production.

In response to this perceived advantage, trade organizations within Europe have advocated trade sanctions against the United States, arguing that there cannot be free and fair trade under World Trade Organization Agreements on Subsidies and Countervailing Measures until the United States implements the Kyoto Protocol. In the past, the EU has used trade sanctions whenever the United States has garnered an unfair competitive advantage by subsidizing exports. Two recent examples—the sales corporation/extraterritorial income and steel import cases—demonstrate that the EU can use trade sanctions to force a change in U.S. behavior.[117]

Greenhouse gas emissions are also regarded by many analysts as the single most important factor in U.S.-EU relations. When President George W. Bush chose to withdraw from the Kyoto Protocol, Margot Wallström, the EU environmental commissioner, called his decision "very worrying." EU spokesperson Annika Östergren went further when she stated that "sometimes people think that the Kyoto Protocol is only about the environment, but it's also about international relations and economic cooperation." Accordingly, clean power technologies could help the United States informally meet obligations under international law. At a minimum, they would improve the reputation and credibility of the managers of the American electric utility sector. If used more widely, renewables could ultimately help U.S. companies avoid trade sanctions and improve relations between the United States and the European Union.[118]

The political controversy over America's greenhouse gas emissions is not limited to the relationship between the United States and the European Union. New coal-fired facilities are a political liability at home as well, where they have become especially difficult to permit and construct in Arkansas, Florida, Kansas, Oklahoma, Oregon, and Texas. For instance, in 2006 the Southwestern Electric Power Company planned to build a 600 MW pulverized coal plant in Hempstead County, Arkansas, but was successfully opposed by local duck-hunting clubs, and Oregon denied PacificCorp approval to charge consumers for the costs of new coal plants built outside of the state. Dallas-based energy corporation TXU intended to build 11 new pulverized coal plants in early 2007, but their announcement resulted in public outrage and a $32 billion private equity buyout of the utility. In June of the same year, Florida enacted a law requiring utilities to consider renewable energy and energy efficiency in the permitting of all new power plants, and in September, Oklahoma regulators denied a subsidiary of American Electric Power permission to build a coal plant in the northern part of the state. In 2008, state environmental officials in Kansas rejected two coal-fired power plants on climate change grounds.[119] Another survey conducted in early 2008 found that at least 48 coal plants were being contested in 29 states.[120] Such rejections, buyouts, mandates, and delays, of course, are costly to utilities and conventional power producers, which could be why the DOE reports that in 2007, power providers added 5,329 MW of wind capacity ($9 billion worth of investment and a 46 percent increase from

2006) but only 1,400 MW of new coal plants.[121] If these trends continue, the coal industry may have to scuttle its plans for expansion discussed in Chapter 1.

Because fossil fuels involve inherent competition over limited commodities, perpetually constrained supply and constantly growing demand create a vicious cycle that increases the value of the fuel and adds additional costs absorbed by ratepayers. Since improved energy efficiency and renewable energy technologies utilize domestic and widely available fuels to produce electricity, they decrease demand on fossil fuels and, therefore, lower prices.

Energy efficiency, for example, has the greatest potential to minimize natural gas prices. Reducing peak electricity use by just 1 percent would immediately save 2 percent of all American natural gas usage and cut its price by 3 to 4 percent.[122]

Several studies have documented that an increase in renewable power production would also decrease costs for electricity generation by offsetting the combustion of natural gas.[123] Because some renewable resources generate the most electricity during periods of peak demand, they can help offset electricity otherwise derived from natural gas–fired "peaking" or reserve generation units. PV, for example, has great value as a reliable source of power during extreme peak loads. Substantial evidence from many peer-reviewed studies demonstrates an excellent correlation between available solar resources and periods of peak demand. In California, for example, an installed PV array with a capacity of 5,000 MW reduces the peak load for that day by about 3,000 MW, cutting in half the number of natural-gas "peakers" needed to ensure reserve capacity.[124]

The value of renewable energy to offset natural gas combustion varies with the projected supply (and thus the price) of natural gas. When demand for natural gas increases (or supply decreases), its price increases and so does the value of the renewable resources used to displace it. Given the historic volatility of the natural gas market, a 1 percent reduction in natural gas demand can reduce the price of natural gas by up to 2.5 percent in the long term.[125] Researchers at LBNL confirmed this inverse relationship between renewable generation and natural gas prices when reviewing the projected effect renewable energy technologies would have on future natural gas prices:

> Each 1 percent reduction in natural gas demand could lead to long-term average wellhead price reductions of 0.8 percent to 2 percent, with some of the models predicting more aggressive reductions. Reductions in the wellhead price will not only have the effect of reducing wholesale and retail electricity rates but will also reduce residential, commercial, and industrial gas bills.[126]

In a 2007 study, UCS assessed the cumulative effect of 20 percent renewable energy penetration on annual electricity prices and found it would save consumers more than $49 billion largely by depressing the price of natural gas used for electricity production and home heating.[127]

UCS is not alone in their findings. In Pennsylvania, where more than 90 percent of electricity comes from coal and nuclear resources, a study conducted by the consultancy Black & Veatch concluded that renewable energy deployment could result in a substantial reduction in fossil fuel consumption, lowering the price of coal and oil and ultimately providing cost savings to ratepayers. The study noted that even a 1 percent reduction in fossil fuel prices would lead to a $140 million reduction in fossil fuel expenditures for the state.[128]

Local Economic Growth

The more capital intensive a product is, the less embodied labor it has. Nuclear and fossil derived electricity are the most capital intense.[129] Power plants create net reductions in regional employment, as ratepayers must reduce expenditures on other goods and services to finance construction. Ralph Cavanagh from the NRDC estimates that, per dollar of capital expended, energy efficiency creates up to four times as many jobs as large central station plants. Moreover, the jobs created by energy efficiency are more dispersed geographically in proportion to the general population, avoiding the socially disruptive effects of energy-related boomtowns.[130]

On the supply side, the United Nations Environment Program (UNEP) assessed the employment impact of various electricity generation technologies and found that renewable energy technologies generate three times as many jobs per MW of installed capacity as fossil fuel–based generation. The wind industry demonstrates the disparity quite clearly. According to a survey by Danish wind energy manufacturers, 17 worker-years are created for every MW of wind energy manufactured, and five worker-years for every MW installed:

> Renewable energy technologies are up to three times more employment-intensive than fossil fuel power options: 188 worker-years are created locally for every MW of small solar electric systems. In Germany, wind power accounted for 1.2 percent of electricity generation in 1998 and the industry employed 15,000 people, compared to nuclear with 33 percent share and about 40,000 jobs, and coal with a 26 percent share and 80,000 jobs. Based on a market share comparison, the potential to create jobs is far greater for wind than for coal and nuclear options.

In the year 2000, the wind energy industry provided more than 85,000 jobs worldwide and UNEP projects that the sector could provide up to 1.8 million jobs by 2020.[131]

Researchers at the University of California at Berkeley found an even larger job creation ratio in the United States. Professor Daniel Kammen, head of UC Berkeley's Renewable and Appropriate Energy Laboratory (RAEL), confirmed that investments in renewable energy technologies would produce as much as 10 times as many American jobs than comparable investments in fossil fuel or nuclear technologies.[132]

Moreover, investments in wind have the potential to create jobs where they are most desperately needed. The Renewable Energy Policy Project used North American Industrial Classification Codes to map the dispersion of manufacturing activity related to the development of wind energy. They found that expansion of the domestic wind industry could create 831,499 jobs in 20 states, and that those states receiving the most investment and most new manufacturing jobs would account for 75 percent of the total U.S. population, and 76 percent of the manufacturing jobs lost in the United States between 2001 and 2004 (see Table 6).

Table 6
Wind Energy Job Creation Potential

The development of a robust domestic wind energy industry would bring thousands of new jobs to the 20 states that have lost the most manufacturing jobs in the past two decades. The states here are ranked by average potential investment.

State	Employees at Potential Companies	Number of New Jobs	Average Investment ($ Billions)
California	102,255	12,717	4.24
Ohio	80,511	11,688	3.90
Texas	60,229	8,943	2.98
Michigan	66,550	8,549	2.85
Illinois	57,304	8,530	2.84
Indiana	53,064	8,317	2.77
Pennsylvania	50,304	7,622	2.54
Wisconsin	48,164	6,956	2.32
New York	47,375	6,549	2.18
South Carolina	20,532	4,964	1.65
North Carolina	30,229	4,661	1.55
Tennessee	28,407	4,233	1.41
Alabama	21,213	3,571	1.19
Georgia	20,898	3,532	1.18
Virginia	20,201	3,386	1.13
Florida	24,008	3,371	1.12
Missouri	23,634	3,234	1.08
Massachusetts	27,955	3,210	1.07
Minnesota	26,131	3,064	1.02
New Jersey	22,535	2,920	0.97

Source: Renewable Energy Policy Project.

However, the regional dispersion of economic benefits is not limited to the wind-manufacturing sector or to the states of the upper Great Plains. In the Southeast, for example, researchers at the University of Tennessee estimated that renewable energy technologies (including biomass generators and incremental hydropower) could create more jobs per MWh for the region than any other type of electricity generation.[133] The American Society of Mechanical Engineers (ASME) estimated that coal and gas facilities provide only 11 job-years per MWh, while renewable technologies can provide as many as 121 job-years per MWh.[134]

Clean power sources also contribute to local economic growth. They make local businesses less dependent on imports from other regions, free up capital for investments outside the energy sector, and serve as an important financial hedge against future energy price spikes. Ninety-three percent of the fuel distributed to electric utilities in the Southeast in 2005 had to be imported from other states and countries. The region lost $7.3 billion of revenue paying for coal (with more than $700 million going to coal from Columbia, Venezuela, and Poland) and $1.1 billion importing uranium over the course of one year.[135] A study conducted in California found that every dollar invested in solar thermal plants resulted in $1.40 of gross state output, where dollars invested in fossil fuels merely transferred revenue out of the region.[136] The state of Arizona estimates that 79.4 cents of every dollar spent on fossil fuel and 56.5 cents spent on electricity are completely diverted from the local economy.[137]

In contrast, researchers studied the 20 most successful DOE energy efficiency programs and found that they saved the nation 5.5 quads of energy worth around $30 billion, but cost consumers and taxpayers just $712 million, creating savings of more than 97 percent.[138] All of these savings went back into the American economy, creating more jobs and stimulating economic productivity. Similarly, the DOE attempted to quantify the benefits of state energy efficiency activities. Their survey, which covered 18 project areas including energy audits, retrofits, procurement, technical assistance, codes and standards, tax credits, and financial incentives, found that while states spent $44.1 million on energy efficiency in 2002, the annual energy savings of these programs exceeded $333 million. Each dollar of energy efficiency funneled $7.22 back into the local economy.[139]

No. 4: The Environment Challenge

Beyond economics, clean power sources yield plentiful environmental benefits related to water conservation, land preservation, and air quality.

Water

Energy efficiency and some renewables do not consume or withdraw water, and the DOE acknowledges that wind power and solar PV could play a key role in averting a "business-as-usual scenario" where "consumption of water in the electric sector could grow substantially." [140] Another DOE report noted that "greater additions of wind to offset fossil, hydropower, and nuclear assets in a generation portfolio will result in a technology that uses no water, offsetting water-dependent technologies." [141] Ed Brown, director of Environmental Programs at the University of Northern Iowa, estimated that a 100 W solar panel would save approximately 2,000 to 3,000 gallons of water over the course of its lifetime. Similarly, Dr. Brown concluded that "billions of gallons of water can be saved every day" through the greater use of renewable energy technologies.[142]

AWEA conducted one of the most comprehensive assessments of renewable energy and water consumption. Their study estimated that wind power uses less than 1/600 as much water per unit of electricity produced as does nuclear power plants, 1/500 as much as coal units, and 1/250 as much as natural gas facilities (small amounts of water are used to clean wind and solar systems).[143] By displacing centralized fossil fuel and nuclear generation, clean power sources such as energy efficiency and renewables can conserve substantial amounts of water that would otherwise be withdrawn and consumed for the production of electricity (see Figure 14).

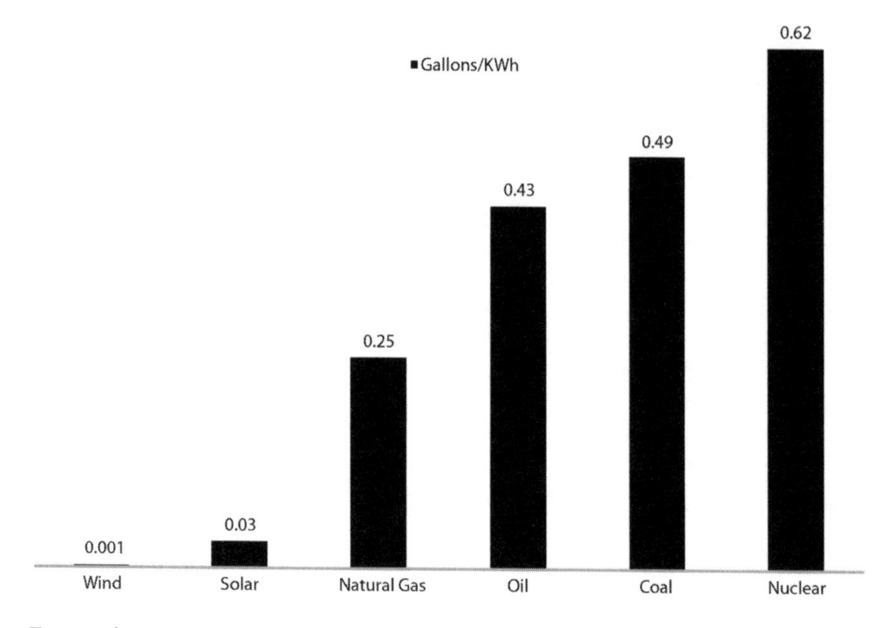

Figure 14
Water Consumption for Conventional and Renewable Power Plants

Shown here in gallons/kWh for power plants in California, renewable generators greatly conserve water when compared to conventional units.
Source: American Wind Energy Association.

Air

Wind turbines also drastically improve air quality. A single 1 MW wind turbine running at only 30 percent of capacity for one year displaces more than 1,500 tons of CO_2, 2.5 tons of sulfur dioxide, 3.2 tons of nitrogen oxides, and 60 pounds of toxic mercury emissions.[144] One study assessing the environmental savings of a 580 MW wind farm located on the Altamont Pass near San Francisco, California, concluded that the turbines displaced hundreds of thousands of tons of air pollutants each year that would have otherwise resulted from fossil fuel combustion.[145] The study estimated that the wind farm will displace more than 24 billion pounds of nitrogen oxides, sulfur dioxides, particulate matter, and CO_2 over the course of its 20-year lifetime—enough to cover the entire city of Oakland in a pile of toxic pollution 40 stories high.[146]

In terms of climate change, and greenhouse gases, the IAEA estimates that when direct and indirect carbon emissions are included, coal plants are about five times more carbon intensive than solar and 140 times more carbon intensive than wind technologies. Natural gas fares little better, at three times as carbon intensive as solar and 20 times as carbon intensive as wind.[147] In the United States, the DOE estimates that every kWh of renewable power avoids the emissions of more than 454 grams of CO_2.[148] According to data compiled by UCS, achieving 20 percent renewables penetration would reduce CO_2 emissions by 434 million metric tons by 2020—the equivalent of taking nearly 71 million automobiles off the road.[149]

An almost identical study published in *Energy Policy* found that biomass facilities were about 10 times cleaner than the best coal technology and that wind, solar panels, and hydroelectric were almost 100 times cleaner than the cleanest coal system.[150] Martin Pehnt from the Institute for Energy and Environmental Research in Heidelberg conducted life-cycle analyses of 15 separate DG and renewable energy technologies and found that all but one, solar PV, emitted much less gCO_2e/kWh (grams of

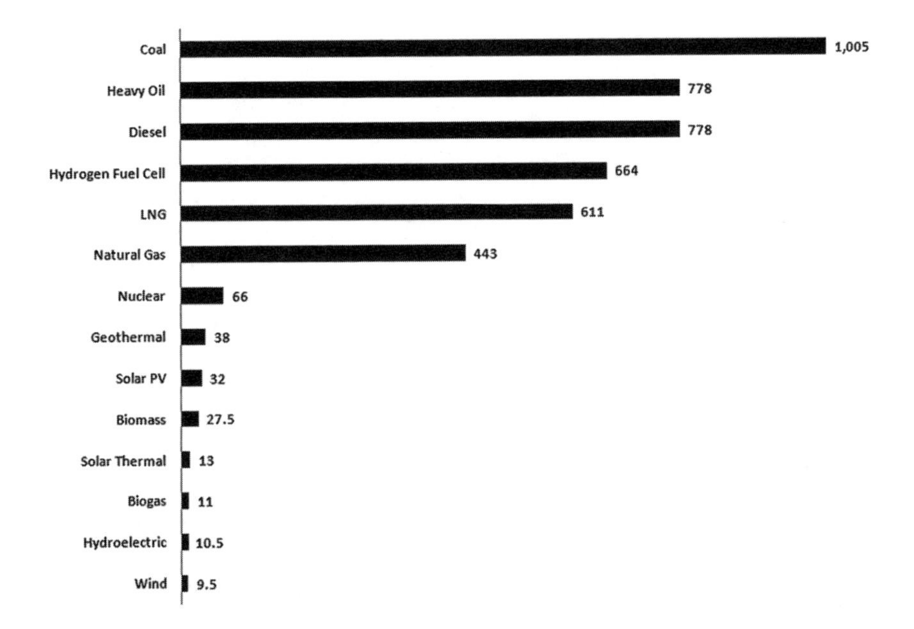

Figure 15
Life-Cycle Greenhouse Gas Emissions for Conventional and Renewable Power Plants (equivalent grams of CO_2/kWh)

Shown here in equivalent grams of carbon dioxide per kWh, renewable power generators are often magnitudes of order better at fighting climate change than nuclear and fossil-fueled units.

Source: Sources for wind, hydroelectric, biogas, solar thermal, biomass, and geothermal estimates taken from Marin Pehnt, "Dynamic Lifecycle Assessment of Renewable Energy Technologies," *Renewable Energy* 31 (2006): 55–71. Diesel, heavy oil, coal, natural gas, and fuel cell estimates taken from Luc Gagnon, Camille Belanger, and Yohji Uchiyama, "Lifecycle Assessment of Electricity Generation Options: The Status of Research in Year 2001," *Energy Policy* 30 (2002): 1267–1278. Solar PV estimates taken from V. M. Fthenakis, H. M. Kim, and M. Alsema, "Emissions from Photovoltaic Life Cycle," *Environmental Science and Technology* 42 (2008): 2168–2174. LNG estimate taken from Christopher Dey and Manfred Lenzen, "Greenhouse Gas Analysis of Electricity Generation Systems," *ANZSES Solar 2000 Conference Proceedings* (Griffith University, Queensland, November 29–December 1), 658–668. Nuclear estimate taken from Benjamin K. Sovacool, "Valuing the Greenhouse Gas Emissions from Nuclear Power: A Critical Survey," *Energy Policy* (forthcoming, 2008).

CO_2 equivalent per kWh) than nuclear plants.[151] In an analysis using updated data, researchers from Brookhaven National Laboratory found that current estimates of the greenhouse gas emissions for a typical solar PV system range from 29 to 35 gCO_2e/kWh[152]—about one-half the equivalent emissions for nuclear power and about 33 times less the equivalent emissions from a coal-fired facility. While it may be unfair to compare base-load sources such as nuclear to intermittent or non-dispatchable sources such as wind and solar PV, if these updated numbers are correct, then renewable energy technologies are two to seven times more effective on a per kWh basis at fighting climate change.

Energy efficiency is even better. Physicist Amory Lovins calculates that every 10 cents spent to buy a single kWh of nuclear electricity could have purchased 1.2 to 1.7 kWh of wind power, 2.2 to 6.5 kWh of building scale CHP, or 10 kWh or more of energy efficiency. Put another way, nuclear power saves as little as half as much carbon dioxide per dollar as wind power and cogeneration, and from several fold to at least tenfold less carbon per dollar than end-use energy efficiency.[153] Researchers Keepin and Kats similarly conclude that each dollar invested in energy efficiency displaces nearly seven times as much CO_2 as a dollar invested in nuclear power.[154] More recently, a 2008 study from the consulting firm McKinsey and Company found that investments in energy efficiency were so profitable that more than 40 percent of greenhouse gas emissions could be abated at negative marginal costs. That is, the things businesses should be doing to fight climate change are those they should be doing anyway to improve manufacturing efficiencies and maximize profits.[155]

Land

Clean power sources also require less land area than conventional generators, and most of the land they occupy is still available for other uses. When configured in large centralized plants and farms, wind and solar technologies use around 10 to 78 square kilometers of land per installed GW per year, but traditional coal-fired plants can use more than 100 square kilometers of land per year to produce the same amount of electricity when using open cut coal mines.[156] A separate study calculating the lifetime fuel cycle and land use impacts of a coal plant confirmed that, in every case, coal plants exceed the land use of comparable clean power sources.[157]

In open and flat terrain, newer large-scale wind plants require about 60 acres per MW of installed capacity, but the amount drops to as little as 2 acres per MW for hilly terrain. While this may sound like a lot, however, only 5 percent (3 acres) or less of this area is actually occupied by turbines, access roads, and other equipment; 95 percent remains free for other compatible uses such as farming or ranching.[158] Alan Nogee from UCS estimates that only a small fraction of contiguous land in the country, ranging from between 0.11 and 0.26 percent, would be needed to supply 20 percent of the nation's electricity from wind energy, and of that land, more than 98 percent would be available for other uses.[159]

At the High Winds Project in Solano, California, eight different landowners host 90 separate 1.8 MW wind turbines that total 162 MW of electricity capacity, but are still able to use about 96 percent of farmland around and between the turbines. Using a conservative figure of 26 acres for each wind turbine, researchers from Oberlin College estimated that 40 square miles could support roughly 38,000 turbines producing 3 to 4 percent of total U.S. electric demand each year. The actual footprint of these turbines would be roughly 10,000 acres, leaving the surrounding 990,000 acres of land either untouched or available for other uses. This figure beats both coal and natural gas in terms of total land use.[160]

When integrated into building structures and facades, solar PV systems can require no new land. The California Exposition Center in Sacramento, California, fully integrates 450 kW of PV into a parking lot. Indeed, NREL concluded that "a world relying on PV would offer a landscape almost indistinguishable from the landscape we know today." [161] The Energy Policy Initiatives Center at the University of San Diego recently estimated that the city could construct 1,726 MW of solar PV relying only on available roof area downtown.[162] In fact, the Worldwatch Institute noted that solar power plants that concentrate sunlight in desert areas require 2,540 acres per billion kWh. On a life-cycle basis, this is less land than a comparable coal or hydropower plant generating the same amount of electricity.[163]

(Many additional environmental benefits are discussed in Chapter 6 when refuting common environmental objections to clean power systems.)

Putting It All Together

Historian Peter Novick once compared compiling statistics to trying to "nail jelly to the wall." The process of collecting and synthesizing data on the costs and benefits of clean and dirty energy sources is complicated and imperfect. This is partly because there are many different types of electricity "prices." There are electricity costs at the bus bar, that is, before it enters the T&D grid; its price at the wholesale level when traded between firms; and its retail price for both residential and industrial customers.[164] Whatever type of price we are talking about for coal-generated electricity, it must also include the money put into air quality, care of people with asthma and other respiratory problems, water shortages, and land reclamation. The cost of nuclear energy must include money for constructing and maintaining permanent waste storage sites, the risk of a catastrophic accident, and the price of decommissioning plants. The price of renewables must include the money that goes into R&D, construction, and maintenance. Not all things can be quantified; not all of those that can be are consistent; comparisons seldom assess costs and benefits across technologies; and most methods suffer from methodological shortcomings. Using three concepts together, however, provides a more accurate picture of the real costs of electricity. While definitely imperfect, marginal costs, levelized costs, and full social pricing are much better at nailing the jelly to the wall.

The concept of the margin or marginal costs separates the past and the present (which we cannot influence) from the future (whose shape we can determine). The marginal cost of a product, or in this case a generator, is the cost of the next power plant to be planned and built in the future. Marginal costs tend to be much higher than historic or current costs because the cheap and easy things have already been done. Engineers cannot put another Hoover Dam where one already is. Confusion is often caused, sometimes deliberately, by comparing the historic or current cost of one alternative with the marginal cost of another. To be fair and consistent, marginal costs must be compared equally among all technologies.[165]

Full life-cycle or levelized costs account for initial capital costs, future fuel costs, future operation and maintenance costs, and decommissioning costs, and average these out over the lifetime of the equipment and the expected electricity it will generate. In short, the levelized cost of electricity (LCOE) refers to the cost over the life of a generator divided by the numbers of kWh it will produce. Using data from the IEA, Cornell University, CEC, NREL, and the Virginia Center for Coal and Energy Reserve (and looking at *marginal* costs), the LCOE for conventional, DG, CHP, and renewable generators is presented in Table 7.

Four clean power sources—energy efficiency, offshore wind, hydroelectric, and landfill gas—are the most competitive technologies *at today's rates and prices.* These findings have even been confirmed by nonpartisan researchers at the United Nations, who estimated that new hydroelectric, geothermal, wind, and biomass plants represented the cheapest sources of power when all subsidies were excluded.[166] The reason is that new wind technologies operating at lower wind speeds and employing stronger materials and solar technologies utilizing plastics, nanostructured materials, and thinner modules have greatly improved efficiency, lowered cost, and enhanced performance.

Today's rates and prices, however, are still biased toward conventional sources because they do not include "externalities" in energy prices. Defined as costs and benefits resulting from an activity that do not accrue to the parties involved in the activity, externalities have won attention in recent decades as a way to make equitable choices of generation equipment.[167] Scientist Russell Lee explains that "externalities are part of the overall *social* cost of producing electric power . . . including the value of any damages to the environment, human health, or infrastructure."[168] The DOE defines externalities as "inadvertent and unaccounted for effects of one or more parties on the welfare of another."[169]

Take the classic example of unregulated pollution from a smokestack. A factory produces items that are priced by taking into account the demand for the products, labor, capital, and other costs, but the damages from the factory's pollution—health and other effects—are true costs borne by society that are unaccounted for in the price of the factory's widgets. These latter costs are commonly referred to as "externalities" because people tend to consume them as by-products of other activities that are "external" to market transactions, and therefore unpriced.[170] One survey of American electricity markets in 2005, for instance, found that clean power technologies were not adequately valued for at least six of the *positive* externalities that they provided, including risk management, environmental performance, investment, reduced resource use, improved public image, and economic spillover effects (see Table 8).[171] Shimon Awerbuch, a financial economist, found that the "risk management" benefits of clean power sources amounted to at least 0.5 ¢/kWh that were

Table 7
Nominal Levelized Cost of Electricity for Different Generators, 2007

Looking at the costs of power production spread over the lifetime of each generator, energy efficiency, offshore wind, hydroelectric, and landfill gas offer the cheapest forms of electricity.

Technology	Nominal LCOE, $2007 (¢/kWh)
Energy Efficiency and DSM	2.5
Offshore Wind	2.6
Hydroelectric	2.8
Biomass (Landfill Gas)	4.1
Advanced Nuclear	4.9
Onshore Wind	5.6
Geothermal	6.4
Integrated Gasification Combined Cycle	6.7
Biomass (Combustion)	6.9
Scrubbed Coal	7.2
Advanced Gas and Oil Combined Cycle	8.2
Gas Oil Combined Cycle	8.5
IGCC with Carbon Capture	8.8
Parabolic Troughs (Solar Thermal)	10.5
Advanced Gas and Oil Combined Cycle with Carbon Capture	12.8
Solar Ponds (Solar Thermal)	18.8
Advanced Combustion Turbine	32.5
Combustion Turbine	35.6
Solar Photovoltaic (panel)	39.0

Source: Data for biomass, nuclear, onshore wind, IGCC, scrubbed goal, advanced gas and oil combined cycle, gas and oil combined cycle, IGCC with carbon capture, Advanced Gas and Oil Combined Cycle with Carbon Capture, Advanced Combustion Turbine, Combustion Turbine, and Solar PV taken from Michael Karmis et al., *A Study of Increased Use of Renewable Energy Resources in Virginia* (Blacksburg, VA: Virginia Center for Coal and Energy Research, 2004). Those estimates assume a 25-year system life; All federal tax incentives and credits as of 2007; Accelerated depreciation with half-year convention (MACRS); A discount rate of 7.0 percent; Current costs and capacity factors (no cost and performance improvements over time); Inflation rate of 2.5 percent per year; Fixed and variable operations and maintenance costs escalated at the inflation rate; Capital costs associated with the connection of centralized systems to the electricity grid are not included; Fixed and variables costs associated with electricity distribution and transmission are not included. Data for offshore wind from W. Short, N. Blair, and D. Heimiller, "Projected Impacts of Federal Policies on U.S. Wind Market Potential," Presentation at the 2004 Global WindPower Conference, Chicago, Illinois, March 29–31, 2004 (Golden, CO: National Renewable Energy Laboratory, NREL/CP-620-36052), p. 5. Data for geothermal from Alyssa Kagel and Karl Gawell, "Promoting Geothermal Energy: Air Emissions Comparison and Externality Analysis," *Electricity Journal* 18, no. 7 (August/September 2005): 90–99. Data for solar thermal from D. R. Mills, P. Le Lievre, and G. L. Morrison, *First Results from Compact Linear Fresnel Reflector Installation* (Sydney, Australia: University of New South Wales, June 2004). Data from hydroelectric from David Pimentel et al., "Renewable Energy: Current and Potential Issues," *Bioscience* 52, no. 12 (2002): 1111–1120. Data for DSM and energy efficiency taken from International Energy Agency, *The Experience with Energy Efficiency Policies and Programs in IEA Countries: Learning from the Critics* (Paris, France: International Energy Agency, August 2005).

Table 8

The Unpriced "Positive" Externalities from Clean Power Sources

One of the reasons clean power technologies sometimes appear so expensive is because many of the benefits they provide are not properly valued in the market.

Risk Management	Environmental Performance	Investment	Reduced Resource Use	Improved Public Image	Economic Spillover Benefits
Hedge against fuel price volatility	Emissions credits	Production tax credit	Reduced water use	Improved relations with stakeholders	Rural revitalization
Hedge against future environmental regulations	Reduced emissions fees	Accelerated depreciation	Lower production costs	Corporate social responsibility	Jobs and employment
Hedge against future carbon tax	Avoided remediation and pollution abatement costs	Local tax base improvements	Reduced energy use and wear and tear on T&D grid		Economic development
Minimization of reliance on futures markets					Avoided environmental costs of fuel extraction and transport
Reduced insurance premiums					

Source: J. E. Pater, *A Framework for Evaluating the Total Value Proposition of Clean Power Technologies* (Golden, CO: National Renewable Energy Laboratory, Technical Report NREL/TP-620-38597, February 2006).

not reflected in traditional electricity markets.[172] Figure 16 illustrates some of the negative externalities associated with the fuel cycle of a typical coal plant.

So, while LCOE is a useful starting point for calculating the costs of electricity generation, it fails to price a host of externalities associated with electricity generation, including:

- Catastrophic risks such as nuclear meltdowns, oil spills, coal mine collapses, natural gas wellhead explosions, and dam breaches;
- An increased probability of wars due to natural resource extraction or the securing of energy supply;
- Public health issues and chronic disease, morbidity, and mortality;
- Worker exposure to toxic substances and occupational accidents and hazards;
- Public deaths and injuries due to coal trucks, barges, and trains;
- Direct land use by power plants, pipelines, and upstream infrastructure;
- The destruction of land by mining operations including acid drainage and resettlement;
- Acid precipitation and its damage to fisheries, crops, forests, and livestock, especially the effects of sulfur dioxide on wheat, barley, oats, rye, peas, and beans and the

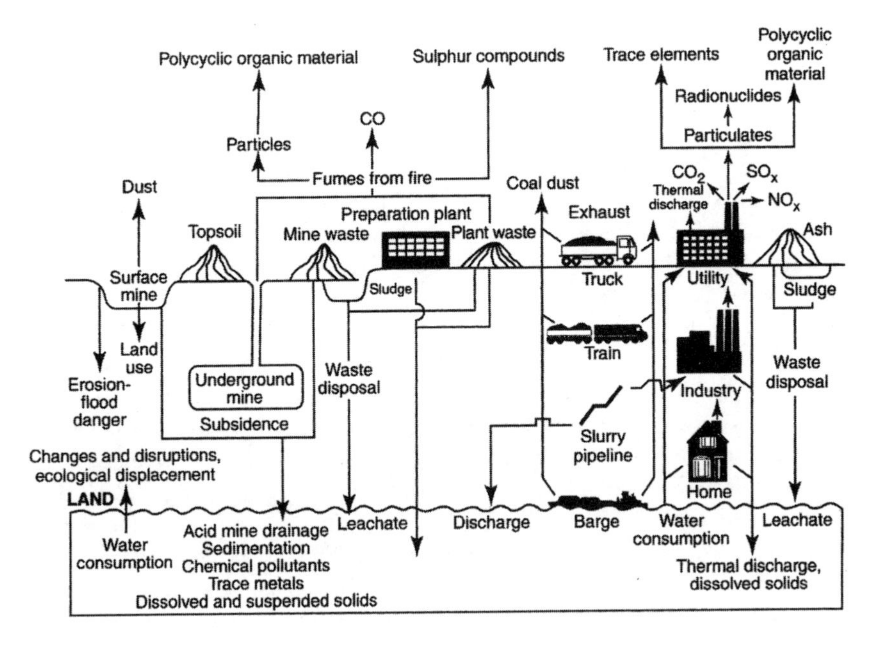

Figure 16
Negative Environmental Externalities of the Coal Fuel Cycle

The environmental damage from coal production and use occurs at almost every point in the fuel cycle.

Source: Oak Ridge National Laboratory.

impacts of acid deposition on other high value crops such as vegetables, fruit, and flowers;

- The effects of water pollution on fisheries and freshwater ecosystems, sensitive to water chemistry, as well as the release of radionuclides, drill cuttings, drilling muds, and oils;
- Consumptive water use, with consequent impacts on agriculture and ecosystems where water is scarce;
- Degradation of cultural icons such as national parks, recreational opportunities, or activities such as fishing or swimming;
- Atmospheric damage to buildings, automobiles, and materials by corrosion and the increased maintenance costs for natural stone, mortar, rendering, zinc, galvanized steel, and paint;
- Continual maintenance of caches of spent nuclear fuel;
- Cumulative environmental damage to ecosystems and biodiversity through species loss and habitat destruction, as well as the ecosystem services provided by wetlands, waterways, different types of forests, grasslands, deserts, tundra, coastal and ocean habitat;
- Changes to the local and regional economic structure through the loss of labor and jobs and transfer of wealth and reductions in GDP;
- Incidence of noise and reduced amenity, aesthetics, and visibility.[173]

While this list is still incomplete, Thomas Sundqvist and Patrik Soderholm analyzed 38 electricity externality studies and 132 estimates for individual generators to determine the extent that positive and negative externalities were not reflected in electricity prices.[174] They found that these costs, when averaged across studies, represented an additional 0.29 to 14.87 ¢/kWh. Taking the median values from Sundqvist and Soderholm and adjusting them to $2007, one gets Table 9, which shows that the eight technologies with the lowest full social costs are energy efficiency, offshore wind, onshore wind, geothermal, hydroelectric, two types of biomass, and solar thermal. When all of their social costs are included, scrubbed coal is 10 times more expensive than energy efficiency; advanced nuclear energy is five times more expensive than offshore wind; and hydroelectric and geothermal is half as much as the most advanced natural gas turbine. Taking just the extra cost associated with scrubbed coal—19.14 ¢/kWh—and multiplying it by coal's generation last year (1,191 billion kWh), the amount is boggling: $228 billion. That is right. Coal generation created $228 billion of additional costs that neither coal producers nor consumers had to pay for in 2006, costs that were instead shifted to society at large.

Three important points need to be made about these numbers. First, by combining LCOE and full cost pricing and looking at marginal costs of generation, the numbers above are a much more accurate assessment of fuel costs than estimates relying on each in isolation.

Second, the costs above already factor in the intermittent nature of some renewable resources such as wind and solar, assigning wind a capacity factor of 35 percent and solar PV a capacity factor of 17 percent.

Third, there are many reasons why Table 9 is extremely conservative:

- When surveying externalities, Sundqvist and Soderholm did not include *any* value for CO_2 and climate change. They explain that their metasurvey found a range of damages so large (from 1.4 ¢/kWh to 700 ¢/kWh) that they decided to exclude climate change externalities.
- In some cases the studies analyzed relied on a "willingness-to-pay" metric to assess damages, but many things (such as clear skies, absolute silence, or a dead child) are impossible to quantify in dollars. Furthermore, virtually none of the studies accounted for the risk of environmental damages—such as tipping points that are crossed as the earth's climate changes, unknown ecological thresholds that are passed, and species extinctions that are fundamentally irreversible—that entail impacts impossible to recover from once they begin.
- Most of the studies surveyed modeled damages associated with a single power plant, and not the combined or cumulative damages from a fleet of power plants or an entire utility system.
- Many studies assumed reference, rather than representative, technologies; that is, they assumed benchmark and state-of-the art technologies instead of those used by utilities in the real world where one-fifth of the nation's power plants are more than 50 years old.[175]
- Almost no studies included the environmental damages associated with T&D; these have environmental impacts ranging from land use conflicts, soil erosion, destruction of forests and natural habitat, as well as audible noise and interference with radio and television reception, deleterious effects on local birds colliding with power lines and towers, electrocutions, the use of chemical herbicides and vegetation management techniques along rights of way, and the human health effects of exposure to electric and magnetic fields (which some researchers claim may contribute to childhood cancer).[176]

Others have made similar estimates of the relative life-cycle costs of conventional and renewable fuels. In one recent study, traditional coal boiler generation technology appeared to produce relatively cheap power—under 5 ¢/kWh over the life of the equipment, which included capital, operating and maintenance costs, and fuel costs—while wind-turbine generators and biomass plants produced power that cost 7.4 ¢/kWh and 8.9 ¢/kWh, respectively (and tended to require larger amounts of land). But when analysts factored in a host of externality costs, coal boiler technology costs rose to almost 17 ¢/kWh, while wind turbines and biomass plants yielded power costing around 10 ¢/kWh.[177]

Researchers from ASE found that if damages to the environment in the form of noxious emissions and impacts on human health resulting from combustion of coal, oil, and natural gas were included in electricity prices, coal would cost 261.8 percent more than it does; oil, 13.4 percent; and natural gas, 0.5 percent. If priced to include the risks from greenhouse gas emissions and climate change, the costs of coal would rise 35 to 70 percent more; oil, 9 to 18 percent more; natural gas, 6 to 12 percent more. The researchers also found that if electricity *was* priced this way, fossil fuel use would decrease 37.7 percent compared to projections; CO_2 emissions would

Table 9
The Full Social Cost of Power Generators, 2007

The table on the left takes the nominal LCOE costs for generators and adds negative and positive externalities to reach the full social cost. The table on the right shows each power generator reordered according to its full social cost.

Technology	Nominal LCOE, \$2007 (¢/kWh)	Nominal External Cost, \$2007 (¢/kWh)	Full Social Cost, \$2007 (¢/kWh)
Energy Efficiency and DSM	2.5	0.0	2.5
Offshore Wind	2.6	0.4	3.0
Hydroelectric	2.8	4.94	7.8
Biomass (Landfill Gas)	4.1	6.7	10.8
Advanced Nuclear	4.9	11.10	16.0
Onshore Wind	5.6	0.4	6.0
Geothermal	6.4	0.7	7.1
Integrated Gasification Combined Cycle	6.7	19.14	25.9
Biomass (Combustion)	6.9	6.7	13.6
Scrubbed Coal	7.2	19.14	26.3
Advanced Gas and Oil Combined Cycle	8.2	11.97	20.2
Gas Oil Combined Cycle	8.5	11.97	20.5
IGCC with Carbon Capture	8.8	19.14	27.9
Parabolic Troughs (Solar Thermal)	10.5	0.9	11.4
Advanced Gas and Oil Combined Cycle with Carbon Capture	12.8	11.97	24.8
Solar Ponds (Solar Thermal)	18.8	0.9	19.7
Advanced Combustion Turbine	32.5	6.46	39.0
Combustion Turbine	35.6	6.46	42.1
Solar Photovoltaic (panel)	39.0	0.9	39.9

decrease 44.1 percent; GDP would improve 7.7 percent; and household wealth would jump 5.5 percent (primarily as the result of improved health).[178]

Professors Daniel Kammen and Sergio Pacca found that if they internalized the cost of mortality and asthma—just two items—into electricity rates, assuming the value of a life was \$5 million (cheap in my opinion), then the annual cost of operation for conventional coal power plants in Illinois, Massachusetts, and Washington was 50 ¢/kWh, almost eight times higher than the average 6.5 ¢/kWh paid by consumers.[179]

Another study found that even when considering best available technology, and only looking at the cost of air emissions and solid and liquid wastes in Europe, damages from conventional generators would cost the entire amount that consumers were paying as the actual market price for electricity. Put another way, "internalizing" the environmental damages from the cleanest conventional technologies

Technology	Full Social Cost, $2007 (¢/kWh)
Energy Efficiency and DSM	2.5
Offshore Wind	3.0
Onshore Wind	6.0
Geothermal	7.1
Hydroelectric	7.8
Biomass (Landfill Gas)	10.8
Parabolic Troughs (Solar Thermal)	11.4
Biomass (Combustion)	13.6
Advanced Nuclear	16.0
Solar Ponds (Solar Thermal)	19.7
Advanced Gas and Oil Combined Cycle	20.2
Gas Oil Combined Cycle	20.5
Advanced Gas and Oil Combined Cycle with Carbon Capture	24.8
Integrated Gasification Combined Cycle	25.9
Scrubbed Coal	26.3
IGCC with Carbon Capture	27.9
Advanced Combustion Turbine	39.0
Solar Photovoltaic (panel)	39.9
Combustion Turbine	42.1

Source: Virginia Center for Coal and Energy Research, California Energy Commission, Cornell University, and Sundqvist and Soderholm. Nominal external costs for IGCC are based on the average given for the separate values for gas and oil.

available would still *double* the price of electricity for consumers.[180] As two follow-up studies documented, if the environmental costs of electricity generation were included in its price, they could very well be equivalent to 1 to 2 percent of the entire GDP for the EU.[181]

The Way Forward

In many ways, the traditional, centralized, and dirty structure of the electric utility industry is becoming antiquated and obsolete. Demand for electricity continues to outpace supply. The fuel used for most electric generators is expensive and located great distances from its point of combustion. Older power plants and the transmission grid continue to deteriorate. And the environmental problems from conventional sources grow daily. Most utilities, argues Shalom Flank, chief technical officer for Pareto Energy Limited, are "in long term stasis" and risk "simply disappearing" unless they change their business strategies.[182] As Vikram Budhraja, former senior vice president at Southern California Edison, confided:

> The biggest problem facing the electric industry is to modernize the current infrastructure. We have power plants that are operating with an average age of thirty to forty years. We really need to replace them, modernize, and upgrade them, as well as build

> additional capacity to meet population increase and economic growth throughout the country. The industry needs to diversify the resource mix with renewables and replace aging power plants. The industry needs to invest in new grid technologies for reliability, market efficiency, and security.[183]

Some in the industry, in other words, seem to realize that it is critical to adapt and respond to many of these current challenges.

But why, despite the potential for clean power systems to provide comparatively abundant, secure, efficient, and cleaner forms of electricity, are these technologies not more widely deployed? Given that in the next few hundred years fossil fuel supplies could completely run out, more than 90 percent of domestically available energy resources are renewable and clean, and the full social cost of power clearly shows that clean energy technologies are cheaper than conventional technologies, why are policymakers not rigorously pursuing them? The benefits of clean energy technologies, backed by more than 250 studies cited in Chapters 1 and 2 of this book, should be as obvious as the color of ink on this page. The fact that they are not tells us that something is terribly, ludicrously amiss. The next four chapters explore this dilemma in detail.

CHAPTER 3

Financial and Market Impediments

Very few people purchase combined cycle natural gas turbines or pulverized coal power plants for personal enjoyment. Game show hosts do not offer such items as prizes, actors do not showcase them on television programs, and politicians do not feature them in lotteries. And yet they are more integral to society than game shows, television, or even politics. Their services may be a matter of life and death, for example, by refrigerating vaccines at hospitals, lighting the workplace, and providing heat for schools.

This chapter shows that, despite the massive and incontrovertible benefits from clean power sources discussed in Chapter 2, a mesh of financial and market impediments prevents residential customers, small and large businesses, industries, and utilities from investing in them. While clean power sources are cheap, they are not free, and they have higher, comparative capital costs (in dollars per kilowatt) than conventional options. These higher capital costs place many smaller, decentralized systems out of the price range of most residential consumers. Entrepreneurs and business owners argue that investing in clean systems is too expensive and deviates from the core mission of their corporate goals. Even some electric utility managers generally shun clean power technologies, using their incumbent market power to block their interconnection to the grid. The market share of large utilities coupled with discriminatory regulations can play a significant role as well.

Basic economic theory explains (and even predicts) many of the impediments to the adoption of clean power technologies. Chapters 1 and 2 demonstrated the existence of externalities associated with all energy systems, but another closely related concept is that of a "public good." Private goods, such as apples and automobiles, are goods that can be owned by individuals. There is a long history of defined property rights associated with private goods, and individuals also can be excluded from owning them.

Public goods, in contrast, are "nonexcludable" and "nonrivalrous." Exclusion from consuming public goods is not readily possible, and consumption of public goods by additional parties does not reduce the quantity of the good available to others. The classic example of a public good is national defense: whether you pay taxes or not, you are still "defended," and no matter how secure you are others can enjoy the same security at no extra cost.

Public goods give rise to the problem of "free riders," people who benefit from the policies or actions of others without shouldering their fair share of the cost. The free rider is the person who refuses to be inoculated against smallpox because, given that everyone else is inoculated, the risk of smallpox to them is less than the risk of harm from inoculation; or the person who refuses to pay for a park even though she uses it, since she believes others will maintain it.[1] Economic theory posits that rational individuals will optimize their use of free public goods (or commons) until that good is no longer of any value. In other words, people will ride for free unless forced by country or conscience to do otherwise.

Markets, the thinking goes, provide optimal and efficient allocation of private goods but not public goods. Such a statement may sound rather dry and theoretical, but it has very real implications for the electric utility sector, especially when positive and negative externalities are thrown into the mix. Importers of LNG and oil, for instance, have little incentive to change the nature of their imports since the national security benefits of doing so are distributed to all companies and importers, including their competitors.[2] States will often require higher smokestacks on fossil-fueled power plants as a way to minimize the environmental harms of air pollution within their state, shifting the pollution instead to a broader geographic area encompassing other states.[3] Officials will locate trash incinerators, coal mines, and landfills near state borders (when possible), so that some of the harms from leakage and waste are transferred to other states, a problem known as "state line syndrome."[4] The state of Indiana recently authorized British Petroleum to release 1,584 pounds of ammonia and 4,925 pounds of suspended solids into Lake Michigan daily from their Whiting refinery, primarily because the benefits from increased output (more fuel, jobs, and tax revenue) accrue within the state while the costs (pollution and waste) are distributed to Illinois, Michigan, and Wisconsin.[5] The ability for existing energy companies to manipulate the production of negative externalities (and degrade public goods) to their advantage also helps explain why federal agencies had to spend $2.8 billion of public money cleaning up abandoned mine sites from 1997 to 2008—the benefits of harvesting those sites has long since accrued to a select number of firms, while its costs are still borne by society.[6] Such situations give rise to what economists refer to as "market failure."

Generally, theorists have determined that markets require at least six components in order to function properly:

- *Perfect information:* all participants in the market must be fully informed as to the quantitative and qualitative characteristics of goods and services (and substitutes to them) and the terms of exchange among them;
- *Transaction costs:* exchange must be instantaneous and costless;
- *Rationality:* consumers must maximize utility and producers maximize profits; economic actors must be able to collect and process all relevant information, hold rational

expectations about prices and products, and make decisions that always promote their self-interest;

- *Perfect competition and openness:* no specific firm or individual can influence any market price by decreasing or increasing supply of goods and services; there must be many buyers and sellers; they must act without collusion; firms cannot use their market power to influence the market themselves; predatory practices by incumbent firms against insurgent firms must be restricted; there must be no barriers to entry and exit;
- *Internalization:* all costs (or positive and negative externalities) associated with exchanges must be borne solely by the participants of the transaction, or internalized in prices so that all assets in the economic system are adequately priced;
- *Excludability:* those involved in the exchange must be able to prevent those not involved from benefiting from it.[7]

All six criteria must be satisfied. The violation of just one constitutes a market failure or market imperfection, a circumstance where "resource allocation will not yield efficient outcomes in the absence of policy intervention."[8]

Economists have recently argued that electricity is one of those cases where market competition (i.e., a greater number of power providers) can actually worsen anticompetitive practices. This is because industries often use storage or inventories to hedge against the risk of product shortages and reduce the risk of interruptions in supply and variations in demand. For electricity, storage is extremely costly. Most electric utilities are obliged to provide electricity at all times under the "duty to serve" law, meaning that short-term demand for electricity is inelastic. However, utilities have their limits. Most facilities that generate electricity have capacity constraints that cannot be breached for significant periods without risk of costly damage. And even if generating units could produce limitless amounts of energy, transmission and distribution systems impose hard constraints on the amount of electricity that can be delivered at any point in time.

The supply of electricity is always tight, and the properties of transmission and distribution mean that an imbalance of supply and demand at any one location can threaten the stability of the entire grid and disrupt delivery to all suppliers. Because there is very little demand elasticity and other supplies are unable to increase their output quickly or easily, even small generators can exert undue influence over the market. For instance, on a hot summer afternoon, when a system operator needs 97 percent of all generators running, a firm that owns just 6 percent of capacity can exercise disproportionate market power. Given extreme inelasticity of supply and demand, any supply shortages, real or imagined, can potentially drive prices up many thousands of times higher than their normal level.[9] Thus, the electricity industry represents a peculiar instance where all parties can exercise disproportionate market power during times of peak demand even when the "market" is operating perfectly.

Because markets seldom achieve perfection, analysts have identified a huge "energy efficiency gap," or the difference between the level of energy efficiency that should be achieved (if everyone were always rational) and the one that actually is (since none of us are).[10] One study identified 60 to 80 separate market failures relating to energy efficiency alone.[11] In their interviews with landlords, residential property managers, commercial property operators, real estate agents, building contractors, and construction trade associations, researchers in California identified a combination of misplaced incentives, misinformation, financing and capital costs, and consumer customs that all impeded the promotion of energy efficiency in residences.[12] Another survey of consumers in California, where such efficient technologies like compact fluorescent light bulbs (CFLs) and energy-efficient clothes washers have penetrated less than 5 percent of the market despite comparatively higher electricity prices, found that lack of information, unavailable capital, split incentives, product unavailability, and transaction costs were to blame.[13]

Much like perfect politicians, perfect electricity markets do not exist. Intentional market distortions, such as subsidies, and unintentional market distortions, such as split incentives, prevent consumers from becoming rationally invested in their energy choices. Chapters 1 and 2 document the negative externalities associated with fossil fuels and the free rider problems associated with the public goods provided by clean power systems. As this chapter details, customers and businesses seeking to invest in clean power systems face a paucity of accurate information and lack clear price signals for making rational choices about them. Homeowners, businesses, and utilities consistently make irrational decisions when they discount future energy savings and make energy purchases. Finally, barriers to entry and the use of predatory practices among incumbent utilities and power providers eliminate the principles of "competition and openness." Electricity consequently meets *none* of the six criteria for functioning markets.

Information Failure

Thomas Jefferson noted that knowledge is like a candle. When you take that candle and light another, the first still continues to glow, even though the second catches fire. The second candle, in turn, can light more candles. The light of the candle can thus be transmitted from one person to the next and not diminish as it goes on.[14] George Bernard Shaw made a similar distinction when he remarked that "if you have an apple and I have an apple and if we exchange these apples then you and I will still each have one apple. But if you have an idea and I have an idea and we exchange these ideas, then each of us will have two ideas." [15]

The provision of information is thus subject to a classic public goods problem because it violates both tenants of nonexcludability and nonrivalrousness. If I generate useful information, it creates a positive externality because it provides knowledge that is valuable to others. Those that have information may have strategic reasons to manipulate its value; self-interest is an incentive for the provision of misinformation by sellers; and the costs of acquiring information may be high enough to inhibit acquisition of sufficient unbiased information needed to overcome well-distributed misinformation.[16]

The production of information is subject to information failure and the adverse selection problem. Nobel Laureate George Akerlof demonstrated that the existence of "lemons" in the automobile market revealed the problems inherent with information distribution; bad cars sell at the same price as good cars since it is impossible for a buyer to know the difference.[17] A supplier of air conditioners will have better information than the buyer, so she can deceive customers, leading to a reluctance of consumers to trust even an honest seller's high efficiency claims. Or, sometimes sellers lack information. Nobel Prize winning economist George Stigler argued that prices change with varying frequency in all markets, and unless a market is completely centralized, no one will know all the prices that various sellers quote at any given time. Price dispersion is indeed the manifestation and measure of ignorance in the market.[18]

On the reception side of information, individuals and firms are limited in their ability to use, store, retrieve, and analyze information. Nobel Laureate Herbert A. Simon hypothesized that human decision makers are only as rational as their limited computational capabilities and incomplete information permit them to be. He characterized this condition as "bounded rationality." [19] Simon argued that in the need to search for decision alternatives, people look for satisfactory choices instead of optimal ones. "Bounded rationality" is an understandable reaction to a complex and uncertain world where humans must calculate consequences, resolve uncertainties, and pursue courses of action that are sufficient, rather than truly self-maximizing. Put another way, most people have selective inattention. Faced with a breadth of information, most of us filter the information we receive and narrow our options. For instance, when searching to buy a new computer, would you call every single store in America or the three or four closest to your home? Would you search every Internet site on Google or just the first ten? Our rationality, in other words, is limited. As anthropologists Mary Douglas and Baron Isherwood put it, "the idea of the rational individual is an impossible abstraction from social life." [20]

Numerous examples of information failure exist in the energy sector. The availability of information itself has a significant aspect on the price of gasoline and diesel. Economists have found that the amount of available information in a given region affects the price dispersion among gasoline stations. In other words, stations set prices in response to changes in the amount of knowledge they think consumers have.[21]

The dissemination of information is difficult in the market for energy-efficient products. Small and medium size firms dominate the residential building sector, with the largest residential builder accounting for less than 1 percent of total market sales. Three-quarters of all residential construction firms build fewer than 25 homes per year, and information on the energy efficiency of new homes is complicated, since new buildings have no prior conditions unless compared in aggregate to a collection of previously constructed homes.[22]

The Southern States Energy Board has found that the challenges inherent with renewables and DG technologies have created an information asymmetry: where stakeholders are familiar with unconventional power sources, the negative experiences are often the best known.[23] The DOE similarly found that since the attractiveness of DG technologies is very site specific, projects must be determined on a case-by-case basis, meaning information about "what works" is hard to locate. In instances where financially attractive DG opportunities exist, they are rarely pursued because there is often a lack of familiarity with DG technologies. This lack of experience, in turn, manifests itself in a dearth of standard data, models of analysis, or standardized practices for incorporating DG units into system planning and operation.[24]

Small scale renewable-energy technologies suffer especially from difficulty in standardization, as access to renewable "fuels" often dictates site-specific requirements. Indeed, this is often identified as one of their strongest advantages: they can be tailored to match any load at any location. However, such flexibility also makes information about them complex and sometimes contradictory. For example, a wind turbine might work best atop a cloudy mountain, whereas a solar PV system reaches optimal performance in a hot and cloudless desert. As a result of these differing experiences, the costs, capacity, need for storage, and rate of payback will differ slightly for installation of almost every clean power technology.

The complexity of building a renewable energy plant in the right size and in the right place makes developing a "standard approach" like the one in use for constructing large fossil-fueled plants extremely difficult. The organizers of the Cape Wind project near Nantucket Sound in Massachusetts, for instance, had to assess not only the effects of their offshore wind turbines on avian species, as well as noise, climate, and maritime security, they also needed to consider the effects on fish species, water quality, marine habitats, commercial and recreational navigation, and telecommunications systems.[25] The need for collecting this information can result in longer delays for the approval of offshore wind projects.

In terms of purchasing energy-efficient technologies, consumers must make complex decisions. Properly assessing energy savings versus higher capital costs involves comparing the time-discounted value of energy savings with the present cost of equipment.[26] Considering that electricity prices increased by 50 percent between the early 1970s and 1982, and then reversed course with prices falling 10 percent between 1982 and 1988, decisions about whether to invest in energy efficiency measures (and what kind) become very hard to make.

One survey of the energy industry found that a host of greenhouse gas–reducing technologies—such as CHP systems, new process schemes, resource efficiency, substitution of materials, recycling, changes in manufacture and design, and fuel switching—remain impeded by the inability to obtain reliable information cheaply.[27]

Consider the case of onshore wind turbines. A pressing problem has emerged related to wind rights contracts for small landowners. Landowners face significant transaction costs related to acquiring accurate information about wind resources on their land, especially since landowners take part in wind rights deals only once, while wind developers take part in a number of deals. Wind developers have been able to exploit such situations by creating contracts that improperly assess site valuation (windiness, proximity to electric power lines, and cost of building access roads), inspection and performance measurements, the right to an independent audit of the developer's books, property taxes, tort liability for damages, and plans for decommissioning and land reclamation once the wind contract expires. Since wind rights contracts typically span 30 years or more, poorly designed or abusive contracts can have lasting consequences for the landowners and their children, as well as subsequent purchasers of their land, and can create significant disincentives for investing in wind technology.[28]

Since information is free, some consumers may wait for others to use clean energy sources and "pay no cost" to see if energy-efficient technologies perform as they should. One study discovered that decisions by industrial managers to purchase more energy-efficient devices were not based on purely rational decisions concerning the superiority of the new technology versus existing options. Instead, managers compared options to the best possible technology they could imagine. In the study, industrial managers were surveyed about their attitudes concerning the purchase of surface mount technology (a new and more efficient method for automated assembly of integrated circuits). The researcher found that a majority of firms decided not to purchase the new technology, even though it provided them with a comparative advantage and improved efficiency, because they anticipated even greater future innovation that had not yet occurred.[29]

Producers face information failure as much as consumers. One study of American utility executives found that they had very little knowledge of customer demand, tastes, or preferences. After the Harvard Business School convened a conference on New England's power needs, for instance, sponsors publicly denounced the "shared ignorance among some of the most informed people in the country about some very fundamental things concerning the nature of electrical energy demand. The managers appeared to know less about electricity demand than other business executives knew about the nature of demand for toothpaste, lifesavers, or beer."[30]

Returns on Investment

By far the most commonly cited financial impediment to clean power technologies is their comparatively higher capital cost (per installed kilowatt). Out of the hundreds of utility executives, state and federal regulators, manufacturers, energy analysts, and economists interviewed for this book, almost 90 percent considered "cost" as the single greatest impediment to DG, CHP, and renewables. Despite all the benefits smaller and distributed systems can offer consumers and communities, their relatively higher installed capital cost make them too expensive for most residential customers and too risky for most utilities and businesses. This is especially true for two of the most common distributed renewable technologies: wind turbines and solar PV.

For wind turbines, the main capital cost is the turbine, although the cost of installing the tower, and the expense of delivery, interconnection, and metering hardware must also be included when calculating the turbine's total "up-front" expenses. Wind hardware designed for individual homes (in the range of 5 to 15 kW) costs between $2,500 and $3,000 per kW installed, yielding an average price of about $27,000 (including batteries and storage mechanisms). Wind energy becomes much more competitive when purchased in bulk. AWEA estimated that the cost of utility-scale wind projects can be as low as $1,000 per kW, but most consumers do not have the ability to purchase multiple turbines.[31] While wind turbines require less maintenance than a coal- or natural-gas plant of similar capacity, the expense of replacement parts for wind turbines can complicate or greatly diminish the rate of return for wind projects. A 2004 DOE report cautioned that many turbines break during construction, and that costly drivetrains tend to wear out more rapidly at low wind speeds.[32] The relative novelty of wind projects in the United States also increases the financing and insurance costs for wind farms, adding a further financial disincentive to invest in wind (though not so much in countries such as Denmark, Germany, and Spain that have greater experience with wind than the United States).

Solar panels tend to be more expensive to manufacture, site, and install than both large centralized fossil fuel generators and nonrenewable forms of DG. PV cells are manufactured from costly multicrystalline materials (the most popular being silicon, but also including gallium-arsenide, copper-indium-diselenide, and cadmium-telluride) in both "thick" and "thin" models. Like wind turbines, they remain somewhat capital-intensive renewable technologies having low operating costs. Solar installation sites must be carefully selected after consideration of several complex factors, including the amount of direct solar radiation, diffuse sky radiation, degree of cloudiness, and air temperature. While these factors do not apply to all solar systems, the complexity associated with solar siting and installation has convinced some policymakers that the calculation of the total cost to produce a kilowatt-hour of PV-generated electricity can be difficult and imprecise.

For example, a typical "stand alone" PV system for home use requires a collector array, controller, inverter, and (usually) battery bank, representing an investment of more than $20,000 for a few kilowatts of power. The IEA estimated that the installation cost of a basic PV system for individual use ranges from $5,000 to $7,000 per peak kW.[33] A similar study undertaken by a U.S. solar cell manufacturer found that utility-scale PV farms often average around $7,000 per peak kW.[34] At even the best rates (i.e., when the cost of purchasing conventionally produced electricity is the most expensive), the capital expense of a PV system is three to seven times as expensive as coal and natural gas facilities (and up to four times as expensive as other renewable technologies such as biomass gasification and wind turbines).

One study found a general aversion against solar PV systems in the housing market for precisely these reasons.[35] Such systems greatly add to the initial cost of purchasing a home. Moreover, when times are good and houses are selling well, builders and real estate agents view it as an indicator that alternative energy technologies are not needed to make sales. When times are bad, they place even more emphasis on minimizing costs and keeping house prices low. Home owners also worry about project delays (and thus rising costs) associated with PV availability, installation scheduling, and utility interconnection. Builders believe that most home buyers are not interested in PV, given its extra cost, and that many may even be opposed to it for concerns of aesthetics, maintenance, or reliability.

Nonprofit groups and utilities also claim that capital costs prevent investment in larger renewable energy resources, typically operated by utilities rather than businesses and residential consumers (such as hydro, ocean, and offshore wind power). The International Rivers Network, a nonprofit organization concerned with preserving the quality of the nation's rivers and streams, argues that both small and large hydroelectric facilities remain "hugely expensive to build and their costs are usually far higher than estimated."[36] After assessing more than 180 dam projects in 27 countries, the *World Commission on Dams Report* concluded that, on average, dams end up costing 56 percent more to build than predicted. Coupled with inflated costs comes underperformance and expensive maintenance needs, as well as costly resettlement programs for displaced people.[37]

While offshore wind turbines include many of the same parts (and costs) of land-based turbines, manufacturers admit that they are more expensive to install, operate, and maintain (although they also have a higher possible capacity, and thus greater potential for making these costs back). A comprehensive study of DG in Europe concluded that the foundations for offshore turbines are understandably more expensive than their land-based counterparts because they must be moored and stabilized to the seabed floor. Additionally, the same study noted that the cost of sea transmission cables typically add more than 20 percent to the costs of an offshore wind project.[38]

Government and electric utility industry analysts believe that the same trends concerning capital costs hold true for most nonrenewable DG technologies. Thomas Petersik, a former analyst for the EIA, put it this way:

> The most significant impediments [to nonrenewable DG systems] are cost, cost, and cost. More specifically, DG technologies experience much higher (in the neighborhood of 50 percent) capital costs per kilowatt, in part because they lack economies of scale; they are much less fuel efficient; and operations and maintenance are much higher . . . As a result, the per kWh costs of DG from, say, microturbines, fuel cells, small wind, solar PV, and others, are significantly more than for central station supply.[39]

Brice Freeman, who models DG project economics for EPRI, agrees and says that first-time capital costs are always "the Joker in the deck . . . because . . . any unexpected costs can sink project economics."[40] And Joe Catina, who manages all North American DG projects for the Ingersoll Rand Company, stated that the biggest issue in selling DG remains capital cost. Catina explains that:

> [r]elative to other forms of central power generation [DG technologies] are expensive in terms of capital costs . . . From the DG generator side, the biggest issue right now is cost. It's all about cost.[41]

In the end, elevated capital costs deter customers, utilities, and businesses from investing in DG and renewables. David Garman, under secretary of energy and former assistant secretary for energy efficiency and renewable energy at the DOE, admits that "the capital cost of most renewable energy systems still places them out of the means of most customers." [42]

For consumers and homeowners, clean power projects cannot meet the extremely rapid rates of return that many individuals set for their investments. Put simply, most people do not understand how money functions. One survey asked 1,004 persons in a national probability sample about the rate of inflation. The average estimate among those who answered was 7.8 percent, and in none of the demographic categories did it drop below 6.0 percent. The surveyors found this extremely troubling, as the actual inflation rate was less than half at 2.9 percent.[43] The researchers also asked members of the sample how much they would have to spend today to get what one dollar bought a year ago, and the average answer was an astounding $1.41. Many people therefore lack an understanding of how investments work.[44]

Implicit discount rates, the rate at which consumers want to recover their investment, are often 20 to 35 percent for home air conditioners and home insulation, above 80 percent for electric water heaters and furnaces, 45 to 300 percent for refrigerators, and 500 to 800 percent for gas water heaters.[45] Under this last discount rate, investments in energy-efficient products would require a payback period of fewer than five months. A survey conducted by the Potomac Electric Company asked consumers about the payback times expected for investments in energy-efficient products, and found that one-third of the respondents were unable to answer the question at all. Of those who answered, more than three-fourths indicated payback periods of three years or less.[46]

Consider an innovative study in the 1980s concerning two refrigerators. Consumers in stores throughout the country were given a choice of two refrigerators that were identical in all respects except two: energy efficiency and price. The energy-efficient model saved 410 kWh per year, more than 25 percent of the standard model's energy usage, and cost just $60 more. The energy-efficient model was cost effective in almost all locations of the country, providing an annual return of investment of about 50 percent in most regions (i.e., it paid for itself in two years). In spite of these favorable and easily observed characteristics, more than half of all purchasers choose the inefficient model.[47]

A second study on consumer adoption of energy-efficient core coils versus commercial ballasts found that even though the energy-efficient coils were commercially available, provided identical amenity to consumers, had a much higher return on investment than most alternatives, resulted in no increased risk to the consumer, and had no detectable hidden costs, such technologies were not adopted because of unrealistic personal discount rates.[48] Three other studies have documented the *same* occurrence with energy star personal computers, color televisions, and magnetic fluorescent ballasts.[49]

Implicit discount rates also (and understandably) vary by income bracket. Those in the lowest bracket often have discount rates for energy efficiency investments between 62 and 89 percent, whereas those in the highest bracket require only 4 to

15 percent.[50] Similarly, in another survey of homeowners who had just purchased residential air conditioning units, those in the lowest income bracket had an average discount rate of 39 percent but those in the highest bracket had an average of just 8.9 percent.[51]

Interrelated with improper discounting is the amount of capital homeowners have to make energy efficiency investments. Once people with fixed incomes invest money in *anything* (the Beatles *White Album,* stock options, a new television, a home improvement), they have already "sunk" their available earnings. Moreover, those who need energy efficiency help the most (i.e., the poorest) often have the least money to invest in it.[52]

All of the impediments—high capital costs, improper discounting, lack of capital—affect businesses as well. Even though businesses can purchase renewable energy systems in larger quantities than residential consumers (driving their price down as production increases), the payback periods for such technologies are substantially longer than most permit. Solar energy offers a prime example. The average American business often looks for a two- to three-year payback on energy investments. Yet a recent *New York Times* article explained that "solar power cannot yet provide that; the average commercial installation is expected to pay for itself in five to nine years." [53] Perpetually low energy prices in some regions of the country only enhance this impediment, as it means PV systems take even longer to pay for themselves.

The same holds true for commercial energy efficiency measures. A survey of 610 commercial customers in Niagara Mohawk's commercial sector found that 16.8 percent used lowest first cost as their decision criterion to select space conditioning or lighting equipment replacement.[54] A survey from Consolidated Edison in New York City found that of 54 electric customers that occupied office buildings, between 80 and 90 percent required simple payback times of three years or less for energy-efficient equipment.[55]

For similar reasons, utilities are reluctant to undertake investments in renewable and DG technologies because they fear such purchases will force them to initially raise their rates to cover the installed costs. Claudia J. Banner, a senior engineer and renewable energy planning lead for American Electric Power, explained that "utilities are usually not receptive to mandated renewable portfolio standards, citing an increase in generation costs that could get passed back to the consumer, resulting in larger utility bills." [56] Theresa M. Jurotich, a consulting engineer in the energy industry, added that the problem is that utilities believe that "renewables combine high capital investment with low output. To recover the capital investment, utilities have to either charge a lot for each kWh generated or obtain long-term financing resulting in a lower annualized cost." [57]

Why are costs so important? In the golden days of regulation, utilities did not care too much about higher costs, since they could pass them on to consumers and still earn their guaranteed return on investment. But in a partially deregulated world, utility managers are conscious of costs so they do not get undersold by new competitors in the market. In an industry as large as the electric utility sector—with trillions of dollars of embedded investment, billions of dollars in sales, and millions of customers—a staggering number of banks, financing firms, manufacturers, and publicly traded utilities have become invested in the system. Such complexity, in turn, has convinced many utility managers to place priority on cost as a matter of maximizing profits and keeping prices low. All members of the electric utility industry, from large, investor-owned utilities to small, rural electric cooperatives, are concerned with keeping prices stable for consumers. Additionally, market restructuring (also called "deregulation") of the electric utility industry occurred in some states in the late 1990s and early 2000. Such restructuring created wholesale markets in which electricity can be bought, traded, and sold like any other commodity. As a result of these changes, power providers have become even more concerned with keeping a financial edge over their competitors.

Split Incentives

When refining his ideas on economics, classical economist Adam Smith classified the "principals" as the owners of joint-stock companies (now the modern corporation) and the "agents" the managers hired by the corporation. Smith warned that since the managers were hired by the ownership but were not owners themselves, their actions would not be optimal at maximizing the welfare of the corporation. Economist Kenneth Arrow noted a similar problem in the relationship between doctors and patients: the physician makes decisions for the patient, but does not have to live with them.[58] These examples have given rise to the "principle-agent" problem in economics, where incentives for making a certain decision are "split" among two or more classes of people. The principal-agent problem comes in many different flavors relating to energy technologies:

- Architects, engineers, and builders design homes that they will not live in;
- Landlords purchase appliances and equipment for tenants that they will not use themselves;
- Industrial procurers select technology for their plants;
- Specialists write product specifications for military purchases;
- Fleet managers select vehicles that others will drive;
- New car buyers determine the pool of vehicles available for all drivers.[59]

A severe misalignment occurs when consumers use technologies selected by others, especially when intermediaries overemphasize up-front cost rather than life-cycle costs. Four sets of split incentives relating to clean power technologies appear to be the most pernicious: distinctions between and among builders and homeowners, landlords and tenants, business investors and electric utility managers.

Architects, engineers, and builders select energy technologies that homeowners and dwellers use. Yet the prevailing fee structures for building design are based on a percentage of capital cost of a project, penalizing engineers and architects for designing efficient but more expensive systems. The pressure to lower first costs is reinforced by banks, lenders, and financers, since the builder is, in effect, building the house for them, and their criteria for selling a loan include keeping the ratio of monthly payments to monthly income low enough to make the loan a reasonable risk.[60] Developers and investors typically want fast, cheap buildings up and running quickly so they can start maximizing returns. The result is that "utilities see buildings as physical structures with energy flowing through them; developers see them as financing structures with money flowing through them."[61]

Building designers are also concerned about liability, and reduce risks by oversizing equipment. As one designer pointed out:

> If building services engineers were to design tables, they would be made of titanium and have six legs, two to spare, just in case; if they were to design cars, these bullet-proof vehicles would have eight wheels (one spare each, just in case) with twin engines, twin steering wheels, and twin seat belts.[62]

Designers, in other words, want things to work; they do not necessarily care if they work inefficiently.

The Office of Technology Assessment (OTA) conducted a series of interviews with building owners, architects, homeowners, engineers, and equipment manufacturers.[63] It found that incentives rarely promoted energy efficiency in the buildings sector and that most contractors believed they would win competitive bidding for projects only if they had the lowest bid, which required specifying lower up-front costs and often required installation of inefficient equipment. Moreover, when contractors were faced with investing $1,000 in insulation or $1,000 in landscaping, most said they would always choose the second, since they believed prospective buyers valued visible, tangible objects more highly.

About one-quarter of commercial building space and residential space in the United States is leased or rented. Tenants have no interest in investing in efficiency improvements, since they do not own the property and may have short-term occupancy; landlords do not because they can pass energy costs onto tenants and retrofits often appear risky and unprofitable. This classic problem of "split incentives" explains why energy consumption and expenditure per unit of floor area are much greater in rented buildings and public housing than in owner-occupied single-family housing.[64]

These problems also occur in universities and office complexes. Energy consultant Wilson Prichett explains that:

> Most office buildings are owned by a company and the spaces are rented out to tenants. The rent for the complex includes the price of utilities, so the people that own the facilities don't care about electricity prices. And the people renting the space don't want to invest in a facility that they don't own. Similarly, universities don't want to spend extra money on fixing the university up and recovering utility savings, because the department that spends the money on the conservation improvements is not the same one that gets the energy savings.[65]

So among certain residences, office buildings, and universities, the market signals between the price and use of electricity are divided and distorted.

The same types of split incentives occur in some of the country's most energy-intensive industries. Many industrial facilities have only one utility meter to measure plant-wide consumption. In these situations, traditional accounting practices treat energy as an overhead cost, which then becomes allocated across departments according to their numbers of workers or square feet of floor space. However, one drawback is that the cost of any one department's energy waste is distributed to all departments. Conversely, any department that undertakes energy savings will have its improvements diluted by the artificial allocation of costs.[66]

In a similar vein, most small and large businesses resist investing in and using clean power sources because these technologies are believed to deviate from each company's core business mission. For the typical business, energy costs are a miniscule fraction of labor costs; therefore, management and capital are drawn to other areas. Even though these businesses obviously use electricity, they seem to have little to no interest in producing power (or promoting *any* type of electricity system, clean or not). Because most nonutility and nonenergy businesses have goals and priorities that have nothing to do with electricity, they tend to be concerned with the promotion of their own corporate strategy. The interests and technologies such businesses promote are not interoperable with the growth of the electric utility system, and so they refuse to invest in energy technologies.

For instance, managers of small businesses remain constrained by limited resources and time, and larger businesses believe that they can best maximize their profits by focusing on nonenergy related issues. Rodney Sobin, former innovative technology manager at the Virginia Department of Environmental Quality, put it this way:

> The potential users of [clean power] tend to be unfamiliar with the benefits such systems offer. The cookie baker is concerned with making a better chocolate chip cookie. The Chicken McNugget manufacturer is concerned with perfecting a better Chicken McNugget. The shopping mall managers are interested in providing retail services. None of them are interested in generating power and becoming a micro or quasi utility.[67]

In other words, most companies do not want to be in the "business of making energy," and would rather use their resources—financial and otherwise—promoting core business activities.

Tom Casten, who has almost three decades of experience selling DG and CHP units to businesses, explained that both large and small manufacturing firms resist undertaking novel energy projects. According to Casten:

> The typical manufacturing enterprise focuses its intellectual and financial resources on core activities—making beer, or steel, or chemicals, etc. The vast bulk of industry will not invest in energy plants unless they are "broken" or, in the best case, when an energy efficiency project will pay back the capital investment in 12 to 18 months. Companies have a much higher investment hurdle rate for core activities than non-core activities, and they employ very few specialists in any non-core activity.[68]

Even though, as Casten concluded, today's successful firms "outsource payroll, security, food service, janitorial services, grounds maintenance, call centers, and other functions," energy is not an important enough of an activity.

Casten's comments parallel conclusions about business practices made by ASE. Chris Russell, the director of industry research at the ASE, noted that energy projects are often resisted by all levels of the business community because they distract personnel from more profitable ventures. Russell elaborated:

> Facilities are thinly staffed, running flat out every day to meet production goals. Therefore distractions aren't welcome. For them, routine is a good thing, and their mantra becomes "that's the way we've always done it." So when you propose energy [projects for] a facility, you are really proposing changes to the way they operate. You have people in operations, finance, procurement, and engineering—all of whom will be impacted by energy management, and all of whom usually have some reason to resist change . . . Decision makers are continually making a tradeoff between risk, time, and money. If you propose an energy efficiency measure that saves X dollars, the facility manager wonders what the additional costs are in terms of risk and time. What labor hours are needed to support energy efficiency efforts? Should they allocate labor hours to making dollars, or saving dimes?[69]

Company employees may also be reluctant to admit the need for more efficient energy technologies, Russell adds, because they believe such admissions become evidence of ineffective job performance.

Professor Stephen J. DeCanio found that a looming possibility of hostile takeover or shareholder revolt exerts profound pressure on business leaders to maximize shareholder wealth, meaning that executives prefer investments with rapid paybacks to enhance their reputations with the owners. His research demonstrates that top management gives low priority to relatively small cost projects, as attention and resources are scarce, and instead focuses all of their social and financial capital on areas deemed crucial to the survival of the firm (such as marketing, improved logistics, or R&D). Managers select projects with higher rates of return, meaning identification of projects does not occur as an unbiased sample, and projects with underestimated costs lock decision makers into spending more capital on those investments and making even less capital available for energy efficiency. Decisions to avoid investing in energy efficiency become self-justifying because once energy management loses priority, historical data needed for comparisons to prove efficiency improvements are working no longer exist.[70]

At the end of the day, energy use continues to be invisible to most companies. Shalom Flank, chief technical officer for Pareto Energy Limited, explained that no single business "can improve their bottom line substantially with alternative energy technologies, because energy inputs usually constitute less than 1 percent of their total cost structure."[71] In other words, companies will often choose to pursue more profitable ventures instead of investing in energy technologies. A systematic study of industrial energy usage undertaken by the ASE found that compressed air leaks were often overlooked because air was believed to be free, even though the same companies needed five horsepower of electricity to generate one horsepower of compressed air. Similarly, the study found that plant operators often assumed that scrap rates were unimportant because scrap can be melted down and reused, not realizing that recycling scrap required energy consumption and associated costs.[72]

Ironically, in the case of DG installations at industries and businesses, the smaller the project, the *less* likely it will be undertaken. Mark C. Hall, senior vice president at Primary Energy, and an advocate of DG technologies, argues that:

> Size matters—not necessarily from a capital cost or efficiency standpoint, but it takes a lot of effort to do a small project as it does a large project, so the tendency is for an organization like ours to focus on the bigger projects because they can support the kinds of efforts needed to get the projects done. In contrast, getting smaller projects done requires such a disproportionate amount of senior management attention, legal attention, and other time and effort that it really burdens those projects with greater and greater costs so that people say it's not worth it.[73]

This attitude helps explain why so many smaller businesses never undertake energy improvements.

As a result of the aforementioned trends, nonenergy related companies almost always choose to do nothing rather than undertake profitable energy investments. According to Casten:

> My son, who is CEO of a company that offers integrated steam turbine generator plants that convert wasted steam pressure drop into electricity, often with over 35 percent returns on investment, analyzed company activities over the last 17 years. Out of 7,000 proposals, all of which had at least 25 percent returns on investment, only around two percent were accepted. Only 1 percent was lost to another competitor. "Mr. Do Nothing" won 97 percent of the time.[74]

Most executives and managers believe that saving energy, it seems, is one of the least important things a business can do.

Interestingly, utilities have a "core" business strategy as well, and it often involves selling as much electricity as possible. Yet electric utilities, because they already serve as the intermediary between consumers and fossil fuel providers, are also the ones charged with promoting energy efficiency. Adam Serchuk, from the Oregon Energy Trust, rightfully noted that this "doesn't make sense" and represents a classic conflict of interest: "utilities have the responsibility of returning profits to their shareholders (and thus trying to get Americans to purchase more electricity), but are also asked by regulators to use energy efficiency programs and renewable resources to get Americans to conserve energy."[75] Toben Galvin, a senior energy analyst with the Vermont Energy Investment Corporation, suggested that "there is complete disinterest on the part of utilities and fossil fuel suppliers to conserve electricity or energy," either through energy efficiency programs or through more efficient generation technologies.[76] Ralph Loomis, senior vice president for Exelon, comments that "utilities are not in the energy efficiency business. It is almost antithetical to say that the people making money selling electricity to people should somehow not be in the business of selling electricity."[77] David Garman, current under secretary of energy for the DOE, put it this way:

> Right now, a lot of investor owned utilities are in the business of selling electrons, and while they have legitimate interests in issues like backup or peaking power, they do not have much of an incentive to switch.[78]

As a result, since electricity sales are seldom decoupled from revenues, most utilities perceive they would lose profits in excess of any savings derided from lower fuel and operating costs gained from clean power sources.[79] The central problem, from a utility perspective, is rate design. Most utilities use sales-based formulas, particularly per-kWh delivery prices, as a method of ratemaking. Rates are often set over long periods of time, meaning changes in sales directly affect profitability. Richard Cowart estimates that under current rate designs, integrated utilities with rates of 8 ¢/kWh would experience a 23 percent decrease in profits from a 5 percent decrease in sales. Wires-only companies do even worse, with the same 5 percent decrease in sales dropping returns on equity by more than 50 percent.[80]

One study found that utilities believed that energy efficiency and DSM practices would hurt them because:

- Selling fewer kWhs means the utility's substantial fixed costs have to be divided among fewer units of energy sold, putting upward pressure on rates;
- Energy conservation programs cost money to implement, further adding to costs;
- Conservative and "true and tested" supply side options are more reliable and cost-effective.[81]

All but two of the states in the United States reward distribution utilities for selling more energy and penalize them for cutting customer's bills.[82] Newly privatized utilities are even more obsessed with short-run profits given the need to improve debt-to-equity ratios, demonstrate competence to shareholders, and send positive signals to the stock market.[83] Also, utilities are especially incentivized to increase sales in situations where they have surplus capacity that is cheap to operate.

Even when utilities determine that clean power sources will help them, fragmented decision making prevents reforms from being implemented. Most fully integrated utilities have separate generation, transmission, and distribution resource planning organizations. Distribution planners are often located in operating divisions physically removed from company headquarters, and respond to outside needs rather than having their technology and budgets driven by strategic planning focused on the distribution system directly. DG systems that may ultimately benefit transmission and distribution centers are not promoted by generation and transmission managers because the benefits do not accrue to them directly.[84]

Predatory Market Power

Finally, clean power sources directly threaten the market share of utilities, energy companies, and other power operators. Joe Catina, from the Ingersoll Rand Company, summed it up nicely by stating that "utilities are by nature monopolistic entities that had a charter to provide adequate energy capacity, with near total control over the production of electricity with little responsibility for cost. Now, they are being asked to lose control over the market yet keep costs low. And many are uncomfortable with the issue."[85] Moreover, most local distribution networks have been designed as radial grids, meant only to send power in one direction (after they have received power from a transmission station). Renewable energy technologies complicate this design pattern because they enable the distribution of power in the opposite direction, away from transmission stations. Ordinary consumers have little ability to arbitrage the market power of utilities. The best obstacle to price discrimination is to buy from other sources, but most consumers cannot do this with electricity. Resale by consumers is implausible, distribution is controlled by central stations and vertically integrated with production, and electricity cannot be stored cost-effectively. Customers have no practical way of reselling the product, meaning they have little ability to hedge against the market power of dominant utilities.[86]

Little incentive exists for utilities and system operators to overcome these challenges; such actors are less concerned about R&D of clean power technologies, and more worried about competing in the restructured marketplace. Fuel cell engineer Shawn Collins argues that "at the moment, [none of the utilities] want to work together because the reward for individually solving the problems and monopolizing the market is so high."[87] Analogously, utilities have actively used their "power of incumbency" to further mold federal and state regulations in favor of large, centralized plants (and disadvantage small, decentralized units). Alex Farrell, a professor of energy resources at the University of California at Berkeley, stated the problem, simply, is that "whether local utilities or large manufacturing firms, once large-scale energy systems are in place, it becomes very difficult to dislodge them."[88] Alan Crane from the National Academies elaborated that perhaps the most significant impediment for clean power systems is an institutional, rather than a technological, dilemma. According to him:

> System and transmission operators don't recognize many renewable and DG technologies as being part of their network because they are not dispatchable . . . It's not in the "right people's" interest to fix interface problems like interconnection and intermittency when such improvements will do little (in their view) to improve control of the overall system. Technically, these technologies can be incorporated into the grid easily. Socially, much institutional resistance remains.[89]

Utility resistance to DG systems is therefore more about institutional culture than it is about technology.

Clean power providers face hurdles when dealing with the administrators of the existing transmission and distribution system, who seek to retain a number of traditional, "time-tested" regulatory and utility practices that have existed for upwards of a century. Seeking to maintain control over a system they (and their predecessors) created, they have made it difficult for new players to play on their turf. Numerous advocates of DG technologies argue that they have personally encountered utility resistance to their technologies. Tom Casten, for example, notes that in his experience:

> Regulated utilities have tremendous bias to increase throughput, and they instinctively fight against companies that provide electricity from alternative sources. This is ultimately the much larger cost of our central generation paradigm approach: that unless the many barriers to competition in retail delivery of power are removed and third party providers are able to capture the true value of DG, then the independent power companies will gravitate towards large remote generation and large industrial generation.[90]

Edward Vine, who directs energy efficiency research at LBNL, commented that:

> The people in control of the power system—oil firms, coal companies, nuclear energy producers—have a substantial amount of investment in traditional technologies. Such institutions, which have become a powerful political force because of their market power, are reluctant to change their existing technologies and see the market share of these technologies dwindle.[91]

And Paul DeCotis, director of energy analysis for New York State Energy Research and Development Authority (NYSERDA), comments that "the reluctance on the utility's part to interconnect and buy generation serve as the most significant impediment to widespread adoption of DG. The more DG there is on a system, the less direct control a utility might have over its system."[92]

A few, independent studies confirmed that system operators have attempted to retain their control over the electric utility system by employing a wide variety of predatory and discriminatory practices. Such efforts typically begin with the imposition of fees to connect to the grid. In many states that have begun restructuring their utility systems, formerly regulated "natural monopoly" power companies have been permitted to charge customers "stranded costs."[93] These costs are intended to cover a "fair return" on generation and transmission investments made by utilities during the era of regulation, when the investments were viewed as serving all users. Put simply, when a customer decides to install an electric generator independent from the utility, he or she arguably removes part of the grid's existing load requirement and "strands" part of the investment the utility made in the power system. But such fees greatly increase the cost of renewable energy systems because customers must pay them in addition to the expense of installing the new technology.

Utilities also require payment of a host of charges on those who use renewable energy systems that run intermittently. Rodney Sobin, former innovative technology manager for the Virginia Department of Environmental Quality, commented that:

> Confronted with the uncertainty of restructuring, most utilities defer to their time-tested mentality of a centralized plant, owned and operated by a regulated utility being a monopoly provider of power. And a strong desire exists among many utilities to preserve that power and control . . . Other utility governance policies are predicated on the one way model of the central utility selling power to the customer rather than the more interactive and fluid grid style management. Thus wires charges, stranded assets, unfair rates of return, and other practices continue to make it difficult for third party power producers—DG or not—to generate electricity.[94]

For example, they may ask for high rates for providing backup power for when the intermittent renewable-energy technologies do not produce power. They may also charge demand fees (a charge that penalizes customers for displacing demand from utilities) that discourage the use of intermittent power systems. A recent study undertaken by NREL found more than 17 different "extraneous" charges associated with the use of dispersed renewable technologies.[95] These types of charges, the senior editor of *Public Utilities Fortnightly* exclaimed, "are a major obstacle to the development of a competitive electricity market."[96]

And, as a particularly insidious example, consider the case of interconnection standards. Naturally distribution and transmission system operators need to ensure that clean power systems do nothing that would imperil their ability to maintain the stability of their network. DG technologies must not interrupt the balanced flow of alternating current to support proper grid synchronization. Renewables cannot endanger the lives of people working on the grid by adding power to it when the workers think it has been de-energized or creating legal liabilities and potential hazards for the public. Utilities and system operators therefore tend to argue that connecting a large amount of renewable or DG capacity to the power grid excessively complicates system management. A survey conducted by IEEE found that most distribution engineers believed that DG technologies would fail to start when needed, that any improvement in the overall system would be imperceptible, and that the use of DG would only mask true load growth, leading to forecasting and expansion problems.[97]

Rather than trying to work with DG manufacturers or local customers to resolve these issues, utilities have instead tended to actively block or impede small scale units from attaching to the grid. To be fair, no mandatory national standard for interconnection currently exists (although FERC does provide "suggestions"). Interconnection procedures that have been implemented, when they exist at all, vary greatly between utilities, municipalities, cities, and states. David Garman admitted that, despite almost a decade of discussion on the issue, "interconnection in this country is still highly inconsistent between states."[98] Solar power entrepreneur Scott Sklar called the situation "totally absurd," and he elaborated that:

> The U.S. still needs a national interconnection standard. Currently, over 20 different states have varying technical rules (loosely based on the IEEE 1547 standard) for interconnection, and still they are different enough so that manufacturers of DG technologies' interconnection equipment have to make different devices for almost every different state market . . . When the telephone industry was deregulated, Congress passed a national interconnection standard for telephones so that, whether plugged into a home in New York or California, they would still work. We need the same thing for DG technologies. A national interconnection standard is absolutely essential to provoke an evolution in interconnection equipment desperately needed for DG systems to take off, and will also allow them to be more compatible with smart meters that will be critical in any building or facility.[99]

Instead, utilities and system operators often charge exorbitant sums of money to even begin the process of interconnecting a DG system.

For example, PJM Interconnection—not an incumbent utility, but an incumbent service operator responsible for one of the large power grids in the Northeast—mandates that customers wishing to interconnect distributed generators to the utility's transmission network conduct an extensive feasibility study. In addition, Section 36.1 of the PJM tariff requires a $10,000 surcharge—regardless of the size, ownership, or location of the connecting generator—for anyone who attempts to interconnect to PJM's transmission system. This fee, although refundable, serves as an especially large disincentive for small-scale power generators. In the Pacific Northwest, some utilities require a $2 million bond to be placed by all parties seeking to connect DG systems in order to protect the grid. Adam Serchuk comments that "utilities pretty much still possess unadulterated control over electricity contract practices, and the negotiation process turns out to be real onerous for mid size businesses."[100] And, as Alex Farrell put it, "the degree to which [clean power] of any variety threatens the monopoly franchise—that is generation owned in a service territory by somebody other than the monopoly franchise—is a direct assault on the business plan for that company, so they fight it tooth and nail."[101]

Despite efforts by NARUC, FERC, and IEEE, the federal government has not yet standardized requirements for interconnecting renewable energy technologies and distributed generators with the grid. Forty-eight states (Texas and Vermont are the only exceptions) have not yet established standard interconnection procedures, making utility interconnection specifications greatly variable and inconsistent. A study assessing the impediments to DG systems in New York State cautioned that the complexity of utility interconnection procedures deters investment in DG systems, and that unclear responsibilities, time delays, and the cost of interconnection equipment (and associated studies) severely impede the adoption of DG systems.[102] Collectively, problems with interconnection prevent DG technologies from being fully preassembled for "plug and play" style installation, and make it difficult for industries or corporations to install DG systems in more than one region (since they then must accommodate competing standards and rules). Standards remain oriented toward large generators, and often suffer from inflated tariffs and standby charges. The hurdles associated with interconnection therefore decrease the economic viability of DG technologies because they interfere with the ability for users to sell power back to the grid (and thus pay off their initial capital investment).

As one case in point, regulated utilities (and unregulated utilities, such as rural cooperatives) often make it difficult for individuals and companies to connect renewable technologies to "their" grids. In some cases, managers of these utilities have employed their formidable resources in attempts to thwart interconnections.

Consider the following narrative told by entrepreneur Tom Casten about the opposition he faced when he tried to deploy eight cogeneration units in New York City. As he explained:

> We were trying to develop on-site combined heat and power plants in NYC, which is served by Consolidated Edison (Con Ed), one of the world's largest utilities. Con Ed is threatened and asked the mayor for a special hearing to consider the health effects of cogeneration. In keeping with the societal mind-set that monopoly utilities are optimal and must be protected, the Mayor's office sets up a hearing, chaired by a Deputy Mayor, but fails to invite us. We are at this time the only active developer of on-site power in NYC. We learn of the meeting three days before, rush together a presentation

> and demand a spot. The hearing is a setup. The chairman of Con Ed, Charles Luce, is in the audience; a Con Ed Senior Vice President is running the show. They have an elaborate presentation made by Dr. Peter Freudenthal, Con Ed's environmental expert. We were proposing to install eight engines in the bottom of a commercial building in downtown New York to generate the building's electricity, and by recycling the waste heat to produce heat and chilled water, achieve double the efficiency of the Con Ed generation fleet. Our project would take one building in Manhattan off of the Con Ed system. Dr. Freudenthal made an elaborate, illustrated presentation that was designed to feel "scientific", showing that the nitrogen oxide from our 8 megawatt plant would exit the 40 story high stack, and miraculously drop right to street level, causing a condition he called "pharyngeal dryness." He didn't explain how this hot, buoyant exhaust was going to descend forty stories. He didn't explain how the exhaust from Con Ed's 2,000 megawatts of Manhattan power plants, whose stacks are not any higher than 40 stories, would obey different physical laws and obediently rise and move out over the Atlantic ocean. But the good doctor claimed that if Cummins Cogeneration Company was allowed to install this relatively tiny plant, New Yorkers would begin to suffer from pharyngeal dryness. Thankfully, we persuaded the panel that this argument was silly, and the panel decided to not act on Con Ed's request to ban cogeneration until further study. Thirty years later, the story is funny, but it is sadly one of a great many experiences of monopoly electric utilities using their considerable resources to fight any generation that does not flow through their wires.[103]

Such a story shows the hurdles CHP entrepreneurs faced in the 1970s when dealing with the administrators of the existing transmission and distribution system, but it is not relegated to the distant past.

More recently, managers of a rural electric cooperative spent seven years trying to stop a family farmer in Iowa from connecting to the power company's distribution lines. The farmer sought to obtain net metering rates from the cooperative under the provisions of Public Utility Regulatory Policy Act of 1978 (PURPA), appealing to Iowa's court system and FERC. Ultimately, FERC ruled in favor of the farmer and scolded the cooperative's managers for deliberately disconnecting the family, for using delaying tactics, and for arguing disingenuously to the courts and to FERC. As a final note in the ruling, the FERC commissioners noted, "We cannot help but note that Midland has used the legal process to thwart efforts to compel it to comply with PURPA for seven years, with a long history of using every means at its disposal to avoid its obligation to purchase from [the farmer's] small wind powered qualified facility."[104]

The upper management of large, centralized utilities is not always at fault for impeding the interconnection of DG energy systems. Shalom Flank, an energy consultant, noted that his experience has shown that:

> Interconnection rules and standards have been unable to overcome the absolute veto power of local utilities. Astonishingly, even when the executives of a utility have joined the [clean power] cause, they cannot overrule their own distribution departments. We've encountered that particular phenomenon multiple times. We've talked about getting some foundation to support the formation of an "Interconnection Strike Force": when a utility raises the usual roadblocks to a [clean power] project—such as delays, endless

> studies, safety objections, gold-plating, or simply outright refusal—we would parachute in with a crackerjack team of lawyers, engineers, and lobbyists to use the existing laws, regulations, and political pressure points to break through their resistance. Countless projects have died because of the lack of such a strike force, doomed by the expense, uncertainty, and delay that the utilities can impose almost effortlessly.[105]

In some instances, even when DG systems gain the support of large utilities, local public utility commissions and systems operators can still manage to thwart the adoption of DG technologies. Put another way: even when DG technologies gain the support of one of the layers in the current electric utility system, the rest of the system continues to act as an impediment. The fees and other efforts to frustrate the goals of DG entrepreneurs also appear based on the desire to achieve other goals, such as the apparent need to remain in control of "their" system.

Utility managers and system operators also publicly argue that the intermittency of clean power sources is a serious obstacle to their wider use in the United States. Wind and solar generation and electricity demand (resulting in electricity supply or load) follow different cycles; load exhibits a distinct diurnal pattern through all seasons, while renewable generation is often affected by large-scale weather events that can have cycles of days or weeks. Hydroelectric facilities are prone to alterations in output based on changes in the water cycle such as seasonal rain and runoff from snowpack. Biomass is subject to the seasonal cycle of plants and rotation of crops. The output of solar PV will vary with the season, time of day, and presence of clouds and rainfall. Wave and ocean energy technologies will vary their output according to tidal movements (every 12 hours and 15 minutes) and the changing gravitational fields of the moon and sun.

These forms of variability serve as a powerful impediment in the minds of many utility executives. One senior vice president at Exelon stated that:

> You need to install 6 MW of wind to get something comparable of 1 MW of a more traditional fossil-fueled generation source. So while the cost per MW of wind may be increasingly competitive, you still have to buy six times as much if you want to deal with the resource adequacy side of the equation.[106]

Since many renewable technologies depend on weather-related phenomena, the director of one research institution noted that:

> Renewable energy sources are not well designed (or designed at all) to provide base-load power . . . To force renewables to provide base-load power is like trying to make a pig fly: you won't succeed and you only make the pig unhappier.[107]

Government analysts also argue that some nonrenewable DG and CHP technologies suffer dispatch problems. Dave Stinton, the CHP program manager at ORNL, comments that microturbines, reciprocating engines, and fuel cells do not match the ideal electrical loads demanded by most utilities. Most cogeneration equipment, Stinton notes, offers the most benefit to businesses that need thermal energy and heat as their primary products, not utilities or companies trying to generate electricity. Moreover, Stinton believes that most CHP systems are not designed to run continually. He clarified that:

> [For small businesses like hotels], there are some times during the day when cogeneration is useful, like the beginning of the day when people are getting up and laundry services are in use. Yet at 1:00 in the afternoon, no one is in the hotel and no one really needs the waste heat. The point is that it's really difficult to find the best application for DG/CHP systems, especially when it may not be optimal to run such systems for 24 or even 12 hours a day.[108]

Again, Stinton's remarks highlight that one of the reasons DG technologies have not gained acceptance among utilities and entrepreneurs is because they do not operate in ways conducive to the large, centralized utility model. Instead, DG technologies may work best when *not* connected to an extensive transmission and distribution system.

The arguments pointing out the intermittency of some clean power sources, however, completely obscure that conventional power systems suffer from similar problems, just of a different degree than clean power sources. Conventional power plants operating on coal, natural gas, and uranium are subject to an immense amount of variability related to construction costs, short-term supply and demand imbalances, long-term supply and demand fluctuations, growing volatility in the price of fuels, and unplanned outages. Contrary to the popular belief, renewable energy technologies can improve each of these sources of variability, as the more renewable systems that get deployed, the more—not less—stable the system becomes. The issue, therefore, is not one of variability or intermittency, *per se,* but how such variability and intermittency can best be managed, predicted, and mitigated. And over the past few years, the advantages of renewables have been empirically proven in many parts of the world where their adoption has not faced the kind of impediments seen in the United States.

Modern wind turbines, geothermal plants, and solar panels have fewer unplanned outages than conventional fossil fuel and nuclear units, with technical reliability often exceeding 97.5 percent. Since each individual turbine achieves such high reliability, any amount of significant wind power in an electricity system would never see all (hundreds of thousands of turbines) down at the same time. In fact, the IEA recently concluded that:

> Initially, it was believed that only a small amount of intermittent capacity was permissible on the grid without compromising system stability. However, with practical experience gathering, for example, in the Western Danish region where over 20 percent of the yearly electricity load is covered with wind energy, this view has been refuted . . . Bigger units of power plants bring with them the need for both greater operational and capacity reserve since outages cause greater disturbances to the system. The higher the technical availability, the lower the probability of unexpected outages and thus the lower the requirements of short-term operational reserve. Wind power plants actually score favorable against both criteria, since they normally employ small individual units (currently up to 5 MW) and have a record of high technical availability.[109]

Wind and solar technologies also improve overall system reliability because they tend to be dispersed and decentralized. It is considered a general principle in electrical and power systems engineering that the larger a system becomes, the less reserve capacity it needs. Demand variations between individual consumers are mitigated by grid interconnection in exactly this manner.

Just like consumers average out each other in electricity demand, individual wind farms average out each other in electricity supply. As one European study concluded:

> Wind power is variable in output but the variability can be predicted to a great extent . . . Variations in wind energy are smoother, because there are hundreds or thousands of units rather than a few large power stations, making it easier for the system operator to predict and manage changes in supply as they appear within the overall system. The system will not notice the shut-down of a 2 MW wind turbine. It will have to respond to the shut-down of a 500 MW coal fired plant or a 1,000 MW nuclear plant instantly.[110]

In other words, the modular and dispersed nature of technologies such as wind (and solar and small biomass facilities) improves overall system reliability because outages, when they occur, can be better managed.

A similar assessment performed by General Electric for the New York ISO investigated a 10 percent wind penetration scenario in New York State, or the addition of about 3,300 MW of nameplate wind capacity to a 33,000 MW peak load system. Researchers also assumed that wind capacity was located across 30 different sites. The study found "no credible single contingency" that led to a significant loss of generation, and that since the system was already designed to handle a loss of 1,200 MW due to the unreliability of conventional generators, it had more than enough resiliency to enable the incorporation of wind.[111] This could be why even though the United States has more than 12,000 MW of installed wind capacity, not a single conventional unit has been installed as a backup generator.[112]

Moreover, the claim that renewable resources cannot displace existing base-load units or provide base-load power is simply false. Hydroelectric, geothermal, and biomass electrical facilities already generate vast quantities of reliable base-load power. Multiple reports within the United States and Europe demonstrate that intermittent renewables such as wind and solar displace a significant share of base-load generation as well. The EIA has determined that widespread use of wind and solar power technologies would lead to lower generation from natural gas and coal facilities.[113] Examinations of fuel generation in several states confirm this finding. NYSERDA looked at load profiles for 2001 and concluded that 65 percent of the energy displaced by wind turbines in New York would have otherwise come from natural gas facilities, 15 percent from coal-fired plants, 10 percent from oil-based generation, and 10 percent from out of state imports of electricity.[114] A more recent study conducted in Virginia found that every new kWh of renewable generation would displace a portfolio of coal, natural gas, and oil facilities, and in Texas renewable energy technologies primarily displace natural gas and coal facilities.[115]

Equally important, but often overlooked, is how solar and wind generation would offset nuclear power in several regions of the United States. Researchers in North Carolina determined that wind and solar technologies would displace facilities relying on nuclear fuels and minimize the environmental impacts associated with the extraction of uranium used to fuel nuclear reactors. In Oregon, the Governor's Renewable Energy Working Group projected that every 50 MW of wind energy would displace approximately 20 MW of base-load capacity, including nuclear power. Environmental groups in Michigan estimated that forcing the state to produce 20 percent of its power from wind would displace the need for more than 640 MW of coal and nuclear power.[116] And in Europe, researchers estimate that a network of wind farms over parts of Europe and Northern Africa could displace about 70 percent of the entire generation portfolio (including base-load nuclear and coal units).[117]

Another study looking at the benefits of interconnecting wind farms at 19 sites located in the midwestern United States found that a maximum of 47 percent of yearly averaged wind power from these interconnected wind farms could be used as reliable, base-load electric power. The authors concluded that almost all parameters from the wind farms showed substantial improvements as the number of interconnected sites increased, including fewer deviations in wind availability, improved reliability, and less need for reserve capacity. They also found no saturation of benefits, meaning that the addition of extra wind turbines always improved the reliability of the system.[118]

In addition, modern wind and solar systems can be connected to pumped hydro and compressed air energy storage facilities, or coupled with biomass generators to smooth out intermittency. Pumped hydro storage systems can improve the ability for wind to replace base-load generation. Bonneville Power Administration, a large federal utility in the Pacific Northwest, uses its existing 7,000 MW hydroelectric and pumped hydro storage network to do just that. Starting in 2005, Bonneville offered a new business service to "soak up" any amount of intermittent output from wind and solar facilities, and sell it as firm output from its hydropower network one week later. Such storage technologies can have greater than 1,000 MW of capacity (depending on location), and operate according to fast response times and relatively low operating costs.[119]

Attaching wind turbines to compressed air energy storage systems can improve their capacity factor to above 70 percent, making them "functionally equivalent to a conventional base-load plant."[120] NREL researcher Paul Denholm found that generating base-load power from a hybrid wind/compressed air energy storage system would add only about 0.7 ¢/kWh to its cost of producing electricity, and that converting natural gas plants to biomass generators to backup wind farms would add only about 0.2 ¢/kWh, two relatively inexpensive options to create completely renewable base-load units. Contrary to utility proclamations stating otherwise, "A base-load wind system," notes Denholm, "can produce a stable, reliable output that can replace a conventional fossil or nuclear base-load plant."[121]

Finally, incumbent energy companies and firms use intellectual property rights (IPR) and collusion to impede new technologies from entering the market. British Petroleum, for example, had to shell out $373 million in fines in 2007 after admitting in court that their traders manipulated parts of the natural gas market at will to drive up propane prices, and that the company intentionally ignored safety regulations requiring them to monitor corrosion on their distribution pipelines to keep prices low (this second mistake resulted in a 4,800 barrel oil spill into the Arctic tundra).[122] Other firms use also patents as intentional tools to impede innovation and competition. Generally, patents must balance two competing goals: giving adequate economic incentives to pioneering inventors, and ensuring that the innovations are followed and the public is benefited. Companies, however, have an incentive to suppress new and innovative technology if it threatens to disrupt profits in a market. While patent suppression has been used by caviar companies to suppress artificial caviar, by Xerox to protect plain paper photocopier technology, by the Automobile Manufacturers Association to suppress air pollution control equipment, and by big tobacco companies to suppress safer and less addicting cigarettes, perhaps two of the most famous examples of patent suppression concern early energy-efficiency technologies.[123]

Patent suppression can be found in the history of the electric lamp industry, born after Thomas Edison patented the first incandescent light bulb in 1880. Competing in the industry required a large investment of fixed capital and specialized plants with high overhead costs. These investments, coupled with an inelastic demand for electric lamps, made competition difficult and favored cartelization. General Electric emerged as the leader in the U.S. lamp industry by entering into several cross-licensing agreements with competitors to divide domestic markets, fix prices, and regulate exports. As a result, incandescent light bulbs with long life spans were intentionally suppressed.[124]

A similar pattern of collusive activity occurred with the development of fluorescent light bulbs. By the 1920s, the basic technology for fluorescent lighting was patented and well known. Yet the leading U.S. manufacturers were determined to saturate the incandescent light market before releasing more efficient lights. The suppression of fluorescent lighting was partially in response to pressure from electric utility companies, which believed that the increased efficiency of fluorescent lighting would lead to reduced demand for electricity and lower profits. The lighting and electric industries were highly dependent on each other, and utilities believed that the fluorescent lamp would negatively affect electricity load. They viewed more efficient lighting as a serious impediment toward electrifying rural areas, since large increases in demand were the primary justification for the expansion of electricity grids. Fluorescent light bulbs were suppressed until 1938, when Sylvania, a new competitor, successfully threatened to become the industry leader in the production of efficient bulbs.[125]

Instances of anticompetitive behavior are not limited to light bulbs or the 1930s. The recent consolidation of the wind energy industry has created incentives to use patents to block entry into markets. Four manufacturers, General Electric, Vestas, Enercon, and Gamesa, are responsible for three-quarters of global wind turbine sales. The Danish company Vestas is so large it made 2,533 wind turbines in 2006, installing one somewhere in the world every five hours, and produced nacelles in 13 countries around the world and blades in seven others. However, such consolidation creates a major disincentive for leading wind manufacturers to license proprietary information to companies that could become competitors. Vestas, for example, licensed its turbine technology to Gamesa, but now must compete with it on the world market.[126] General Electric has actively used its patent on variable speed wind turbines to block Enercon (a European manufacturer) and Mitsubishi (a Japanese manufacturer) from entering the American wind market.

Thus, the existing energy policy landscape, far from creating incentives for consumers and businesses to automatically invest in cost-effective clean power technologies and utilities to optimize the efficiency of the electric utility system, remains prone to multiple and interrelated market failures. Consumers lack capital and knowledge about energy efficiency investments; builders, homeowners, and businesses remain uninterested in energy projects; and utilities and energy companies wield their market power to retain control over the existing state of affairs.

CHAPTER 4

POLITICAL AND REGULATORY OBSTACLES

Electricity and politics have always been intertwined. In 1959, President Richard Nixon held up the American standard of electrical living against Nikita Khrushchev in their famous "kitchen debate." Standing in front of an exhibit of an American kitchen in Moscow replete with electrical appliances, surrounded by Russian schoolchildren, Nixon drew out the meaning of the exhibition and answered the charges that only the "rich" could afford to live in a kitchen like this.[1] Electricity, it seems, played a key role in distinguishing the merits of American and Soviet life-styles.

Yet, after digging through archives at the DOE for three years and spending almost another decade writing his study, historian John G. Clark remarked that one thing above all stood out in his assessment of American energy policy. "US energy policy," he concluded, "can be characterized as minimalist, uncoordinated, and based on the erroneous principle that correct policy guaranteed the lowest possible price."[2]

How can electricity be both central to American identity and yet punctuated by haphazard and inconsistent government policy? As a partial answer, this chapter explores the political and regulatory impediments to clean power systems and begins by noting that the strong political support, after the energy crisis of the 1970s, inflated expectations among the public that the use of renewable energy resources would grow rapidly. Yet a number of unforeseen events occurred: the Reagan administration shifted the energy policy of the country, fossil fuel prices fell in the 1980s, and conventional technologies continued to improve. Voters and politicians became disillusioned with clean power, and relinquished whatever social capital they achieved after the energy crises to utility managers and system operators.

After the 1970s, when the country shifted completely back into the fossil fuel paradigm, inconsistent political support for tax credits created great uncertainty regarding clean power technologies, deterring both public and private investment. Well-intentioned interventions from state governments, often in the form of mandatory targets and standards, further contribute to the uncertainty. Public and private firms continue to underfund R&D in clean power systems, and an overall hubris and faith in big technological projects complicates projects and can impede innovation.

Flawed Expectations

Many of the benefits offered by clean power technologies, elaborated in Chapter 2, convinced people, including President Jimmy Carter, to zealously push for their adoption in the 1970s. As a result of the Organization of Petroleum Exporting Countries (OPEC) oil embargo of 1973, President Carter made energy policy his first major initiative. Among a set of five laws proposed by Carter and passed by Congress (albeit in greatly diluted form), the Public Utility Regulatory Policy Act of 1978 (PURPA) had the most far-reaching—and least intended—consequences for small-scale energy systems. PURPA spurred creation of radical technologies, began the process of deregulation, and challenged the control held by power company managers.

Superficially, PURPA appeared to pose little threat to utility companies. But one obscure portion of it offered incentives for the use of efficient CHP plants. A related provision spurred research on environmentally preferable technologies that used water, wind, or solar power to produce electricity. More successful than anyone originally anticipated, PURPA established the first production tax credits for renewable energy systems such as wind turbines and solar panels.

While Carter's support was instrumental in making the public more aware of the benefits from comparatively cleaner energy, it also catalyzed some people to make unrealistic projections about how fast such energy systems could take off. Throughout the energy crisis of the 1970s, policy experts predicted that renewable energy systems would be in widespread use by the turn of the century. President Jimmy Carter privately confided to his friends that he expected renewable energy systems to reach 10 percent of national electricity capacity by 1985. James D. Conant, one of the two overseers of the Manhattan Project and president of Harvard University, thought that by the end of the century solar energy would be the "dominating factor in the production of industrial power." [3] In 1977, Denis Hayes, director of the Solar Energy Research Institute, predicted that:

> By the year 2000, renewable energy sources could provide 40 percent of the global energy budget; by 2025, humanity could obtain 75 percent of its energy from solar resources . . . Every essential feature of the proposed solar transition has already proven technically viable.[4]

Earl T. Hayes, former chief scientist at the U.S. Bureau of Mines, and Michael C. Noland from the Midwest Research Institute similarly thought that solar panels would furnish 10 to 20 percent of all American energy needs by 2000.[5]

As Joe Loper from ASE put it, "The projections [for renewable energy systems] were too optimistic." [6] Michael Karmis, director of the Virginia Center for Coal and Energy Research, commented that the rising political furor over renewable energy resources convinced policymakers and the public to overestimate the technological breakthroughs needed to transition to a renewable based electricity sector. Karmis concluded that "there was an overestimation of the potential for renewable energy systems and an underestimation of the time, resources, research, and development needed to make them work."[7]

A number of unforeseen events shifted the electric utility system back in favor of fossil fuels. Many researchers believed natural gas supplies would run out and that oil prices would rise between $30 and $60 per barrel (in 1970s dollars). However, OPEC strategically decided to increase its supply, driving the price of oil down in 1974. A few years later, natural gas markets were deregulated with the Natural Gas Policy Act of 1978, lowering the cost of natural gas. Around the same time, the Staggers Act of 1980 restructured the railroads, further pushing the price of coal downward. Brian Castelli, the executive vice president for the ASE and a former program officer at the DOE during the 1970s, explains that:

> A big reason renewables never met expected potential is because the target price for technologies that compete with renewable and distributed energy systems continued to drop relative and faster than the prices for most renewable technologies. For example, renewable systems competed against coal fired generation, and coal just became cheaper and cheaper. No matter how much we cut down the cost of PV and wind, the coal price continued to fall faster.[8]

When devising their original economic estimates about the future competitiveness of renewable energy systems, most analysts assumed that fossil fuel prices would continually escalate, rather than gradually decline.

The declining cost of fossil fuel was coupled with improvements in the technical proficiency of fossil-fueled generators. Adam Serchuk, senior program officer at the Energy Trust of Oregon, explained that "established technologies didn't stay as they were built. Natural gas combined cycle turbines, for example, became more efficient, they didn't just stand still as a dead and dying technology."[9] And Chuck Goldman, program director for renewable energy LBNL, argued that "unforeseen improvements in traditional electrical generation technology also ended up distorting the initial estimates and cost targets for renewable systems."[10] Additionally, it took much longer than expected to bring down the capital costs of renewable energy systems. Scott Sklar, the president of a solar panel manufacturing company, notes that, while it was expected that renewables would develop themselves almost by "immaculate conception," it instead took decades to scale up the manufacturing and delivery mechanisms needed to make renewable energy systems economically competitive.[11]

Furthermore, on top of having to compete with more natural gas turbines using cheaper fuels, renewables had to compete with cost effective efficiency measures and overcome utility resistance. Chuck Goldman from LBNL added that:

> In relative terms, with energy efficiency, you can capture a sizeable amount of the resource at a cost considerably less than conventional supply-side options. And for renewables, it's only been in the last couple of years where some renewable technologies are cost competitive with traditional supply-side technologies. On the supply-side,

> fossil-fuels ended up being cheaper, and on the demand side, energy efficiency practices ended up being cheaper. Both impeded a wider use of renewable energy systems.[12]

Rather than being pursued as part of a synergistic or independent strategy, renewables had to compete directly against other clean power technologies for funding and political commitment.

The confluence of these events—lower fossil fuel costs, more efficient fossil fuel generators, and comparatively higher renewable energy costs—greatly challenged the accuracy of predictions made about renewables in the 1970s. Ironically, the very same enthusiasm that engendered so much support for renewable energy systems also contributed to their "downfall." Marilyn Brown, former director of engineering science and technology at ORNL, explains that such premature enthusiasm turned off many people from renewables altogether, and that:

> The policies of the 1970s did not catalyze the renewable energy industry. The technologies, simply put, were not ready for prime time, and when they failed they gained a blemished reputation. Solar technology (for instance) wasn't ready, but political factors associated with the energy crisis pushed it ahead anyway. Renewable energy wasn't ready to be incentivized, and the whole market disintegrated. This set the industry back instead of accelerating it.[13]

Intermittent political support for renewable energy systems did more than just hurt government programs. When the American renewable market collapsed, it poisoned the country's intellectual consciousness against alternative energy systems. The country's physical landscape was littered with images of broken down wind and solar farms, and its business landscape was haunted by memories of bankrupt American renewable energy manufacturers. One study noted that "the specter of failed American wind projects during the 1970s and 1980s has effectively kept wind power from becoming a major player in the U.S. electric industry."[14] James Gallagher, director of the Office of Electricity and Environment for the New York State Department of Public Service, commented that "the early failures or hurdles with renewables turned people off from the technologies . . . people developed a bad taste in their mouth."[15] Thus, renewables were paradoxically a victim of their own success: public favor quickly turned to either apathy or resistance once the expectations for renewable energy failed to materialize.

Inconsistent Incentives

Moreover, public disinterestedness and contempt for clean power systems enabled utilities and interest groups to further fight against their adoption. The most obvious element of this antipathy concerns the inconsistent political support for renewable energy systems. Unlike subsidies and incentives for fossil-fueled technologies, policies aimed at encouraging clean power technologies have changed frequently, greatly discouraging widespread adoption of the technologies. Since all policy is politically and ideologically motivated, inconsistency should not seem unexpected. Joe Loper, vice president for research and analysis at ASE, remarked that:

> It is going to take one hell of a crisis to get serious about energy issues in the U.S. Politicians have some special interests driving them, but they, and their inability to confront energy issues seriously, challenge entrenched attitudes about consumption, and really raise awareness about energy issues, are really a reflection of society at large.[16]

Ken Tohinaka of the Vermont Energy Investment Corporation went so far as to claim that "I don't think there ever has been a true national energy policy in this country . . . the U.S. model tends to operate through annual appropriations which are routinely cut or adopted only at the last possible moment, making it hard to invest in or plan long-term energy efficiency practices." [17]

To be fair, some politicians have tried actively to promote cleaner technologies. Because of his political (and spiritual) beliefs that espoused personal sacrifice and changes in fundamental values concerning consumption, President Carter advocated legislation, regulations, and tax credits to spur energy-efficiency and renewable-energy technologies. However, these incentives expired in 1986 under the Reagan administration. In a symbol of his free-market fanaticism, Ronald Reagan removed the solar collectors from the roof of the White House and took more substantive measures to end federal programs and tax credits that encouraged efficiency and alternative energy technologies.[18] David Baylon, an energy consultant, argues that the end of the Carter administration signified the end of the country's focus on clean power:

> When people in the 1970s were saying they thought renewables were the way of the future, the U.S. government was pumping in over $1 billion into energy R&D. Since the 1980s, we've probably put a $1 billion all together into R&D.[19]

Reagan, in direct contrast to Carter, reverted back to what energy analyst Jan Harris refers to as the "subsidize oil and gas" and the "dig, dam, drill" approach to energy policy.[20] Such a shift in national priority had far reaching consequences in promoting fossil fuel technologies, and disadvantaging clean power technologies.

For instance, the transition from the Carter administration to the Reagan administration did more than financially endanger government R&D for distributed and renewable energy resources: it drove some people out of the renewable and small-scale energy industry altogether. Sam Fleming, an energy consultant with 30 years of experience building renewable energy projects, explained it this way:

> Many people in government had high expectations for renewables, including things that went to Congress for demonstration and commercialization programs, concentrating solar energy projects in the Mojave Desert, various kinds of incentives, power purchase agreements, and tax credits under PURPA which translated into a flurry of activity for renewables. However, in the early 1980s the government quickly removed key incentives, including accelerated depreciation, and several renewable energy projects had to be abandoned mid-way through construction and companies went bankrupt.[21]

Wilson Prichett, a consultant who has managed more than 300 renewable energy projects in the United States, argued that President Reagan discouraged everyone, not just policymakers in the government, from working on energy systems. As a result of Reagan's policies, Prichett recalled that:

> By 1982 or 1983, most everyone had left the industry. All those thousands of people around the country working on ways to convert agricultural waste to fuels and different solar and wind designs just stopped doing it. There was no political encouragement. In fact, there was complete discouragement.[22]

And Tommy Thompson, a researcher for the state of Virginia's energy department, explained that "when we were doing solar technologies back in the 1970s, it was only because of the incentives being given that made the projects viable. The incentives dried up, and then the renewable industry dried up." [23]

As another sign of its potential bias toward fossil fuels and large generators, the federal government has been equally inconsistent in the promotion of tax credits for renewable energy resources. The Energy Policy Act of 1992, signed by Reagan's successor, George H. W. Bush, provided a production tax credit for certain renewable energy technologies. But those credits expired in 1999, and environmental advocates worked diligently to win Congressional approval for their reinstatement, often on an annual basis. When Congress failed to restore the credits before the end of 2001, investment in wind turbine projects dropped precipitously. Developers installed only 410 MW of new wind turbines in 2002, down from about 1,600 MW in 2001 and 2003. The conflicting policies for wind turbines in the United States have created boom and bust cycles within the industry, making it all but impossible to obtain financing for projects. The trend is set to continue, with the U.S. Congress failing to renew tax credits for renewables and clean power in July 2008 despite extending the tax credit for nuclear power plants to 2030.

While government tax credits, in theory, are supposed to foster the development of novel technologies, in practice such tax credits have sometimes been tragic for

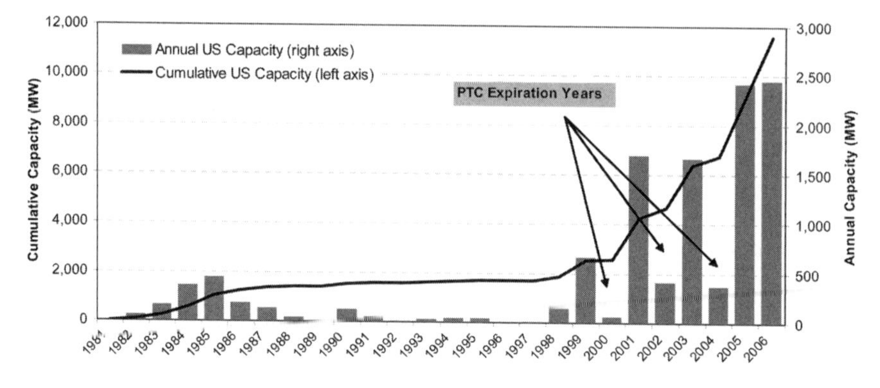

Figure 17
U.S. Wind Power Capacity, 1981 to 2006

The inconsistent U.S. federal production tax credit has been devastating for American wind developers. Note the effect the expiring credit had on wind installments in 2000, 2002, and 2004.

Source: Ryan Wiser, "Wind Power and the Production Tax Credit: An Overview of Research Results," *Testimony Before the Senate Finance Committee,* March 29, 2007.

investors. Larry Papay, the chair of the California Council on Science and Technology, saw a problem in at least two areas:

> Many of the well-intentioned incentives put in place for renewables have been the wrong ones, such as investment tax credits rather than production tax credits thereby guaranteeing the technologies actually get into operations. And then, even when inventive programs can be the "right" ones, they tend to expire on a yearly or bi-yearly basis, meaning that developers are reluctant to commit large sums of money to renewable projects.[24]

Ralph D. Badinelli, a professor of business information technology at Virginia Tech, notes that certainty is the single most important factor in determining whether companies and manufacturers will choose to invest in certain energy technologies. As Badinelli explains:

> The more uncertainty there is about the future, the higher the risk premiums placed on investment returns. Furthermore, the risk of investments in alternative energy generation systems is amplified by the long terms of these investments: generators usually are considered capital investments with decades-long lifetimes.[25]

Adam Serchuk from the Oregon Energy Trust believes that "the mere uncertainty of the tax credit itself creates a host of problems . . . banks often refuse to finance DG projects, insurers refuse to protect them, and local employees resist them."[26] Chris Namovicz from the EIA admits that tax credits have created "significant uncertainty regarding investment" in renewable resources.[27]

The problem, moreover, is not limited just to production tax credits. Many of the laws relating to clean power are internally inconsistent: federal R&D on renewable energy systems has focused on centralized, large-scale, and utility owned technologies, whereas legislation such as PURPA has advanced decentralized, small-scale, and independently owned technologies. The 1935 Public Utility Holding Company Act (PL 74-333), 1980 Wind Energy Systems Act (PL 96-345), 1984 Renewable Energy Industry Development Act (PL 98-370), and provisions of the Energy Policy Act of 1992 (PL 102-486) were allowed to expire or never fully implemented. The federal government placed a moratorium in 2008 on new applications for the construction of solar collectors on federal lands and has refused to expedite the cumbersome, multiyear permitting process for offshore wind turbines, but has no such constraints on oil and gas exploration and production.[28]

Consequently, the variability of policy relating to clean power technologies serves as a serious impediment. Entrepreneurs seeking investment from individuals and institutions often require consistent conditions upon which to make decisions. Forecasts of profitability usually require data concerning tax credits, depreciation schedules, cash flows, and the like, well into the future. When policymakers frequently change the factors that go into these financial calculations, they insert an extra level of uncertainty into the decision-making process. Donald Aiken has stated that "an effort to promote renewables has to be sustained, orderly, substantial, predictable, credible, and ramped."[29] In the United States, formal policy has tended to vary for clean technologies on each of those criteria at the same time it has remained

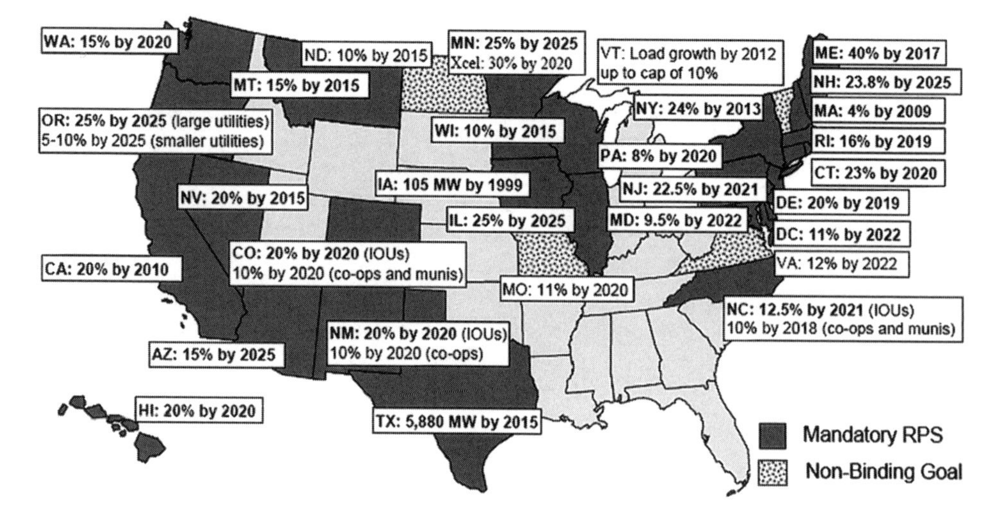

Map 15
States with Renewable Portfolio Standards, 2008

As of March 2008, more than half of the states have implemented some form of renewable portfolio standard in an attempt to promote renewable power within their jurisdictions.
Source: Lawrence Berkeley National Laboratory.

consistent for fossil-fueled generators. The result has been inconsistent government mandates, an increase in electricity rates for consumers, less research and development in innovative energy technologies, and a worsening of uncompetitive practices in the electric utility industry.

Varying State Standards

The states, on the other hand, have taken the lead in promoting clean power systems. Ever since Iowa and Minnesota mandated that utilities purchase renewable energy in 1985 and 1994, no fewer than 28 states and the District of Columbia have implemented some form of mandatory standard (often called a "renewable portfolio standard") forcing power providers to use renewable energy resources. Collectively these states have launched hundreds of millions of dollars in renewable energy projects, the most aggressive states being California and Colorado (20 percent by 2010), New York (24 percent by 2013), and Nevada (20 percent by 2015).

Despite the immense progress individual states have made in promoting clean power, however, state contributions remain constrained by the design and inconsistency of their differing statutes. Contrary to enabling a well-lubricated national renewable energy market, inconsistencies between states over what counts as renewable energy, when it has to come online, how large it has to be, where it must be delivered, and how it may be traded clog the renewable energy market like coffee grounds in a sink. Implementing agencies and stakeholders must grapple with inconsistent state goals, and investors must interpret competing and often arbitrary statutes. If America's interstate highway system were structured like its renewable energy market, drivers would be forced to change engines, tire pressure, and fuel mixture every time they crossed state lines.

To pick just a few prominent examples, Wisconsin set its target at 2.2 percent by 2011, while Rhode Island chose 16 percent by 2020. In Maine, fuel cells and high efficiency cogeneration units count as "renewables," while the standard in Pennsylvania includes coal gasification and fossil-fueled DG technologies.[30] Iowa, Minnesota, and Texas set their purchase requirements based on installed capacity, whereas other states set them relative to electricity sales.[31] Minnesota and Iowa have voluntary standards with no penalties, whereas Massachusetts, Connecticut, Rhode Island, and Pennsylvania all levy different noncompliance fees.[32] The result is a renewable energy market that deters investment, complicates compliance, discourages interstate cooperation, and encourages tedious and expensive litigation.[33]

Using individual states as a crucible for innovations in electricity generation and marketing may have made sense when limits were placed on the size and geographic scope of utility holding companies. But now, multistate utilities must struggle with competing statutes. For instance, American Electric Power serves more than 5 million power customers in 11 different states. The utility operates in Texas, where it must meet the state's mandatory renewables target; Virginia, where it is encouraged to meet the state's voluntary renewables target; Indiana, which is considering a renewable target; and Tennessee, which has none at all. In the Pacific Northwest, state statutes demand that PacifiCorp generate 25 percent of its electricity from renewable resources in Oregon by 2025, 20 percent in California by 2010, and 15 percent in Washington by 2020. Xcel Energy must similarly juggle differing statutes in Colorado and New Mexico (20 percent by 2020), Minnesota (30 percent by 2025), Wisconsin (10 percent by 2015), and Texas (a proportion of 5,880 additional megawatts).

Underfunded R&D

Both the public and private sectors continue to underfund R&D on clean power technologies, especially compared to conventional systems.

Adam Segal, senior fellow at the Council on Foreign Relations, notes that while federal R&D in technology reached $132 billion in 2005, it continues to be concentrated in the fields of defense, homeland security, and the space program.[34] Federal funding for R&D in energy efficiency, DG, and renewables has declined as a portion of real gross domestic product by more than 70 percent from 1994 to 2004.[35] After funding for such technologies peaked in 1980 at $1.3 billion, it declined to $560 million in 1980 and just under $140 million in 1990, after which funding stabilized at about $200 million (see Figure 18). As Kurt Yeager, former president of EPRI, remarked, "today the U.S. invests at a lower rate than [any of] its major international competitors."[36]

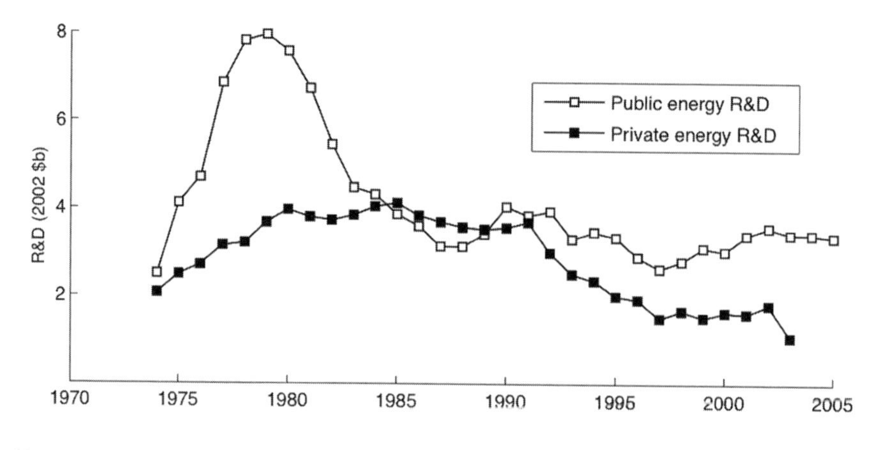

Figure 18
Public and Private Spending on Energy R&D, 1974 to 2005

Depicted in billions of $2002, both public and private spending on energy R&D has sharply declined over the past two decades.

Source: Gregory F. Nemet and D. M. Kammen, "US Energy Research and Development: Declining Investment, Increasing Need, and the Feasibility of Expansion," *Energy Policy* 35, no. 1 (2007): 746–755.

Recent events in the energy sector, such as the restructuring of the electric utility industry and increased competition, make it unlikely that large utilities and firms will compensate for public underinvestment. Restructuring of the electric utility industry and the repeal of the Public Utilities Holding Company Act of 1935 (PUHCA) increased the incentive for companies and firms to invest in short-term technologies with rapid financial returns. A 1999 Office of Science and Technology Policy report cautioned that:

> Privatization, deregulation, and restructuring of energy industries . . . can lead to neglect of the ways the composition and operation of energy systems affect the wider public interest (including meeting the basic needs of the poor, as well as addressing other macroeconomic, environmental, and international security needs).[37]

A gap is thus emerging between what the electricity industry does and what society's interests require.

Increased competition means that electricity firms are likely to make investments only in short-term projects that have better discount rates, lower risk, and perceived better financial return to the company's investors. It also means that, in order to prepare for competition, many utilities cut investment in discretionary spending, and have less of an incentive to collaborate on research.[38]

During the late 1990s, large utilities drastically cut R&D spending on clean power technologies to prepare for restructuring and impending competition. A 1998 study conducted by the GAO concluded that "increased competition from restructuring was cited as the primary reason for the biggest cutbacks in research to date by utilities in California, New York, and Florida." [39] The revocation of

PUHCA could accelerate this trend, since the financial consolidation of utilities and holding companies will likely convince utilities to shift the focus of their R&D from collaborative projects benefiting society to proprietary R&D giving their affiliates a competitive edge. Empirically, the GAO has noted that when utilities must consolidate, the result is "slowing technology development, sacrificing future prosperity to meet short-term goals, and failing to meet national energy goals."[40]

What about smaller firms? MIT economists Nancy Rose and Paul Joskow found a difference between the *size* of utilities and their willingness to adopt new technologies. They surveyed 144 electric utilities that built high-pressure conventional and supercritical coal-fired steam turbine generating units between 1950 and 1980. The authors note that the "willingness to innovate" was affected by a diverse array of factors such as perceived economies of scale in learning and using the innovation, firm participation in complementary R&D activity, and risk aversion.[41] Smaller firms and government owned utilities, they conclude, adopted newer technologies much more slowly than larger firms and investor owned utilities. What accounted for this difference?

Rose and Joscow argued that fuel costs were the most significant determinant of expected cost savings from supercritical and conventional high pressure units, since both technologies reduce operating costs by improving fuel efficiency (typically between 2 and 5 percent). Larger utilities have larger units that consume more fuel, meaning they have an incentive to adopt fuel-efficient technologies sooner. Larger utilities and firms are more likely to possess internal engineering, design, and maintenance staff that are interested in adopting new technologies, and capable of adopting them before substantial operating experience has accrued.

Smaller firms have smaller portfolios of generating units, meaning the impact of a mistake on cost of service, system performance, and profitability cannot be easily absorbed. They often own fewer plants than larger firms, so when lower costs are correlated with operating more plants, they have less incentive to favor early adoption. Furthermore, if new technologies are scale augmenting, they may be more attractive to larger utilities that can economically add capacity in large chunks. Larger utilities build new generating units of all types more frequently than smaller ones, as a result of the relationship between size, growth rates, and the lumpiness of installation.

What this means for energy R&D is that no one will do it. The government is focused on researching other areas. Small firms are less likely to adopt new technologies and conduct R&D programs because they are more risk averse and have fewer resources, whereas large firms are reluctant to undertake R&D because they are more focused on short-term gains and dealing with competitors. As a result, energy R&D *intensity,* expenditures for energy R&D as a percentage of the utility's total sales for one year, among energy companies averages less than three-tenths of 1 percent, compared to an average industrial benchmark of 3.1 percent. Power companies conduct less R&D than manufacturers of dog food.[42] R&D investments made by all energy companies declined 50 percent between 1991 and 2003.[43]

Hubris

Paul Gilman, former executive assistant to the Secretary of Energy, believes that a final problem is the way that American policymakers tackle energy problems. Gilman argued that:

> Lots of the energy legislation that was drafted in the 1970s was rather naïve and vague. It said that we should achieve x percentage of our generation from y source by this year, and there were lots of goals and laudable things . . . That tends to be the context in which Americans approach technological problems: we set lofty and daring goals. We were not smart, doing the market analysis of what is doable, and integrating that with an understanding of the energy sectors—transportation, buildings, and so on—and taking an integrated look at the overall picture.[44]

As a result, Gilman concludes that politicians went about promoting renewable energy systems "in the wrong way." What exactly did they do wrong?

Energy R&D efforts in the United States have been largely characterized by centralized management of research, little product development between steps, selective evaluation, competition between researchers, consolidation of information, corporate ownership, and episodic government policy. American efforts to design wind turbines are most telling, where researchers and managers pursued a "top-down" wind energy R&D strategy, starting with how to determine the optimum design and size for wind turbines, and then heading directly toward the final turbine, focusing on the rotor diameter, blade tip speed, and height that would guarantee the lowest end price.[45]

In other words, American designers started with a "high-tech" approach emphasizing aerodynamic efficiency and then attempted to build all of its components at once. Engineers became insulated from the hands-on problems encountered in construction and maintenance, and the extent of interaction between producers and suppliers was poor, since most relationships were one-time, short-term contracts motivated by profits at the expense of efforts to foster knowledge sharing and interactive learning. An adherence to gigantism (large, multi-MW machines) further limited the interaction points from which producers could learn from each other and weakened the linkages between government agencies (such as NREL) and industry. American designers also sought the lightest type of construction materials and disregarded the structural dynamics of the blades, resulting in an excessive rate of fatigue fractures.

Standardization and evaluation techniques were selective, with less emphasis placed on comparative testing and standards based on generic engineering science that failed to evolve with "hands-on" knowledge from users. Both NREL and its predecessor, the Solar Energy Research Institute (SERI), had the word "research" in their titles but not the word "testing." Researchers at such laboratories focused on finding high-tech designs based on fundamental scientific principles and created abstract models too theoretical to provide the kinds of systematic testing the industry needed. NREL, for example, took three years to develop advanced wind turbine blades while researchers in other countries (such as Denmark) used the same amount of time to complete an entire turbine development cycle, introducing new models every two to three years.

Ownership was different in the United States as well, with most developers keen to exploit subsidies and tax credits that were not dependent on how well the wind turbines actually worked. A fundamental separation of ownership from usage began, and ownership was highly concentrated in a few competitive firms. While more than 30 institutions, laboratories, and universities received funding for wind projects from the DOE in fiscal year (FY) 1979, nearly 90 percent of the funds went to only 8 of them, mostly large aerospace companies and defense contractors such as Boeing, General Electric, and Raytheon.

The approach proved counterproductive. In California, wind technology grew too rapidly in the 1980s, where a combination of federal and state investment tax credits brought significant but unreliable and unproductive capacity online. American research programs were built on the mistaken premise that the design of wind turbines should be based on aerospace models. This meant that blade throw, damage by lightning strikes, and problems with bugs and ice on turbine parts (things not experienced by airplanes in the same manner as wind turbines) became difficult challenges. Consequently, American turbines began failing in large numbers in the mid-1980s. Several factors created a backlash against wind power, resulting in serious damage to the industry. Hyped up performance projections led to false expectations and were enhanced by negative images from the media, which portrayed them as tax scams. As one research director lamented, the only "feedback" his company received was from lawyers and lawsuits. There was also little pressure to develop independent mechanisms for information disclosure and evaluation.[46]

Such problems were not unique to wind R&D. Rather than building a solar R&D program aimed at taking advantage of its "distributed, decentralized, and democratic" nature, American government planners emphasized large, centralized solar stations. In promoting solar in much the same way as planners promoted nuclear and fossil-fueled power stations, early R&D programs ignored smaller and distributed applications and attempted to create solar technologies "in the image of nuclear power."[47] Massive solar engineering projects, often designed by aerospace companies, dominated the government's early solar R&D program, attempting to harness solar power to produce utility-scale electricity rather than small-scale technologies for individuals or communities.

An adherence to large, centralized, top-down projects has even been documented in early research on the electric vehicle. When Congress enacted the Electric and Hybrid Vehicle Research, Development, and Demonstration Act of 1976 (PL 94-413), government planners wanted an electric vehicle with the same performance characteristics as an internal combustion automobile as soon as possible. What should have been an incremental (and possibly more viable) research strategy aimed at scaling up small improvements was instead scrapped for an ambitious program that attempted to refine all vehicle components at once. After the project failed,

Victor Wouk, who worked with the program in 1974, noted that "we learned from the program that technology advances cannot be coerced by legislation."[48]

Thus, a bigger is better ideology, strong belief in technical efficiency, hierarchical management, centralization, and search for economies of scale in gigantism have tended to furnish overconfidence in the potency of American energy R&D (when it has been funded).

CHAPTER 5

CULTURAL AND BEHAVIORAL BARRIERS

The writer Aldous Huxley recounted how during one of his summers, he journeyed to a little valley in England once visited as a child. In his childhood, Huxley recalled an area of delightful grassy glades, but when Huxley returned it had become overgrown with unsightly brush. The rabbits keeping the grasses neatly trimmed had succumbed to myxomatosis, a disease deliberately introduced by local farmers to reduce rabbit-linked destruction of crops. Huxley lamented such unnatural interference with nature, but even he was unaware that the rabbit itself had been brought as a domestic animal to England in 1176, ostensibly to improve the protein diet of the citizenry.[1] The story reminds us that the world we live in is partially a construction, and that the things we take for granted as "natural" may not always be. Most of the time, we are completely unaware of our cultural and human built surroundings—much like water to a fish, the technological world we inhabit is just part of the environment we swim through.[2]

In much the same way, this chapter shows how the apparent disconnection between how electricity is made and how it is socially perceived perpetuates public apathy and misinformation about it. This misunderstanding pervades not just ordinary customers but also the builders of homes, facilities managers, zoning officers, and local officials. Many Americans believe that they are entitled to cheap and abundant sources of electricity, but they lack the necessary understanding of what needs to occur so that they continue to have access to such a supply. As a result, clean power systems are often opposed not because they are a poor alternative to fossil fuels, but because people simply do not comprehend why such technologies may be needed. Far from becoming an unintentional side effect of the "hub and spoke" model of electricity supply, some evidence suggests that public "ignorance" toward electricity is an important, and planned, part of the existing technological system. The lack of public interest in the electricity sector allows utilities and system operators to maintain control over their system and extract stable profits.

Public Apathy and Misunderstanding

To explain how American consumers perceive electricity, it is necessary to explore briefly a few basic facts about the product itself. The president of the American Educational Institute remarked that:

> Electricity is difficult, really impossible, to visualize. I can hold a pound of coal, or a 16 ounce jar of oil or gas, in my hand. A few of us could hold a pork belly. But no one I know could hold a kilowatt-hour of electricity . . . electricity is different, it's an abstraction, not a commodity, but a phenomenon.[3]

Ever since the 1880s, commentators have been speaking about how electricity is unique among all industrial enterprises.

Electricity, while bought, sold, and traded as a commodity on the American market, is really unlike any other item. A unique disjuncture occurs between its appearance at the point of consumption (something clean and abundant) and its point of creation (something typically dirty and manufactured). Electric utility systems are even fundamentally different from other large infrastructures such as air-traffic control centers, natural gas pipelines, and long-distance telephone and communications networks because of the need for continuous and near instantaneous supply, high reliability, and the passive nature of the transmission and distribution network. As former Secretary of Energy Hazel O'Leary used to say, "Electricity is just another commodity in the same way that oxygen is just another gas." [4] Economist E. F. Schumacher agrees that electricity is "not 'just another commodity' but the precondition of all commodities, a basic factor equal with air, water, and earth." [5]

Those using electricity always need a secure and constant supply, yet it cannot be easily saved in large quantities, meaning no surplus can ever be effectively maintained without generating more of it. It is impossible to stockpile electricity without expensive storage technologies, and the product must be produced and consumed almost simultaneously. Of course, batteries can store electricity, but such storage systems are still uneconomical for amassing large amounts of it, and options like pumped storage require the availability of significant sources of water.

Moreover, there are no feasible substitutes for electricity. Demand can be unresponsive to price, fluctuating from season to season and day to day. The T&D system has few "control valves" or "booster pumps" to regulate electrical flows on individual lines, meaning that control over the system is limited to adjusting generation output and to opening and closing switches to add or remove entire transmission lines from service.[6] Because electricity can only be transported through an extensive T&D network with limited ability to route power, the actual physical delivery patterns can never match the contractual arrangements for its sale, since it flows through the entire grid.[7]

Before the invention of large, expansive electric utility networks, people tended to more actively harness energy inside their home. When the primary fuel for energy in Europe was wood, the consequences of its use were immediate and local. Pollution shrouded cities wherever households combusted wood in large quantities, and forests were felled faster than they regenerated. Over time, expansive woodlands whittled away, and the search for substitute fuels intensified.

Yet in the modern electric utility sector, the tendency for power plants to be spatially isolated from population centers dilutes the aggregate impacts of electricity. Utility managers started situating large plants built in the 1940s and 1950s outside cities. Urban expansion depleted the amount of property available for land-intensive electricity generators, and residents living in American cities became more aware of air pollution and environmental problems associated with energy production. Planners located nuclear plants outside cities as a safety measure as well. Furthermore, since water is essential for the transfer of heat in the electrical generation process, it made sense to place plants near its adequate availability, resulting in a tendency for power plants to concentrate near rivers, lakes, seas, and oceans.

The physical "removal" of power stations from most cities and neighborhoods also "removes" them from the minds of most Americans, and contributes to public apathy and misunderstanding. Geographer Martin J. Pasqualetti argued that:

> An out of sight, out of mind pattern misleads the public by suggesting that the environmental costs of electricity are less than they actually are . . . As distance, technology, and our urbanized lifestyle came to cushion us from the direct environmental costs of energy, we become increasingly less aware and eventually less tolerant of the intrusions of energy development on our personal space.[8]

In other words, electricity places unique demands on natural resources, the environment, and the marketplace, yet its delivery segregates these impacts from the population to deliver a product seemingly pure, invisible, clean, and cheap.

Some of this has to do with the "physical" invisibility of energy technologies. When energy use is visible, people think that more is being depleted. This is why people tend to overestimate the energy consumed by household lighting (which is literally visible) but underestimate the larger amount used by water heaters (where consumption occurs without human intervention). A survey of 400 homes in Michigan found that the average resident believed, wrongly, that they could save twice as much money reducing lighting than using less hot water.[9] Building contractors report that it is easier to sell a new home with visible solar collectors on the roof than with passive solar design, added insulation, and other less visible features, even though the latter actions would save energy more cost effectively. Moreover, insulation in walls, flame-retention heads on oil burners, aluminum on automobile bodies, and extra windings on electric motors all save energy without being visible. Because people cannot easily see them, they are less likely to believe these alterations save energy.

Interviews with families and their decision making processes show that direct household energy consumption is based on the *nondecisions* of families. Families do not make explicit decisions to consume energy at some level or up to some dollar amount. Rather, they engage in activities of their choice to meet particular goals and consume energy in the process. Almost never do families decide how many gallons of fuel oil to burn in the next month or kWh of electricity to use, although family decisions about every other commodity from sugar to shoes and molasses to movie tickets are consciously planned. Thus, heating and electricity bills are the consequences of the family's life-style rather than the other way around.[10]

Most people do not even count "electricity" as technology. Historian James C. Williams argues that people know that technology and technological systems are the tools with which they interact in their everyday lives. But once technological landscapes are in place, people fold them so completely into their psyches that those very landscapes become almost invisible.[11] In today's culture, most people conceive of technology only as the latest high tech items, such as new and rapidly developed electronics. Inventions of far larger historical significance, such as electricity, no longer "count" as technology. Sociologist Paul N. Edwards remarked that:

> The most salient characteristic of technology in the modern (industrial and postindustrial) world is the degree to which most technology is not salient for most people, most of the time . . . The fact is that mature technological systems—cars, roads, municipal water supplies, sewers, telephones, railroads, weather forecasting, buildings, even computers in the majority of their uses—reside in a naturalized background, as ordinary and unremarkable to us as trees, daylight, and dirt.[12]

The unique attributes of electricity, segregation between its production and consumption, and physical and mental "invisibility" of energy technologies serves as a serious impediment to the adoption of clean power technologies. Once electric power becomes part of people's lives, they rarely think about how it is produced, how it gets to them, and how to save it.

Despite three serious blackouts in the past five years, plus the devastating effects of Hurricane Katrina on the American energy sector, electricity problems remain insignificant in the minds of many Americans. A recent 2006 *Wall Street Journal*/NBC Poll found that while the public recognized social security, health care, the Iraq war, and unemployment as important issues, neither "energy" nor "electricity" made the list.[13] Toben Galvin of the Vermont Energy Investment Corporation noted that people could take a number of simple, easy steps (such as weatherizing their homes, insulating their boilers, switching off their appliances and lights) to lower their energy bills, but they do not. Instead, Galvin concluded that the "lack of significant, sustained, national political support prevents energy efficiency mechanisms from being even partially effective throughout most areas of the country." [14] The benefits of energy efficiency are often meaningless to people because they are not directly observable, and when they are, their price tends to be bundled with other purchases.[15] The cost of a better insulated home, for instance, is already buried in its base price.

As one result of public apathy, misinformation concerning electricity and energy flourishes. A comprehensive 1978 study undertaken by Southern California Edison, surveying thousands of consumers, asked them, "Where does electricity come from?" Most people said, "Out of the plug in the wall," while others even said "lightning" and "static electricity." One of the authors of the study concluded that "people in this country have no idea how electricity is generated or transmitted." [16]

More recently, 41 percent of respondents in a Kentucky survey identified "coal," "oil," and "iron" as "renewable resources" in 2004. The study also found that the number of Kentuckians answering that "solar" and "trees" were "renewable resources" dropped between 1999 and 2004 (from 61 percent to 55 percent).[17] A 2006 survey of American electricity consumers found that four-fifths were unable to name a single source of renewable energy (even including "dams" and "hydroelectric" generators).[18] While not directly related to electricity, a separate study found that nearly 70 percent of flexible fuel vehicle owners (people who purchased automobiles that could run on gasoline and/or ethanol) were unaware that they were driving one.[19] In another poll, only 39 percent of those interviewed had ever heard of "greenhouse gas emissions," and of those who had, more than half did not know where they came from.[20]

More interestingly, it seems that some consumers prefer to remain uninformed about electricity and energy policy. In a survey during the 1973 oil crisis, researchers queried Iowa residents about the gravity of the energy situation and what personal actions they would take to resolve it. More than 90 percent said that they planned to "talk with friends" about energy problems, but only 15 percent said they would "write to politicians or officials" and less than 3 percent said they would "join a group" or "read publications" to learn more about energy policy.[21] Few respondents sought to influence public decision making on energy and even fewer intended to learn more about the situation. One survey of more than 3,000 residents in Washington before and after the energy crisis found a trend toward increased support for personal security programs such as retirement benefits, health and medical care, and social security, but an actual *decline* in concern for energy policy, social justice, environmental quality, and the public good.[22]

The research suggests that the massive media attention thrown at the "energy crisis" increased public apathy about it, with some respondents saying they were "tired of hearing about energy and the environment" as important issues. Another survey conducted in 1976 found that despite the crisis, energy ranked low on the list of national priorities after unemployment, economic growth and trade, inflation, crime, medical services, fire and police protection, and education.[23] The same study found that Americans were as concerned with noise pollution from energy facilities as they were about nuclear waste, and they were much less concerned about particulate matter and acid mine waste.

Similarly, most people never connect their own energy consumption with some of the environmental problems faced in the production of electricity. Paul Gilman argued, "most people still don't conceive of their homes as places of energy consumption."[24] Consumers often oppose clean power technologies not because they believe these technologies are poor alternatives to fossil fuels, but because they do not realize that new plants of any type appear necessary to provide additional electricity. Rodney Sobin explained that:

> Most Americans don't have any idea what has to happen so that electricity is delivered to their home. Thus they oppose wind turbines because they may ruin, in their mind, the

> view of a mountain top, but they never actually consider that the alternative to that turbine is more smokestacks, cooling towers, and even fly-ash and acid rain, which ultimately end up hurting the mountain much more. People don't campaign against birds hitting windows, cars, or cell phone towers, but they will campaign against birds hitting wind turbines. I think it demonstrates a lack of understanding of context and tradeoffs.[25]

People also tend not to connect their energy use with their energy prices. Professor Alex Farrell comments that "people don't make the connection between inefficient energy use and higher prices that result from it." [26] Perhaps people would object as strongly to plans to build traditional power plants as well, if only they started to think about where power originates and how it gets to their premises.

In addition, consumers are less forbearing and broad-minded about new energy technologies. Historian David Nye argued that strong appeals of tradition and familiarity explain the consumer choices made by many Americans regarding technology. According to Nye:

> The energy systems a society adopts create the structures that underlie personal expectations and assumptions about what is normal and possible . . . Each person lives within an envelope of such natural assumptions about how fast and far one can go in a day, about how much work one can do, about what tools are available, about how that work fits into the community, and so forth. These assumptions together form the habitual perception of a sustaining environment that is taken for granted as always there.[27]

Such technological environments appear natural because they have been there since the beginning of an individual's historical consciousness. A child, Nye noted, born into a world with automobiles and airplanes takes them for granted and learns to see the world naturally at hundreds of kilometers an hour.

Yet the familiarity consumers express toward the old often impedes the use of the new. Before the modern era of industrialization and electrification, utilities and corporations had to convince the public of the durability of their products. However, such actors "did not present their products as being revolutionary, but rather as natural parts of the home, neither unfamiliar nor sensational, but rather safe, familiar, and comfortable." [28] From the 1890s to 1910s, product styling imitated the fixtures and appliances being replaced. General Electric designed their first electric lights to look like gas-fired streetlights. Early appliances such as electric stoves were made to look like gas ranges and coal stoves. Manufacturers designed the first electric vehicles, more prevalent at the time than diesel- or gasoline-powered ones, to look like horse-and-buggy carriages. Denver Gas and Electric Company spoke of their new natural gas unit as "the invisible furnaceman" to emphasize its similarity with solid fuel furnaces that required constant stoking.[29] Utilities, manufacturers, and marketers learned that people tend to resist technologies they perceive as untested, radical, or different.

Many clean power systems have also been around for centuries, but advocates seem not to have learned to frame them as compatible with and similar to existing technologies. Fuelwood, for instance, has provided humans with fire for around 350,000 years. The Romans used human power in treadmills and hand-operated water pumps. Wind energy powered the sea vessels that ended up colonizing North and South America. Solar reflectors, windmills, and passive solar architecture are millennia old; flat plate collectors, solar furnaces, and heliostats are more than two centuries old; and PV and solar heat engines are more than one century old. At the 1878 World's Fair in Paris, France, one engineer even displayed a solar engine powering a printing press.

Before the 1900s, even the United States relied significantly on renewable sources of energy. Wind systems met about one-quarter of all American nontransportation energy needs for most of the 1800s. About one-third of the houses in Pasadena, California, used solar water heaters in 1897. In 1899, Charles Brush of Cleveland, Ohio, built the first wind machine to generate electricity, and one year later, historians estimate that the country at large possessed more than 6 million small wind machines. An American engineer helped design a solar thermal power plant in Meadi, Egypt, that produced enough steam to move a 100-horsepower engine in 1912. During World War I, government researchers carried out a program on the development of small windmills to power aircraft radios.[30]

Yet many of those promoting clean power technologies present them as novel and excessively revolutionary technologies. As one historical example of when the radical nature of a new energy system prevented its adoption, consider the tale of Aramis. The French transportation system Regie Autonome des Transports Parisiens (known as Aramis) was developed from 1969 to 1980 and offered a major advance in personal mass transit: the efficiency of a subway with the individuality of the automobile. Even though its opponents acknowledged that it was a superior form of transport, consumers ultimately rejected it because it was too radical. Aramis would have innovated virtually every aspect of the transportation system simultaneously, from motors, casing, tracks, and computational support to automation, speed, doors, signal systems, and passenger behavior. Because politicians and people saw it as revolutionary rather than complementary, they summarily abandoned it.[31] Aramis is an example of how tradition will always play a secondary but significant role in shaping cultural attitudes and decisions regarding technology even while notions of familiarity and personal identity are constantly shifting.

In fact, familiarity with only the new can sometimes *erase* accurate conceptions of the past. In her interviews with nuclear engineers, Gabrielle Hecht discovered that the creation of newer and more advanced reactors erased previous versions of that technology.[32] More specifically, Hecht found that French engineers working on nuclear power have chosen to forget the struggles over gas-graphite technology, the role of G-2 reactors in manufacturing French nuclear weapons, and the shift to light water reactors. This erasure was so complete that one engineer told Hecht that no nuclear program had existed in France before the 1970s.

Familiarity also determines which energy technologies public utility commissions and local regulators choose to adopt. Since most Americans refuse to take an active interest in electricity, utilities make decisions for people. The disjuncture between how electricity is made and how it is used, combined with the immense technical skill required to manage power plants and an elaborate T&D grid, has supported the rise of a professionalized elite that desires to maintain control over as much of the utility system as possible. Jeff Jones of the EIA explained that "there may also be a certain comfort level with the more 'traditional' technologies. Suppliers may have a tendency to do things that they have done before. They have considerable experience building coal, gas, and nuclear plants; they haven't built as many wind plants."[33] Utilities also do not want to dedicate the time or resources needed to learn about new technologies. For example, an anonymous, high-ranking executive for a large power trading company put it this way:

> [Clean power] is not a market [power generators] are interested in. A typical company, for example, sells power in 50 MW blocks. This is considered a modest size relative to the available generating capacity. It is equivalent to the output of a few small generators, or a fraction of a single medium to large generator. Staff sizes and the capital investment required to create power in this manner are very low compared to alternative methods.[34]

In other words, utilities believe that they would have to employ more staff to run several distributed units than they would for one large and centralized unit.

The same resistance toward the unfamiliar tends to hold true for local officials. When asked to comment about the most significant impediments toward DG systems, a former senior DOE official responded that "fire marshals and zoning regulators" need better education on how stationary electricity sources such as fuel cells, microturbines, and reciprocating engines work. Currently, the official concluded, local officials may "know what a chiller or furnace is, but they don't know how a fuel cell operates. You have a great deal of problems stemming from institutional issues, and they can be a killer."[35]

Consumption and Abundance

As a related impediment, consumers believe that they are entitled to more energy-intensive standards of living, and utilities believe it is their duty to give it to them at the lowest cost possible. For most of the past five decades, Americans have come to place faith in unbridled material progress linked to visions of a high-technology society, a success associated with new cars, large modern homes, and the accumulation of energy-intensive appliances. Historian Martin V. Melosi observed how an overarching theme of such visions is the abundance of cheap, reliable energy. Over the course of the country's development, historically plentiful sources of natural resources (including fossil fuels) enabled the transition from labor-intensive jobs to capital-intensive ones, and provided sources of vast commercial wealth. Americans, Melosi concluded, have become "endowed with an abundance of domestic sources of energy and having access to foreign sources," and continue to expect supplies to be simultaneously "never-ending and cheap."[36]

Abundance affects the way Americans use energy, how businesses develop and market it, and how government establishes policies about it. While bestowing many benefits, the array of energy sources also poses a problem of choice. For those building America, the luxury of choice seemed preferable to the necessity to choose. The explanation for why we consume so much energy is located in history, and rooted in attitudes concerning the natural world, urbanization, industrialization, patterns of consumption, and our modern postindustrial society.[37]

To the Puritans who landed at Plymouth Rock, the New World landscape was Satan's territory, a "hideous wilderness inhabited by the unredeemed and fit chiefly for conquest" in the name of civilization, Christianity, and progress.[38] The paramount concern of the colonists was to keep the "candle of civilization" flickering on the "edge of moral and physical darkness." [39] The European settlers viewed the land as a source of commodities, as "raw material awaiting transformation." [40] Although individual views varied, most believed in expansionism, the Biblical injunction to be fruitful and multiply, and a divine providence in developing the land. Whereas the original inhabitants of North America saw a complicated, reciprocal relationship with nature, the colonists saw natural resources as something to be used, transformed, and integrated into the economy. As Lynn White put it, "Despite Copernicus, all the cosmos rotates around our little globe. Despite Darwin, we are *not,* in our hearts, part of the natural process. We are superior to nature, contemptuous of it, willing to use it for our slightest whim." [41]

The historical antagonism toward nature expanded westward, as settlers faced many challenges including primitive and dangerous forms of transportation, hostile Indians, and the absence of physical or social security. Natural resources, so highly valued now, were viewed as obstacles rather than assets: forests needed to be cleared, marshes drained, rivers controlled, wildlife suppressed, and sod burned before farming could occur. Resources had little value and were thought inexhaustible.[42]

The early nineteenth century witnessed the emergence of an "industrious revolution," where people placed immense value on work.[43] Households increased both the supply of marketed commodities and labor, and the demand for goods in the marketplace. More family members worked, they worked harder, and some had slaves to further increase output. The emphasis during this time was on minimizing waste and putting everything to industrious use. The newfound wealth from such industriousness enabled people to purchase goods from outside their immediate areas. The purchasing power of every class increased, and population expansion created a market without trade restrictions and of a size unmatched anywhere in the world. By 1865, new economies of production and distribution made it feasible to operate larger factories and to concentrate them in advantageous locations. This precipitated the rise of the corporation and displaced people from human welfare. Strangers could purchase shares in a company they had never seen run by employees they had never met as investors became interested more in profits and balance sheets instead of the public good.

Near the turn of the twentieth century, Fordism, industrialization, the discovery of electricity, and the invention of the assembly line integrated a number of ideas that would alter the way people worked. Labor was subdivided into separate, smaller tasks, so that no one person put together an entire artifact; the concept of "interchangeable parts" created a need for reliable and predictable standards; factory owners grouped machines not by type (e.g., all lathes in once place) but according to their function; and most conspicuously, once the machines were in place, something was needed to move parts along them, hence the invention of the continuous moving belt. These changes had far reaching effects not just on production but on consumption. They dissolved bonds between workers and management, making people almost as interchangeable as the parts they worked on. A flood of mass produced goods altered the way Americans conceived of themselves and each other.

It was shortly after the rise of Fordism that leisure began to emerge as an alternative to work. Amusement parks, cinemas, fairs, dance halls, and vaudeville shows perpetuated their own set of values: not work, but play; not sobriety, but frivolity; not saving, but spending. This created a cycle as factory output soared and lower and middle class people became consumers of more varied goods, necessitating more factories and further increase in output. "The new mass-market commodities seemed luxurious, contrasting sharply with what had been available to the previous generation" noted David Nye. "As the phonograph, ready-made clothing, domestic lighting, the automobile, and other novelties became available, each of them had a symbolic resonance. In acquiring them, the consumer was experiencing something unprecedented."[44]

The advertising agency, department store, and mail-order catalog emerged out of the corporate need to call attention to new products. Marketplace transactions were no longer face-to-face as human aspects of commerce became more anonymous. Products proliferated and brand names became more important. Consumers could no longer rely on personal acquaintance with shopkeepers to provide quality. Corporations such as Nabisco, American Tobacco, and Gillette emerged to promote brands, backed by guarantees and distinctive packaging. Department stores sold goods in one centralized location and offered them at fixed prices. The simultaneous manifestation of purchases on credit, wide selection, money-back guarantees, and competitive prices made department stores "palaces of consumption" and shopping itself pleasurable and theatrical.

During this time, three key changes in consumption patterns emerged:

- People based their replacement of items no longer on need or when something wore out, but on style and image or when an item was fashionable.
- People started realizing that selling one item of better quality or fashion would often persuade consumers to buy a whole new array of objects (a concept later known as the "Diderot effect").
- To sell more goods some manufacturers built obsolescence into their products.

So what does all of this have to do with energy policy? In the late 1920s, before the Great Depression, indulgence and excess began to replace work and frugality. The notion of the "workplace" changed as laborers migrated from the country to the city, requiring more energy-intensive workplaces such as factories and offices. These new workers necessitated extra industrial and commercial space that had to be heated and plumbed with hot water and serviced with appropriate quantities of process steam, direct heat, and electricity.[45] A rapid growth in the proportion of women in the workforce caused per capita income to rise, increasing energy consumption much faster than population growth.[46] Urbanization gave way to suburbanization and the "death of the walking city" as consumption patterns became more dispersed.

Americans, in short, entered a postindustrial age of "consumption communities" where they made themselves known through the products they purchased. Today, Americans dispose of some 220 billion tons of solid waste per year, more than 20 times the average person's body weight every day, and it is estimated that less than 1 percent of all materials mobilized to serve the market are actually made into products still in use six months after sale.[47] Each of these historical themes—hostility toward nature, industriousness, urbanization, industrialization, and the creation of leisure and mass consumption—exerts immense influence on how Americans perceive the environment and consume energy. Each is also connected, so that seemingly isolated experiences such as shopping for groceries, turning on an appliance, listening to music, or checking email at work all shape personal and cultural identity and demand continuous forms of usable energy.

Historian and philosopher Langdon Winner noted that the most fundamental shift in American culture has been a transition of values. Most of us now embrace the idea that democratic freedom lies not in moderation, simplicity, self-restraint, and the activities of citizenship, but rather in the enjoyment of material abundance, backed by unquestionably positive and progressive technologies constituting the essence of democratic freedom for the ways they render the conveniences of modern life and demonstrate human ingenuity.[48] The subtle and powerful lure of this model is that the form of energy technology adopted does not matter. If society possesses a cornucopia, why worry about its shape? The appropriate response is jubilant celebration, applauding the abundance provided by modern industry, rejoicing in the accomplishments that industrialization has offered. We have shifted, in other words, from a system premised on self-restraint and moderation to the idea that freedom and social well-being are best achieved through sheer abundance and limitless consumption.

These values, far from floating "out there" in the minds of people, become a powerful and socializing force built into the physical environment. One extensive, 2-year study involving a national survey of 1,500 households and 125 public utilities found that all households have circumscribed choices concerning the energy-related features of their homes. The architectural design, selection of appliances, and structure of built-in equipment are already there.[49] Central heating and cooling systems allow people to move freely from one room to another without thinking about energy, but remove the option of saving fuel by closing off unused rooms. Architects design apartment and office buildings with windows that cannot open for safety, but make it impossible to rely on natural heating and cooling.

The result is profligate and often careless use of energy and electricity.

We harness electricity to burn neon signs in the daytime, run electric toothbrushes and can openers, and power movable sidewalks.[50] With just 4.6 percent of the world population, Americans use 26 percent of all energy consumed worldwide and

23 percent of all electricity. We have less than one-twentieth of the global population but use more than one-fourth of the world's energy. On a per capita basis, Americans consume 2.5 times more energy than Western Europe, 8 times more than Latin America, and 10 times more than China.[51] To put these numbers into perspective, consider that the combustion of a single pound of coal is equivalent to the work of 300 horses, meaning that the force set free in the burning of 300 pounds of coal is equivalent to the work of one able-bodied person for a lifetime. We use more than 200 times the amount of energy per person today than we did in 1850.[52] Historian David Nye commented that "the average household of 1970 commanded more energy than a small town in the Colonial period. The color television alone, which the family watched 4 hours a day, consumed 1.4 kW an hour—more energy than a team of horses could provide in a week."[53]

Psychological Resistance

Strong albeit subtle psychological factors encourage wasteful electricity consumption as well. "Economists," mused John Kenneth Galbraith, "are economical, among other things, of ideas; most make those of their graduate days last a lifetime." While not all of us are economists, we do tend to behave in similar fashion, holding onto our conventions and habits, letting them exert great influence on our experiences, doing things "as we have always done them" rather than embracing change. Indeed, psychological factors such as comfort, freedom, control, trust, social status, ritual, and habit deeply shape American attitudes toward the consumption of energy. Psychologists and economists, for instance, have observed that:

- People have an aversion to loss, and will sometimes ask for a higher price to replace a good they own than the cost of attaining a new one (conflicting with the view that the willingness of consumers to exchange goods is invariant to the good that they own).[54]
- People hold a strong preference for the status quo, and once familiar with a particular energy product, it attains higher value (conflicting with the view that people will invest in changing their life cycles if it maximizes self-interest).
- People frequently require faster rates of return for shorter payback periods than for longer ones, and for smaller investments than for bigger ones (conflicting with the view that discount rates for energy efficiency investments will be consistent within the same person).[55]

One longitudinal study of American couples found that, contrary to expected primacy of concerns about cost, "comfort" was the single most important determinant of their energy use, an attitude so consistent that neither the location of those surveyed nor the year they were contacted changed the answer.[56] In other words, the most inelastic and consistent component of energy use, one that did not change with availability of energy, was thermal comfort, making it the "most significant predictor" of actual energy consumption. The logical extension of this view is that "rather than being exhorted to make sacrifices for energy conservation, people should be told of the ways they can save energy and be comfortable at the same time." [57]

Freedom and control appear to be almost as significant in influencing energy choices. People will resist energy technologies that impede their freedom or appear to diminish their control.[58] An army experiment utilized a device to save fuel that created physical resistance when drivers tried to accelerate cars or trucks too rapidly. Drivers soundly rejected the mechanism, and in about 10 percent of the cases dismantled and disconnected it themselves.[59] Researchers at the Center for Energy and Environmental Studies at Princeton University applied this principle to energy efficiency. There, researchers studied personal resistance to the installation of automatic thermostats, and found that people disliked them since they did not have enough control over temperature settings. The Princeton team redesigned thermostats so that residents could temporarily override the system, and the simple modification made them so much more attractive that they reduced household electricity consumption by 19.4 percent in the summer and natural gas consumption by 31.3 percent in the winter.[60] Other studies have found that control plays an important role in antilittering messages. Those making explicit external pressure against littering tend to be much less effective than those appealing to social or normative standards. ("Don't mess with Texas!" proved more successful than "Please don't litter.")[61] Indeed, the need for freedom is such a powerful factor that one psychologist has gone so far as to claim that the need to control one's personal environment "is an intrinsic necessity of life itself." [62]

Issues of trust also play their part. One study sent a pamphlet describing how to save energy in home air conditioning to 1,000 households in New York City. Half of the households received information in the mail from a local utility, the other half from a state regulatory agency. The following month, households that had received pamphlets from the regulatory agency used around 8 percent less electricity than those that had received identical information from the utility. The researchers concluded that utilities were perceived as untrustworthy, and therefore the information they provided was ignored, implying that some organizations may be unable to influence consumers no matter how accurate or important their information is.[63] A similar study conducted in Texas after the oil crisis found that just 14 percent of those surveyed thought that they could trust government sources of information concerning energy and that almost 60 percent thought the crisis was a "phony issue" created by the government to draw attention away from more serious problems.[64] A 1974 sample of 1,069 Los Angeles residents found that only 6 percent of those surveyed reported that they believed the energy crisis was real and significant.[65]

A more recent national survey of 1,011 adults found that, when asked about whom they trust as a source of information, 79 percent said "scientists" or "teachers," whereas only 3 percent said "the government." [66] Other surveys have shown two-thirds to four-fifths of Americans do not have a great deal of confidence in the government as a source of information relating to the use of energy.[67] This lack of trust contributes to less enthusiasm, more grudging acceptance, and a feeling of helplessness concerning energy problems. In response, most people become disinterested and, ironically, perpetuate the very system they mistrust.

Social stigmatization molds energy beliefs as well. In her surveys on the acceptance of energy-efficiency practices among builders in Utah, consultant Shelly Strand noted that most people lack "interest and knowledge" about even basic energy-efficiency practices. One of the more troubling conclusions reached by Strand concerned the widespread public stigma in Utah against more efficient energy devices such as evaporative coolers, weatherization techniques, and geothermal heat pumps. Strand found that most builders perceived the more efficient technologies as "low-class" and "trashy" because, years earlier, a successful energy program had distributed these technologies primarily to low-income areas.[68] Energy efficiency techniques and practices were functionally "stigmatized" among more affluent members of the Pacific Northwest.

Once values concerning familiarity, energy, and consumption are formed, whatever they may be, they tend to be very difficult to alter, especially when transmitted between generations. Psychologists have found that people tend to rationalize all of the decisions they have made afterwards, emphasizing the positive aspects of chosen alternatives and the negative aspects of unchosen options. As time goes by, individuals come to increasingly view the selected option as clearly superior to all other alternatives.[69] Moreover, the greater the commitment in terms of cost, effort, or irrevocability, the stronger and more permanent the effect. People tend to remember the plausible arguments favoring their own position and the implausible arguments opposing their position, serving the need for self-justification rather than objective fact seeking.[70]

An assessment of the offshore wind power controversy in Cape Cod, for instance, found that many opponents evaluated the project's environmental effects based on readily accessed factors such as scale, familiarity, and permanence. However, researchers discovered these opponents frequently overestimated the likely environmental impacts of offshore wind and underestimated the consequences from more familiar (and damaging) activity such as trawling or industrial pollution.[71] The study also found that supporters of the wind project committed similar errors. They characterized their opponents as primarily objecting to the project for aesthetic concerns and coming from only wealthy property owners and boat operators. In doing so, they overlooked how sentiment against the project had broader roots extending beyond aesthetics into the belief that humans should not interfere with the oceans as a matter of respect and value. Neither side, the researchers concluded, accurately portrayed the other.

A follow-up study of the controversy over Cape Wind found that opponents also greatly underestimated the environmental value of offshore wind farms. A 2007 survey of 500 local residents living in Nantucket Sound, Massachusetts, found that 72 percent felt the project would have negative impacts on aesthetics and that 42 percent strongly opposed the Cape Wind project for environmental reasons, even though the Cape Wind project would displace about 1.5 GWh of more polluting fossil-fueled capacity.[72]

When applied to the consumption of electricity, these psychological factors mean that people resist change because they are committed to what they have been doing. They justify that inertia by downgrading information that implies that change is needed, partially explaining the failure of people to install energy-efficient appliances. Human behavior also suggests that change may be brought about by a process that uses small commitments to energy-saving action and then moves under its own momentum toward more significant efforts. This strategy clashes, unfortunately, with the idea presently being propagated by advocates emphasizing the revolutionary aspects of clean power systems.

Even the simplest energy-efficient devices and practices, such as plugging leaks, changing furnace air filters, insulating water heaters, and maintaining pilot lights on gas stove tops are impeded by an interconnected mesh of psychological impediments. These include relative advantage, or the degree to which energy-efficient technologies appear superior to prior innovations fulfilling the same needs; risk, or the degree to which economic, physical, functional, and psychological burdens are perceived to be associated with the innovation; compatibility, or the degree to which the innovation is consistent with the values, experiences, and infrastructure of potential adopters; complexity, the degree to which energy efficiency is easy to understand and use; divisibility, the degree to which it can be tried on a limited basis; and communicability, the degree to which results of innovation can be disseminated easily and effectively.[73]

John Warren, director of the Division of Energy for the State of Virginia, noted that one of the biggest problems he faces is that:

> People tend to look at things very narrowly and hear the things that they want to. If you oppose wind turbines and you have a strong attitude towards that, you will read and absorb the information that opposes wind turbines as well. And you will be in a position to defend the arguments that they kill too many birds and bats, and even if information may contradict that, you will believe the information that you want to hear. And so it's hard to sway the public.[74]

In truth, Americans have become so complacent that the energy system, as currently configured, makes crises a natural and recurring by-product.[75]

The confluence of these factors—utility preference for large, geographically distributed power plants, subtle yet deeply engrained attitudes favoring familiar technologies, American attitudes concerning their standard of living, and psychological resistance to change—result in an almost reckless consumption of electricity. Ironically, perhaps, people have become accustomed to low-priced electricity, and they often consume it indiscriminately, creating demand for construction of new power plants. Consequently, Americans' preferences for sprawling growth, automobiles, individually controlled climates, and huge electricity consumption impose conditions on their future energy choices. Nye noted that, "Americans have built energy dependence into their zoning and their architecture . . . they think it natural to demand the largest per capita share of the world's energy supply."[76] Increased consumption is often further encouraged by utilities (who make money selling power) and wires companies (who make money based on throughput, the amount of electricity traveling through their network). So, in other words, Americans simultaneously create a demand for more electricity, but they frequently oppose construction of new generation facilities (including renewable technologies) because they do not realize that they contribute to the necessity of new plants.

As we will see in Chapter 6, these subtle and often hidden impediments explain why people find clean power systems unattractive and aesthetically displeasing.

CHAPTER 6

Aesthetic and Environmental Challenges

Forester and philosopher Aldo Leopold once wrote that promoting sustainable development "is a job not of building roads into lovely country, but of building receptivity into the still unlovely human mind." His comment underscores that the true challenge for any energy system, conventional or alternative, is to win the "hearts and minds" of the people relying on it. Clean power sources often save or generate electricity more cost effectively, through nondepletable renewable fuels and with minimal environmental damage. Then why does a loose collection of concerned farmers, property owners, environmental scientists, and activists attack systems for harming the environment and ruining the aesthetic beauty of the land?

As an answer, this chapter begins by discussing the genesis of environmental ethics in the United States, or how some Americans came to place importance on the protection of the environment and preservation of species, ecosystems, and the biosphere. American concern toward the environment peaked in the 1970s after environmental crises precipitated a growing social frustration with the perceived mounting challenges of industrial pollution, population growth, and resource scarcity. These challenges convinced voters and politicians to acquiesce to a social movement that had been arguing for legislative actions to protect the country's environment. As a direct consequence of such action, and perhaps incongruously, clean power systems have become challenged on ethical and environmental grounds and are frequently opposed by local communities and environmentalists for aesthetic reasons.

Closer inspection, however, reveals that much of this antagonism has little to do with clean power technologies themselves. A slow and subtle shift has made energy production and use less visible in modern society. Over the past few thousand years, humans have transitioned away from utilizing ambient energy (natural energy flows moderated with simple technologies such as clothing and shelter) to fuel energy (usable energy stored in a form that can be stockpiled, transported, and released in concentrated form whenever and wherever humans desire it).[1] In the 1800s, for example, most homes were heated by wood stoves, meaning that the person heating the home was also the one chopping and stacking wood. This form of heat and energy production was very labor-intensive, since wood had to be retrieved and loaded frequently. Next came the coal-fired furnace, which displaced some of this labor by having coal delivered by a company, but users still had to shovel coal into a furnace, and the daily ritual of loading coal made the householder aware of dwindling fuel supplies. The oil-fired furnace, which came next, no longer required daily tending and relied on storage tanks so that users did not have to closely monitor fuel supply. Finally, the modern-day natural gas or electric furnace removed almost all human involvement in the production of heat apart from flicking a switch, and fuel collection, distribution, and delivery (along with energy use) became effortless and invisible.[2]

The transition from an energy system prioritizing human labor and direct interaction to one more convenient and indirect means that clean energy services and power technologies are more taken for granted, less understood, and, as we shall see, more likely to be opposed when they suddenly become visible again. In the end, the aesthetic and environmental impediments facing clean power systems—or, in fact, all energy systems—serve as important reminders that technologies must not only work in the technical sense. In order to succeed, technologies must also fit within people's values, expectations, and ethical codes.

The Emergence of Environmental Ethics

For most of the country's history, Americans have tended to place their own personal needs above that of the environment. Nearly 200 years of cheap fuels, industrial growth, abundant natural resources, and an environment that could seemingly absorb pollution convinced many people that they were, in fact, entitled to dominate nature. These classical ideas created a worldview that correlated energy consumption with economic growth, convinced many Americans that they are somehow entitled to consume as much energy as possible, and promoted the notion that technology can overcome all resource constraints.

Historically, such a worldview can be connected to New England Puritan ideals. The original pilgrims found their natural surroundings strange and threatening, and their writings often referred to nature as an "enemy to be subjugated."[3] Forests were cleared, wilderness declared an obstacle to progress, and the advancing frontier was conceived only as a terrain to be conquered through manifest destiny. To assist them in their conquest over nature, settlers placed their faith in technology, their trust in the experts, and their confidence in the idea that American ingenuity could solve all problems. Mastery over nature went hand-in-hand with other American ideals including the right to own land and property, individualism, independence, and self-reliance. And these ideals were threaded together into an overall belief in progress. Such progress was to be accomplished through technological development and more energy consumption.[4]

However, in the late eighteenth and nineteenth centuries, the emergence of the Romantic Movement began to challenge this worldview. The work of William Blake, William Wordsworth, and Johann Wolfgang von Goethe (among others) argued that nature should not be viewed as an impersonal machine, but an organic process with which humanity is united. Other writers associated wilderness with the sublime, wild, and untouched landscape. The transcendentalists in New England referred in similar terms to the sacred dimension of nature. Henry Thoreau held that nature was a source of inspiration, vitality, and spiritual renewal, writing that "in wilderness is the preservation of the World." [5] During the 1870s John Muir circulated the philosophy of wilderness preservation, and the works of George Perkins Marsh (1864) and Charles Darwin (1859) described how species of plants and animals are part of nature in continuity with other forms of life, including humans.[6]

A few decades later, during the Great Depression and the Dust Bowl in the 1930s, many Americans witnessed perhaps the country's "worst ecological disaster" firsthand, further inspiring a conservation ethic.[7] Congress created the Soil Conservation Service (the future Natural Resources Conservation Service), Grazing Service (future Bureau of Land Management), and Civilian Conservation Corps to educate Americans about the value of preserving natural resources, a trend later disrupted by World War II and industrial wartime production.

Romanticist, transcendentalist, and conservationist ideals still greatly influenced later twentieth-century work arguing in favor of protecting wilderness and the environment. In his 1947 introduction to *A Sand County Almanac,* Aldo Leopold directly challenged American destruction of the natural world and likened society to a hypochondriac, so obsessed with its own "economic health" that its people have lost the "capacity to remain healthy, the whole so greedy for more bathtubs that it has lost the stability necessary to build them, or even turn off the tap." [8] The fundamental flaw with such a strategy, Leopold argued, was that it transformed the landscape from something humbling and natural into an economic symbol degradable and exploitable; a land that entails privileges but not obligations. Likewise, Charles S. Elton's 1958 *The Ecology of Invasions by Animals and Plants* argued in favor of an environmental ethic. Elton suggested that protecting the environment could be justified from a very practical desire to preserve the land, crops, forests, water, and fisheries that are needed to sustain human life.

The nation's environmental consciousness transformed dramatically during the 1960s. Rachel Carson's 1962 *Silent Spring* documented the terrifying threat from uncontrolled use of pesticides, and concluded that the challenges posed by threats could only be addressed by a sustained, coordinated, and a thoroughly ecological worldview. Around the time Carson's text was hitting bookshelves, American politics became more nationalized, a phenomenon encouraged by the growth of the national media. The Santa Barbara oil spill, Cuyahoga River fire, and other environmental events became national events broadcast at a national level. While numerous environmental disasters had occurred in the preceding decades, stories of such events were not distributed as widely nor in as visceral a manner. The 1960s thus witnessed the start of national environmental legislation. The 1960 Multiple-Use Sustained-Yield Act directed federal agencies to better manage national forests. The 1964 Wilderness Act established federal lands as wilderness areas. The 1966 National Historic Preservation Act created a network of protected parks and areas. The 1966 Department of Transportation Act called for more efficient automobiles. And the 1968 Wild and Scenic Rivers Act aimed to clean up some of the country's rivers and streams.

These early acts laid the foundation for modern environmental ethics, advocated in turn by an organic environmental social movement. Even though, as a whole, the idea of "environmental ethics" is much more variegated than presented here, it advances at least four basic themes: humans are dependent on nonhuman forms of nature; pollution of air, water, and land is clearly detrimental to human life; limits should be set on the exploitation and use of natural resources; and humans have a duty to preserve the biosphere for future generations. While certainly supported on strong spiritual and ethical grounds, the concept of an environmental ethic also has its connections to advances in ecology, conservation biology, and evolutionary biology. These sciences developed the idea that species exist as part of an ecosystem; that humans are interdependent with all members of a biotic community; that biological diversity is needed for ecological balance and stability; and that finite limits exist for population growth and the capacity of the environment to provide resources.

The force of the environmental movement on society and policy was most profound in the United States during the 1970s. The federal government implemented copious federal statutes, including the Clean Air Act of 1970; National Environmental Policy Act of 1970; Federal Insecticide, Fungicide, and Rodenticide Act of 1972; Coastal Zone Management Act of 1972; Federal Water Pollution Control Act of 1972; Endangered Species Act of 1973; Safe Drinking Water Act of 1974; Resource Conservation and Recovery Act of 1976; Toxic Substances Control Act of 1976; Clean Water Act Amendments of 1977; Comprehensive Environmental Response, Compensation, and Liability Act of 1980 (commonly known as "Superfund"); and the Alaska National Interest Lands Conservation Act of 1980. According to historian Robert Nash, it was during this time that "environmentalism changed from a religion to a profession" and moved from a "blue-jean-and-granola style of conservation evident at the time of the first Earth Day" to a sophisticated and lasting social movement.[9] While laws aimed at protecting the environment have certainly been passed since, it was during this period that the conceptual and political groundwork was laid to support the wider environmental movement.

Today, about three decades later, environmentalists continue to exert influence on state and federal policy. A 2006 article in *The Economist* concluded that "America's environmentalist movement has undeniable political clout and a huge reservoir of public sympathy." [10] A similar Yale University survey found that three-fifths of registered American voters considered the environment to be an important consideration, and two-fifths believed that the federal government should do more to protect it.[11]

Environmental Objections to Clean Power

Given the importance that many Americans place on environmental protection, it may come as no surprise that some have come to believe that, as part of preserving nature, clean power systems should be opposed because they harm species and destroy ecosystems. Clean power sources, while cleaner than alternatives, are not without environmental consequences.

For wind energy, the most vociferous environmental concern relates to the death of birds and bats resulting from collisions with wind turbine blades, an issue termed "avian mortality." Onshore and offshore wind turbines present direct and indirect hazards to birds and other avian species. Birds can directly smash into a turbine blade when they are fixated on perching or hunting and pass through its rotor plane; they can strike its support structure; they can hit part of its tower; or they can collide with its associated T&D lines. These risks are exacerbated when turbines are placed on ridges and upwind slopes, close to migration routes, and when there are periods of poor visibility such as fog, rain, and night. Indirectly, wind farms can physically alter natural habitats, the quantity and quality of prey, and the availability of nesting sites.[12] A 1992 CEC study estimated than more than 1,766 bats and 4,721 wild birds (including more than 40 species) died every year at the Altamont Pass Wind Resource Area, where about 5,400 wind turbines operate.[13] Some wind farms operating in Tennessee, Pennsylvania, West Virginia, and the Midwest have been known to kill about 48 bats per turbine annually.[14]

Photograph 2
Sign Against Wind Energy near Maxwelton, West Virginia

A farmer in West Virginia publicly announces his or her opposition to wind energy.

Some environmentalists argue that effective and large wind farms are sometimes highly land intensive. The DOE notes that large-scale utility wind turbines usually require one acre of land per turbine.[15] When these big machines are built in densely forested areas or ecosystems rich in flora and fauna, they can fragment large tracts of habitat. At the Mountaineer Wind Energy Center in West Virginia, more than 40 acres of forest were bulldozed and 150 acres of forest interior were lost to erect eight turbines. Similarly, 350 acres of forest habitat were destroyed to construct 20 wind turbines at a Meyersdale, Pennsylvania, wind farm.

Other people find wind turbines visually unattractive, especially in significant tourist or recreational destinations where the human-built turbines impose obtrusively on the natural environment. The regions in the United States with the most offshore wind potential include areas along the eastern seaboard, coastlines highly valued for their fisheries, aesthetics, and recreational activities. A recent article in the *Boston College Environmental Affairs Law Review* noted that for many people, "fears of three hundred foot spinning turbines and blinking navigational lights blanketing the horizon have caused an uproar that threatens to drown out wind power's loudest advocates."[16] And onshore, older wind turbines from the 1970s sometimes created interference with radio, TV, and other electromagnetic transmissions. While recent improvements in turbine technology eliminated these problems, blade noise from 1970s prototypes, induced by low-frequency aerodynamic sounds generated by the interaction of turbine blades and the tower, could often be heard up to one kilometer away.[17]

Consequently, many citizens campaign aggressively against wind farms, as they have in Virginia's Highland County.[18] One survey of 1,200 residents in Watauga County, North Carolina, found that 64 percent believed that wind turbines would "harm mountain views." The community even stated that they would pay as much as $724,000 per year to have wind farms sited somewhere else.[19]

Environmental advocates sometimes express concern about bioelectricity. While biomass combustion has the advantage of not releasing any net CO_2 into the atmosphere (and thus contributes little to the global inventory of greenhouse gases), it releases measurable levels of PM, NO_x, and SO_2. These air pollution issues parallel aesthetic concerns about land use, smell, and traffic congestion. The use of agricultural wastes, forest residues, and energy crops such as sugar, legumes, and vineyard grain to generate electricity, when harvested improperly, can strip local ecosystems of needed nutrients and minerals. Widespread use of these crops can contribute to habitat destruction and deforestation. Furthermore, the combustion of biomass has been reported to release foul odors near some plants, and they can contribute to traffic congestion when large amounts of fuel must be delivered by trucks.

For geothermal electricity, plants can emit small amounts of hydrogen sulfide and CO_2 along with toxic sludge containing sulfur, silica compounds, arsenic, and mercury (depending on the type of plant).[20] Geothermal systems require water during drilling and fracturing processes, and are ill-suited for desert areas or regions with low levels of water. Extra land may also be required for the disposal of waste salts from geothermal brines, and contamination of groundwater and freshwater can occur if plants are poorly designed.

Concerning hydroelectric facilities, the most extensively debated and complex problems relate to habitat and ecosystem destruction, emissions from reservoirs, water quality, and sedimentation. All these concerns arise because of the dam's role as a physical barrier interrupting water flows for lakes, rivers, and streams. Consequently, dams can drastically disrupt the movement of species and change upstream and downstream habitats. Such barriers also result in modified habitats with environments more conducive to invasive plant, fish, snail, insect, and animal species, all of which may overwhelm local ecosystems. To maintain an adequate supply of energy resources in reserve, most dams impound water in extensive reservoirs. However, these reservoirs can emit greenhouse gases from rotting vegetation.[21]

DG and CHP technologies use fossil fuels but smaller smokestacks than conventional facilities, meaning they may appear to have few environmental benefits. Most emissions control efforts in the electric utility industry have been aimed at large, central station generators greater than 25 MW in capacity, so that current regulations and codes, with the exception of product standards, leave smaller DG and CHP technologies unregulated.

One study conducted for NYSERDA, which projected that New York will add 4,736 MW of DG by 2020, estimated that NO_x emissions could increase by as much 140 percent relative to the base case (because of the displacement of central station units that have NO_x controls with DG technologies that do not).[22] A similar study in Brazil compared the emissions from a 1,000 MW combined cycle power plant to 1,000 small diesel power plants (each ~100 kW) in operation from 1990 to 1999. The study found that because most of the small generators were exempt from environmental regulations, they emitted much more pollution: about 50 times more PM, 826 times more SO_2, 1.3 times more CO_2, and 4 times more NO_x.[23]

It has long been known that the higher the smokestack of a power plant, the farther it disperses its pollutants. Since DG technologies generate electricity close to the end user, their exhaust systems concentrate pollution near the point of combustion, rather than dispersing it over a larger area. The University of California Energy Institute modeled the plume of air pollutants for five DG technologies and compared it to that of traditional centralized generators.[24] The study concluded that the pollution emitted by DG systems, inhaled by downwind populations, could be more than an order of magnitude greater for each of the five DG technologies when compared to the existing generators. The difference was a consequence of the closer proximity of DG sources to densely populated areas.

Contextualizing Environmental Objections to Clean Power

While pointing out the environmental concerns of clean power sources is important, the call for balance and clarity is counterproductive if it obscures the truth that they are much better for the environment than conventional sources. As Harvard professor John Holdren and Lawrence Livermore National Laboratory researcher Robert J. Budnitz pointed out more than 30 years ago, "no existing or proposed energy technology is so free of environmental liabilities as to resolve satisfactorily the central dilemma between energy's role in creating and enhancing prosperity and its role in undermining it through environmental and social impacts."[25] One researcher even jokingly commented that modern resistance toward *any* energy project is so strong that "not in my backyard," or NIMBY, is rapidly turning into "build absolutely nothing anywhere near anything," or BANANA.[26] In assessing the environmental benefits of alternative energy systems, it is important to remember that *all* sources of electricity supply have environmental concerns.

Chapter 2 details the numerous environmental advantages—less water consumption, fewer harmful emissions, land conservation, more climate-change fighting potential—that clean power sources have over conventional systems. Even legitimate concerns about avian mortality, land use, bioelectricity, geothermal energy, hydroelectric power, and DG/CHP plants are not as environmentally damaging as they may appear.

While the avian mortality issue should certainly be taken seriously, several facts make bird deaths unique to older wind sites. Altamont Pass, for example, is located near bird migration routes and has terrain, such as craggy landscapes and various canyons, making it ideal for birds of prey. One of the most comprehensive studies ever undertaken, a 3-1/2-year study including fatality searches at 1,536 turbines across Altamont Pass, concluded that a majority of fatalities were during the first few years of operation, and that birds became aware of operating wind turbines and took measures to avoid them.[27] Outdated turbine designs are also prevalent at these older sites. It takes 15 Altamont turbines to produce as much electricity as one modern turbine, and early turbines were mounted on towers at the same level as bird flight paths (60 to 80 feet in height).[28]

Today, newer facilities can produce the same amount of electricity with fewer turbines, and turbines are mounted on towers that typically avoid birds at a height of 200 to 260 feet. Death rates of all flying animals have decreased in recent years as wind power entrepreneurs have installed larger turbine blades that turn more slowly, and have used advanced thermal monitoring and radar tracking to site turbines more carefully. Developers commonly avoid placing wind farms in areas of high nesting or seasonal density of birds, remove potential perches on lattice towers, and utilize micrositing to position turbines in ways that minimize intersection with flight paths.

Ecologists have documented a relative absence of avian mortality at newer turbines. Researchers monitoring the 216 turbines at the Buffalo Ridge wind farm in Minnesota, for instance, observed almost 12,000 sightings of 188 species of birds over the course of their study, but documented only 29 bird fatalities (a risk of less than one-thousandth of 1 percent).[29] Rigorous observation of a 22-turbine wind farm in Wales documented that it has killed *no* birds, and researchers found a shift in bird activity to a neighboring area.[30] A series of weekly checks at a much larger 256-turbine wind farm in Spain documented only 106 avian mortalities over the course of a year.[31] Aerial surveys, radar monitoring, and video surveillance of

Table 10
The Environmental Concerns of Different Power Plants

Every type of power system exerts some sort of damage on the natural environment, although it must be noted that the damage from clean power sources is much less than the degradation inflicted by conventional and nuclear units.

Source	Environmental Concerns
Oil	Global climate change, particulate matter, acid rain, oil spills, rig and tanker accidents, occupational hazards, water use
Natural gas	Global climate change, methane leakage, accidents and explosions, occupational hazards, water use
Coal	Global climate change, particulate matter, acid rain, mining, ground water pollution, health effects on miners, water use
Nuclear power	Radioactivity from accidents, waste disposal, terrorism, proliferation of nuclear weapons, environmental pollution from tailings, health effects on uranium miners, water use
Biomass	Effect on landscape, groundwater pollution from fertilizers, competition with food production, water use
Hydroelectric power	Displacement of populations, effects on rivers and groundwater, visual intrusion, seismic accidents, downstream affects on ecosystems and agriculture, methane emissions from reservoirs
Wind power	Visual intrusion on sensitive landscapes, noise, avian mortality
Tidal power	Visual intrusion and destruction of ocean habitat, alteration of water flow
Geothermal energy	Release of some polluting gases, groundwater pollution, seismic effects
Solar energy	Sequestration of large land areas for centralized facilities, use of toxic materials in manufacturing, visual intrusion

Source: Based on Table 13.1 from Godfrey Boyle, "Assessing the Environmental and Health Impacts of Energy Use," *Energy Systems and Sustainability* (Oxford: Oxford University Press, 2003), 520; and Table 3.1 from Russell Lee, "Environmental Impacts of Energy Use," in *Energy: Science, Policy, and the Pursuit of Sustainability,* ed. Robert Bent, Lloyd Orr, and Randall Baker (Washington, DC: Island Press, 2002), 81–86.

offshore wind farms in Denmark revealed that the risk of a collision between a bird and a turbine was less than 1 out of 30,000.[32]

To put the issue in greater perspective, the absolute number of avian deaths from onshore and offshore turbines is incredibly low compared to other sources. Millions of birds die annually when they strike tall stationary communications towers, get run over by automobiles, or fall victim to stalking cats. After surveying wind development in California, Colorado, Iowa, Minnesota, New Mexico, Oklahoma, Oregon, Texas, Washington, and Wyoming (the 10 states with more than 90 percent of total installed wind power capacity), the U.S. Government Accountability Office (GAO) calculated that building windows are by far the largest source of bird mortality, accounting for 97 million to 976 million deaths per year. Attacks from domestic and feral cats accounted for 110 million deaths; poisoning from pesticides comes next at 72 million; and collisions with communication towers are next at 4 to 50 million.[33] Another study projected that glass windows kill 100–900 million birds per year; transmission lines to conventional power plants, 175 million; hunting, more than 100 million; house cats, 100 million; cars and trucks, 50 to 100 million; and agriculture, 67 million.[34] In neither study did "wind turbines" even make the list.

As Chapter 2 documents, clean power sources use less land than conventional generators, keep most of the land they use available for other activities, and in some cases directly improve habitats. The land employed for wind turbines, unlike the property needed for a coal plant or nuclear facility, can still be used for farming, ranching, and foresting. Solar panels can be easily integrated into building facades and existing land uses. Yet because almost all of the fuel cycle impacts for a wind plant are in one location (whereas the fuel cycle impacts of a coal plant are spread out over a number of different and unconnected locations), the apparent impact of some clean power sources (misleadingly) seems larger.

Considering bioelectricity, dedicated biomass electrical plants release no net CO_2 emissions into the atmosphere (as long as they avoid combusting fossilized fuel) and produce fewer toxic gases. One study conducted by the Center for Energy Policy and Technology found that combined cycle biomass gasification plants produce one-twentieth the amount of pollutants emitted by coal-fired power plants (and one-tenth the pollution of equivalent natural gas plants).[35] Landfill capture generators and anaerobic digesters harness methane and other noxious gases from landfills and transform them into electricity. This does not just produce useful energy, but also displaces greenhouse gases that would otherwise escape into the airshed.

High-yield food crops leech nutrients from the soil, but the cultivation of biomass crops on degraded lands can help stabilize soil quality, improve fertility, reduce erosion, and improve ecosystem health. Perennial energy crops improve land cover and enable plants to form an extensive root system, adding to the organic matter content of the soil. Agricultural researchers in Iowa, for instance, discovered that planting grasses or poplar trees in buffers along waterways captured runoff from corn fields, making streams cleaner.[36] Prairie grasses, with their deep roots, build up topsoil and put nitrogen into the ground, and twigs and leaves decompose in the field after harvesting, enhancing soil nutrient composition (switch grass grows without irrigation and is harvested with a low-labor process similar to mowing the lawn).[37] Biomass crops can also create better wildlife habitats, since they frequently utilize native plants that attract a greater variety of birds and small animals, and poplar trees, sugar beets, and other crops can be grown on land unsuitable for food production.

Geothermal plants also have immense air quality benefits. A typical plant using hot water and steam to generate electricity emits about 1 percent of the SO_2, less than 1 percent of the NO_x, and 5 percent of the CO_2 emitted by a coal-fired power plant of equal size. Its airborne emissions are "essentially nonexistent" because geothermal gases are not released into the atmosphere during normal operation.[38] Another study calculated that the geothermal plants currently in operation throughout the United States avoid 32,000 tons of NO_x,78,000 tons of SO_2, 17,000 tons of PM, and 16 million tons of CO_2 emissions every single year.[39]

All forms of hydroelectric generation combust no fuel, meaning they produce little to no air pollution in comparison with fossil fuel plants. Luc Gagnon and Joop F. van de Vate conducted a full life-cycle assessment of hydroelectric facilities, and focused on the activities related to the building of dams, dykes, and power stations; decaying biomass from flooded land (where plant decomposition produces methane and CO_2); and the thermal backup power needed when seasonal changes cause hydroelectric plants to run at partial capacity. The study found that typical emissions of greenhouse gases for hydropower were still 30 to 60 times less than those from equally sized fossil-fueled stations.[40]

Even though they utilize fossil fuels, DG and CHP systems are also considerably better for the environment. One study analyzed air emissions from five technologies—gas engines, diesel engines, gas turbines, microturbines, and fuel cells—and then compared them with the emissions from centralized combined cycle gas and coal steam turbines. The study found that centralized combined cycle gas turbines performed the best in electricity-only applications, but that the efficiency advantages of CHP systems improved as the heat requirements increase. Thus, when credited for heat energy, the authors conclude "this finding stands in clear contrast to the concerns of some analysts that use of . . . DG technologies will necessarily result in higher emissions."[41]

One of the most comprehensive assessments of DG and air emissions comes from a study conducted by ORNL.[42] The study assumed that utilities operated DG in two strategies: a peaking strategy and a base-load strategy. The study found that DG technologies tend to displace more polluting sources and estimated that 313 peaking DG units in the PJM region would displace oil steam turbines, gas combustion turbines, natural gas combined cycle turbines, and coal facilities. To the extent that cogeneration is used to improve overall system efficiency and cleaner fuels are substituted for more polluting fuels, the base-load DG strategy in the PJM region was projected to reduce NO_x emissions by as much as 7.3 lb/MWh, SO_2 by as much as 13 lb/MWh, and CO_2 by as much as 1,500 lb/MWh.

These environmental advantages are not merely theoretical. A 94 MW CHP plant in East Chicago, Indiana, managed by Primary Energy Incorporated, recycles waste heat recovered from coke batteries, supplying one-fourth of Mittal Steel's total electric requirements and 85 percent of its process steam needs. That plant alone displaces an average 13,000 tons of NO_x, 15,500 tons of SO_2, and 5 million tons of CO_2 emissions per year. A similar 161 MW CHP facility operated by U.S. Steel Corporation in Gary, Indiana, uses captured blast furnace gas to provide more than 40 percent of the electric and all of the process steam needs of the facility. The company estimates that they save more than $1 million in energy costs and displace 3,000 tons of NO_x, 3,300 tons of SO_2, and 1.1 million tons of CO_2 every year.[43]

This could be why, in practice, between 1991 and 1997 CHP penetration in the Netherlands reduced annual emissions of CO_2 by between 4.4 million and 6.7 million tons, or 2.4 to 3.8 percent of national CO_2 emissions. CHP was so successful that it became "the major tool of distribution utilities in meeting industry CO_2 reduction targets."[44] A British analysis estimated that domestic DG technologies reduced CO_2 emissions in the United Kingdom by 41 percent in 1999; a similar report on the Danish power system observed that widespread use of CHP throughout their country cut emissions by 30 percent from 1998 to 2001.[45]

Lastly, claims that clean power sources should be rejected for aesthetic and even environmental reasons naturalizes the harms from the current power system and, in turn, indirectly promotes what the Reverend Benjamin Chavis Jr. called "environmental racism."[46] Economist Kenneth E. Boulding referred to it as the "milk and cream" problem: both within and between communities, a certain proportion reaps the benefits of cleaner power sources while another, often larger proportion becomes worse off, leaving them disorganized and polluted.[47] Dirty infrastructure can sometimes create "national sacrifice zones" that condemn poorer communities to suffer disproportionately.[48] Eric M. Uslaner, professor of Government and Politics at the University of Maryland, termed this as "shale barrel politics," the reverse of "pork barrel politics." Pork barrel politics involves the distribution of benefits to congressional constituents, whereas shale barrel politics involves the distribution of environmental costs that are wholly destructive to minorities.[49]

Whatever one decides to call it, there are at least three ways in which the current American power system promotes class- and race-based discrimination: most of those working in energy-related jobs with greater occupational hazards (such as coal mines or refineries) tend to be near the poverty line; poorer families have less capital to invest in energy efficiency and thus live in homes that consume more electricity; and lower income families live in neighborhoods in closer proximity to conventional power plants, T&D lines, nuclear reactors, municipal landfills, trash incinerators, pipelines, abandoned toxic dumps, and nuclear waste repositories (and thus are more exposed to the life-endangering pollution that they bring).[50]

The "NIMBY" attitude emanating from opponents to clean power do not eliminate hazards completely, but instead redistribute them. The risk from dirty power is not reduced but instead shifted to less affluent populations. People of color (African, Hispanic, and Native Americans) must then bear "a disproportionate share of the nation's noxious risks and environmental hazards" as the consequences of energy production move "from white, affluent suburbs to neighborhoods of those without clout."[51] Paul Mohai and Bunyan Bryant conducted a meta-analysis of studies documenting the spatial distribution of pollution and found "clear and unequivocal evidence that income and racial biases in the distribution of environmental hazards exist."[52] A similar assessment of environmental pollution across 2,083 counties found that "toxic releases increase as a function of [minorities in] the population."[53]

Sadly, the existing configuration of the electricity industry reinforces these inequities since people living in poverty pay proportionally more for power, meaning they are less likely to accumulate the wealth needed to make investments to escape their poverty, and the deleterious health effects from power plant related pollution are more likely to impact household members. The United Nations has warned that energy pollution has an often ignored class dimension: infant mortality rates are more than 5 times higher among the poor, the proportion of children below the age of five who are malnourished is 8 times higher, and maternal mortality rates are 14 times higher.[54]

Electricity has racial, geographic, and age-based dimensions along with its classism. African Americans consume more fish in larger portions than other Americans, meaning that they have a higher exposure to mercury poisoning from power plants.[55] More than two-thirds of all African Americans live within 30 miles of a coal-fired power plant. They are rushed to the emergency room for asthma attacks at more than 4 times the national average, and have children 3 times as likely to be hospitalized for treatment of asthma.[56] About half of African American children have unacceptable levels of mercury and lead in the bloodstream compared to 16 percent of the general population, and nationwide studies demonstrate that the air in communities of color contain higher levels of PM, carbon monoxide, ozone, and SO_2.[57]

Pollution is also concentrated among certain locations and age groups. Ohio has the most polluted air of any state in the nation, and the 1.4 million people living there in Cuyahoga County face a cancer risk more than 100 times the goal established by the Clean Air Act.[58] One of the most comprehensive studies ever undertaken on environmental externalities, a $3 million, 3-year study by ORNL and Resources for the Future, found that power plant pollution was primarily responsible for increased mortality among the elderly, the very young, and individuals with preexisting respiratory disease.[59] It also found that the effects of airborne pollutants from power plants on human health were two orders of magnitude greater in the Southeast.

A Clash of Values

So why is opposition to clean power sometimes so fierce? Three reasons stand out: the symbolic aspects of clean power technologies, differing interests and values of energy users, and lack of a coherent clean energy industry or lobby.

First, clean power systems possess symbolic meaning. Opposition to the siting of new power plants can occur because such technologies inflame preexisting social conflicts that have little (and sometimes nothing) to do with electricity. Rural residents, for example, often resent urban developers who wish to build electricity projects in their midst. Others oppose new generators because they feel that they have been excluded from the policy making, permitting, or siting process. In other cases, rural residents want clean power projects for their own use, as a vehicle for economic development, and resent what seems like meddling by urban residents intent on preserving the countryside for its scenic and recreational value. In this way, clean power technologies become more than simply an electricity generator: they symbolize a method of organizing the landscape, a system of ownership and control, and a personal ethic or a reflection of attitudes.[60]

Much of this conflict has to do with the immobility of renewable resources. Wind moves but windy locations do not. Wind and sunlight differ from coal and conventional fuels because they cannot be extracted and transported for use at a distant site. For wind farms to be successful, turbines can only be installed where sufficient wind resources exist. Thus, the site-specific nature of wind invites conflict with existing or planned land uses. The landscape itself can shape public attitudes toward renewables, as some landscapes are more valued than others. Place turbines in sensitive areas, perhaps along the coast or in a national park, and prepare for social uproar. Place them out of view or in low value areas such as sanitary landfills, and opposition diminishes.[61]

Yet opposition to clean power technologies for aesthetic reasons is far from uniform. In some cases, opposition to conventional energy infrastructure (such as T&D lines) can turn to broad public support if the infrastructure is justified by the need to interconnect clean power technologies (such as wind turbines). In 2003, for example, Xcel Energy received approval from the Minnesota PUC to site 178 miles of new transmission lines and four new substations to facilitate a tripling in size of its Buffalo Ridge wind farm. Early in the process, Xcel justified the new transmission as critical to expanding wind power generation at Buffalo Ridge, whose transmission lines were already fully subscribed. In a remarkable reversal of norms, local stakeholders accused the company of not proposing an adequate amount of new transmission and not working to build it fast enough. One senior environmental consultant noted how local landowners and advocates perceived environmental and economic benefits from clean power and that perception translated into overwhelming support for Xcel's transmission upgrades:

> The combination of expanded use of renewable energy and the associated influx of potential economic gain in rural, primarily agricultural, regions have led to unprecedented support of the transmission line projects. Environmental groups view the increased use of a renewable energy source as a positive step and recognize the need for additional transmission capacity to support siting of renewable generation facilities.[62]

Buffalo Ridge offers a case study for how other utilities can win public approval for T&D expansions by highlighting that they will serve renewable generators.

The broader lesson, however, seems to be that clean power sources, because of their symbolic nature, will always elicit some type of social reaction. Owing to the immobility of renewable resources such as wind and sunlight, total mitigation of public opposition is almost impossible. Because social and environmental values change from place to place and from time to time, generic solutions are few and elusive. Since nothing can make clean power technologies invisible, little will make them more acceptable to those perceiving land-use interference. There is no escaping the essence of wind turbines, for instance, as they will always be "spinning, pulsing, exoskeletal contraptions that attract the eye."[63]

Second, electricity users have heterogeneous interests, values, and worldviews. Social scientist Paul Stern and psychologist Elliot Aronson commented that "energy" possesses at least five distinct meanings in contemporary society.[64] The *scientific* view of physicists and engineers frames energy as a property of heat, motion, and electrical potential, measurable in joules and BTUs. According to this view, energy can be neither produced nor consumed, quantity is always conserved, quality is always declining, and correct policy is a matter of understanding thermodynamics and physics.

The *economic* view sees energy as a commodity, or a collection of commodities such as electricity, coal, oil, and natural gas, traded on the market. This view emphasizes the value of choice for consumers and producers and assumes the marketplace allocates choices efficiently. According to this view, when prices rise, fuel substitutes will be found, and inequities arise only through irrational behavior. Correct policy is a matter of analyzing transactions between buyers and sellers and minimizing the external costs of these transactions.

The *ecological view* rejects framing energy as scientific or economic, and instead classifies energy resources as renewable or nonrenewable, clean or polluting, and inexhaustible or depletable to emphasize their environmental context. This view prioritizes the values of sustainability, frugality, and future choice. Correct policy is a matter of recognizing that energy resources are finite and interdependent and that present use engenders significant costs to future generations.

The *social welfare* view sees energy services as a social necessity. This view suggests that people have a fundamental right to energy for home heating, cooling, lighting, cooking, transportation, and essential purposes. The central value here is one of equity, and correct policy becomes a matter of distributing energy services to all social classes.

The *energy security* view focuses on the geographical location of energy resources, political stability of producing and consuming countries, and availability of fuel substitutes. This view sees energy supply as a key component of national security, and correct policy becomes a matter of maintaining economic vitality and military strength.[65]

Each of these views differs in its conception of energy, diagnosis of what counts as energy problems, and prescription for them. Multiple surveys of American attitudes about energy during the energy crises of the 1970s identified an even broader collection of incommensurable views, values, and interests (see Table 11). Those who believe in growth, for instance, see an expanding economy underpinning constant improvements in average living standards, upward social and economic mobility, and a general aura of progress. Those who advocate limits see restricted capacity of the environment to absorb pollution from energy processes, adverse effects on human health and safety, and a growing risk of full-fledged catastrophe in a complex, centralized, and independent world.[66] Elements of these views are incompatible. "To conserve energy in a growth-oriented economy," remarks Norman Metzger from the National Research Council, "is like letting the cat into the pigeon

Table 11

American Approaches to Energy Policy and Explanations for the 1973 Energy Crisis

Americans offered at least five different explanations for the 1973 energy crisis, and not one group viewed the crisis in the same way.

Approach	Proponents	Cause of the Energy Crisis	Assumptions	Policy Response
Free Market or "Economics" Approach	Major oil and gas companies, academic experts, politicians from oil producing states, conservative political organizations and individuals	The energy problem was not the consequence of imminent depletion of domestic or foreign reserves, but stemmed from government policy errors exacerbated by the cartel-like actions of oil producing nations	(a) Free market mechanisms achieve the most efficient allocation of energy resources; (b) supply and demand of energy can be adequately responsive to price changes; (c) economic growth and secure supplies of energy are instrumental to a free enterprise system	The government must end interventions and rely instead on unfettered market forces, accommodating consumer's growing demands for energy services

Energy Independence, Machiavellian, or "Energy Security" Approach	Nuclear industry, private power utilities, Americans for Energy Independence, groups within the AFL-CIO	The energy crisis was the result of rapid depletion of world and domestic oil and gas reserves	(a) A group of unreliable and hostile foreign producers had the power to raise fuel prices and interrupt the flow of energy; (b) a strong supply orientation was the only way to achieve high levels of employment, social mobility, and living standards; (c) energy conservation could not take the place of energy development; (d) the market works best in business-government partnerships	The government should launch large supply-side development projects that involve economic costs and risks that private industry is unwilling or unable to accept, and remove stringent environmental restrictions impeding the mining, processing, and transportation of fuel resources and construction of electric power facilities
Liberal New Deal or "We Need More Science and Technology" Approach	Prominent liberal Democrats, most major unions, and consumer organizations	Depletion was not to blame, but instead shortages were the result of OPEC policies in partnership with inept government policy	(a) Economic concentration has resulted in political influence far disproportionate to what is proper; (b) supply and demand of energy will continue to be inelastic to price increases; (c) no preference should be expressed for particular energy sources	The government should initiate massive programs to achieve energy independence and allow the United States to confront OPEC from a position of strength, and a crash energy R&D program is needed

Liberal Conservationist, Malthusian, or "Ecological" Approach	Environmentalists and consumer and public interest organizations	The energy crisis was the result of overconsumption of energy and rapid depletion of natural resources	(a) Future energy production entails immense financial, political, and environmental costs; (b) energy corporations and government regulators are complicit in the crisis; (c) American energy use is wasteful and unsustainable	The government should promote massive energy efficiency and conservation programs and attempt to change America's demand for energy resources
Orthodox Leftist or "Social Need" Approach	U.S. Communist Party, some elements of the labor movement, some intellectuals	The energy crisis was not the result of depletion of resources, but the activities of energy companies and the pro-corporate policies of government	(a) The market cannot allocate resources in an optimal and fair manner; (b) socialism can better manage energy systems; (c) the energy problem was most unfair to the lowest and poorest classes	No limit on energy use is needed, but control over energy corporations, resources, and services should be distributed equally to everyone

Source: Don E. Kash and Robert W. Rycroft, *U.S. Energy Policy: Crisis and Complacency* (New York: University of Oklahoma Press, 1984); Benny Temkin, "State, Ecology and Independence: Policy Responses to the Energy Crisis in the U.S.," *British Journal of Political Science* 13, no. 4 (October 1983): 441–462; William D. Smith, "Shortage Amid Plenty," *Proceedings of the Academy of Political Science* 31, no. 2 (December 1973): 41–50.

coop."[67] A similar tension exists between treating electricity as a commodity and regarding it as a public service. If a commodity, it would make sense for electricity companies to choose their customers carefully and focus only on distributing power where they can maximize profits, even if it meant the exclusion of poor and rural areas. If a public service, then electricity companies should have an obligation to supply everyone regardless of cost.[68]

Extensive sociological and psychological research has found that values and perceptions vary greatly by political affiliation and class. One study investigated energy policy voting trends in Congress and found that party affiliation was more important to policymakers than expectation of economic benefits for constituents.[69] Republican policymakers, the authors noted, are more likely to weigh big-business interests and promote the free market in their energy policies, whereas Democrats focus on mitigating the power of large energy companies and promoting social welfare.

In terms of class, a 1974 survey queried American residents about the severity of the energy situation and found that the upper class placed significantly greater importance on environmental protection (70 percent compared to 37 percent), whereas the lower class put greater emphasis on maintaining low energy prices (40 percent compared to 11 percent).[70] This is because low income families often spend about 14.5 percent of their income on energy, whereas middle and upper income families spend about 3 percent.[71] Counterintuitively, the relationship between energy consumption and percentage of family income spent for energy is reversed: poorer families use less electricity but have to expend a bigger share of their income to pay for it.[72]

These contradicting values impede progress toward any type of coherent energy policy, and become especially important for clean power technologies. Studies have suggested that the most powerful predictor of the intention to purchase clean power or promote energy efficiency is who the respondent blames for energy shortages. If people believe their own consumption is wasteful and accept personal responsibility, they are likely to change their attitudes and values. But if they are able to blame companies, politicians, foreign countries, and other consumers, they will do nothing.[73]

Third and finally, unlike the well-organized and clearly defined fossil fuel and nuclear lobbies, clean power firms often compete directly with one another to gain funding and public credibility. Energy efficiency efforts may have the weakest constituency, since saved electricity is never itemized on bills, energy-efficient technologies often look just like inefficient ones, and savings are spread across millions of separate electricity users rather than one concentrated whole. Most clean power projects attract few "ribbon-cutters or rent-seekers," and constituencies are scattered and weak. Professional engineering societies, such as the American Society of Heating, Refrigeration, and Air-Conditioning Engineers, as well as the Illuminating Engineering Society, ASME, and Association of Energy Engineers, tend to consolidate traditional practices rather than seek innovation, and their committee and bureaucratic structure is not well-suited to fast moving changes in technology.[74]

CHAPTER 7

The Big Four Policy Mechanisms

John Kenneth Galbraith once iterated that "in a state of bliss, there is no need for a Ministry of Bliss." [1] His comment subtly implies that the government should constantly strive to create environments where it is not needed. It also suggests that government intervention has its limits, and should be used only sparingly. This may be why policymakers have not "fixed" electricity markets to include externalities, overcome split incentives, extend accurate price signals to consumers, and change individual behavior.

But as this chapter shows, four big policy mechanisms can do much to address the problems impeding clean power. To respond to the economic impediments among businesses and utilities, clean power technologies should be made mandatory through a national feed-in tariff. To overcome political and regulatory barriers, subsidies for fossil fuel and nuclear technologies should be immediately repealed. To address behavioral and cultural obstacles, regulators should price electricity accurately. And to minimize aesthetic and environmental concerns, a national systems benefit charge should be implemented to educate consumers and property owners, protect low-income families from rising electricity prices, and promote energy efficiency and DSM programs.

No. 1: Make Clean Power Mandatory

Since electric utilities and businesses are risk-averse, are unfamiliar with clean power technologies, face split incentives even when they decide to promote them, and underinvest in R&D, clean power should be made mandatory by implementing a national *feed-in tariff* and guaranteeing clean power suppliers access to the grid (it should be noted that FIT advocates have recently discussed renaming the mechanism "Renewable Energy Payments" to avoid the term "tariff" in their campaigning).

Feed-in tariffs (FITs) force utilities to purchase renewable power by setting a fixed price above market rates (say, 20 ¢/kWh) that they have to pay all suppliers. FITs obligate electric utilities to purchase the electricity from renewable energy resources in their service area at a tariff determined by the public authorities and guaranteed for a specified period of time (usually about 15 to 20 years).[2] Germany stands as the paradigmatic example of effective FIT regulation. The country implemented its Electricity Feed-In Law in 1991 in order to create a market for renewable electricity by offering providers a fixed but attractive price for the recovery of generation costs.[3]

FITs have many advantages over other mechanisms. Rather than leaving renewable power prices to the market, FITs ensure a stable investment stream for project developers, as the profitability of projects is guaranteed.[4] This type of pricing system makes it exceptionally easy for developers to obtain bank financing for investments in renewable energy.[5] Suppliers also get paid immediately, rather than having to wait for the sale of renewable energy credits or reimbursement of tax credits.[6] Generators and power providers are likely to put pressure on equipment producers for lower prices and on developers for the best available locations, shifting competition from electricity prices to equipment prices.[7]

While California, Illinois, Michigan, Minnesota, and Rhode Island have discussed or proposed enacting FITs at the state level, there are strong reasons federal legislation is required. Making clean power mandatory through a FIT at the national level—offering renewable producers a fixed price but differentiating it for technologies (i.e., wind receives something like 10 ¢/kWh but solar 35 ¢/kWh), guaranteeing access to the grid, and creating strict penalties for noncompliance—provides three immediate benefits not captured by state action.

First, a federal FIT is needed to sufficiently and rapidly promote clean power sources. Despite the progress made by state programs, the deployment of renewable resources has stayed relatively the same over the past decade. Almost ten years ago, renewable energy technologies constituted about 2 percent of the country's electricity supply (excluding large hydroelectric facilities), and today they still provide less than 3 percent. Projections suggest that the contribution of renewable resources is unlikely to exceed 4 percent by 2030 without some type of national legislation.

Experience in Canada and Germany implies that a national FIT is the best way to encourage quick expansion of renewable power. The FIT program in Ontario, Canada, started in November 2006 and provided a fixed rate of 11 ¢/kWh for small-scale hydroelectric, biomass, and wind projects and 42 ¢/kWh for solar PV facilities, set in 20-year contracts with guaranteed access to the grid. In just 15 months, the FIT has signed more than 655 MW of wind, 316 MW of solar PV, 66 MW of hydroelectric, and 67 MW of biomass capacity.[8] The Canadian FIT was so successful it exceeded its ten year anticipated target of 1,000 MW in less than two years, with more than 1,300 MW of contracts fulfilled by the end of June 2008.[9] The FIT in Germany, which enables solar PV systems to sell power at 49 €¢/kWh (Eurocents per kWh) and onshore wind producers to sell at 8 €¢/kWh, increased renewable power deployment from 3.5 percent of national capacity in 2005 to 6.7 percent in 2007—an increase of 91 percent in just two years.[10] Under the FIT in its current form, German policymakers expect renewable electricity supply to grow to 15.5 percent of gross electricity production by 2010 and 27 percent by 2020.[11] German utilities that buy power at these higher rates do pass extra costs back to consumers, but by 2007 the national FIT had increased household electricity bills by only €3 per month.

Second, despite the extra initial cost, a national FIT policy would quickly depress electricity prices. The German Federal Ministry of Environment estimates that while their FIT cost consumers $3.2 billion in higher electricity rates in 2007, it *saved* them $3.5 billion in depressed fossil fuel costs.[12] In Spain, where a similar national FIT saw the deployment of 26.7 TWh of wind energy in 2007, the policy cost consumers about $1 billion but depressed the market prices of fossil fuels by 0.6 €¢/kWh, saving utilities (and thus consumers) $1.7 billion in avoided costs (for a net savings of more than €640 million). The FIT also lowered the average delivered cost of wind energy to 3.8 €¢/kWh, and if Spanish policymakers achieve their goal of 20,000 MW of wind by 2010, they expect the FIT to produce a net savings of €2.3 billion per year.[13] Compare these trends to France and the United Kingdom, still heavily reliant on fossil fuels and nuclear power, where electricity prices for some electric utilities have risen 96 percent cumulatively from 2003 to 2008 and are set to rise 40 percent more by the end of 2009.

Third, a national FIT would create harmonization, consistency, and predictability for financers, investors, manufacturers, and producers. As Chapter 4 noted, differing state standards for clean power heighten barriers to interstate trade, and federal uniformity would help manufacturers and industry by providing a consistent and predictable statutory environment. This industry, in turns, brings new and high paying jobs. Economists in Germany, for instance, have credited their FIT with creating at least 157,000 jobs in renewable energy manufacturing and installation.[14] Q-Cells, based in Wolfen, Germany, overtook Sharp to become the world's largest manufacturer of PV cells in 2008 precisely because of the German FIT, which encouraged the solar industry there to grow by a factor of 4 from 2000 to 2006.[15] Enercon, the leading German wind energy manufacturer, expects employment in the domestic renewables industry to increase to 710,000 by 2030, matching the number of jobs offered by the German automobile industry.[16] Correspondingly, the FIT has created 188,682 new jobs in Spain, which has enabled the country to become the third largest manufacturer of wind turbines.[17]

No. 2: Eliminate Subsidies

Governments subsidize energy technologies to protect domestic industries (and employment in them) and to develop a technological lead over other countries.[18] In the United States, these subsidies have mostly included direct financial transfer, preferential tax treatment, trade restrictions, public funding, and direct regulation.[19]

The five largest energy subsidies in order of magnitude per year are accelerated depreciation of energy related capital stock (at $16.6 billion); R&D funding for the DOE ($9.1 billion); assumption of legal risk for nuclear plants ($4.9 billion); maintenance of the strategic petroleum reserve ($3.6 billion); and the general investment tax credit ($3.5 billion).[20] These five subsidies account for a total of half of all government expenditures on energy.

Table 12
Major American Energy Subsidies, 2007

The federal government still continues to massively subsidize energy technologies, with at least 18 different types of subsidies still on the books for FY 2007.

Direct Financial Transfer	Preferential Tax Treatment	Trade restrictions	Public Funding	Direct Regulation
Grants to producers	Rebates	Quotas	Direct investment in infrastructure	Demand guarantees
Grants to consumers	Exemptions	Technical restrictions	Public R&D	Mandated deployment rates
Low-interest or preferential loans	Sales taxes	Trade embargoes		Price control
	Producer levies			Market access restrictions
	Tariffs			
	Accelerated depreciation			

Source: Douglas N. Koplow, *Federal Energy Subsidies: Energy, Environmental, and Fiscal Impacts* (Washington, DC: Alliance to Save Energy, April 1993); Doug Koplow and John Dernbach, "Federal Fossil Fuel Subsidies and Greenhouse Gas Emissions: A Case Study of Increasing Transparency for Fiscal Policy," *Annual Review of Energy & Environment* 26 (2001): 361–389; Doug Koplow, "Subsidies to Energy Industries," *Encyclopedia of Energy* 5 (2004): 749–765.

Because they spread government benefits unevenly, however, *immediate repeal of existing subsidies* would bring about four drastic and important changes.

First and most important, removal of subsidies would send market signals to consumers and encourage more rational use and valuation of power resources. Subsidies actively discourage consumers from seeking cleaner alternatives, encourage the overconsumption of resources and thus higher electricity use, and lead to capacity developments and consumer patterns in excess of true needs. Federal funds used to develop intercoastal waterways and deepwater channels, for example, make it cheaper to deliver coal and oil to markets that would otherwise suffer higher fuel transportation costs. One study found that conventional subsidies distorted price signals for electricity by at least 11 percent.[21] It is both "good economics" and "fair" to adopt policies that force consumers to pay for the true costs that their consumption imposes on society.[22]

Second, elimination of subsidies would improve competition in the electricity industry, eliminating the unfair advantage given to nuclear and fossil fuel technologies.[23] From the moment the federal government formally began funding energy R&D in 1882 (by supporting coal research by the USGS), it has heavily focused on promoting oil and gas development at the expense of alternative and unconventional technologies.[24] Since then, the pattern of subsidization has favored mature, conventional energy sources by a ratio of more than eight to one. End-use energy efficiency has received only $1 worth of subsidies for every $35 spent on forms of conventional supply.[25]

Looking closely at the numbers, conventional sources have received almost 90 percent of all subsidies for the past six decades. From 1943 to 1999, for instance, federal subsidies for nuclear power totaled $144.5 billion, more than 25 times the cumulative spending on wind and solar ($4.4 billion for solar thermal and PV and only $1.3 billion for wind).[26] In 1973, before the energy crisis, the federal government awarded 93 percent of its subsidies to fossil energy but only 6 percent to energy efficiency and renewables. Between 1978 and 1995, federal subsidies for clean coal exceeded $12 billion, nearly two-and-a-half times the total funding for wind.[27] Even in fiscal year 1979, when subsidies for clean power peaked at $1.5 billion, subsidies for fossil fuels were greater at $1.9 billion and more than 58 percent of the DOE R&D budget was focused on nuclear power.[28] Nuclear power development received subsidies worth $15.30 per kWh between 1947 and 1961, which compares with subsidies worth only $7.19 per kWh for solar and 46 cents per kWh for wind between 1975 and 1989. In its first 15 years, nuclear and wind produced about the same amount of energy—2.6 billion kWh for nuclear and 1.9 billion kWh for wind—but nuclear subsidies outweighed wind subsidies by more than a factor of 40, receiving $39.4 billion compared to wind's $900 million. Taken as a whole, from 1947 to 2000 cumulative subsidies for nuclear power amounted to $1,411 per U.S. household, compared to just $11 per household for wind. (And, amazingly, these figures *underestimate* subsidies for nuclear power because they exclude price guarantees, discovery and production bonuses for uranium miners, accelerated depreciation, tax exemptions for industrial development bonds, investment tax credits, and all non-DOE R&D on fusion, high energy physics, and metallurgy).[29]

What about recently, astute readers may inquire? The GAO notes that in 2004, not much had changed, with fossil energy receiving 86 percent of government subsidies, nuclear energy 8 percent, and renewables and energy efficiency only 6 percent.[30] An examination of current federal electricity subsidies over the past six years shows that government policymakers remain heavily committed to supporting conventional sources. Consider two of the more prominent types of subsidies: R&D appropriations and tax credits. During 2002 to 2007, nuclear power received 54 percent of all DOE related R&D subsidies ($6.2 billion out of $11.5 billion), and the amount given to nuclear significantly increased by 59 percent over the same period, from $775 million in 2002 to $1.2 billion in 2007. Fossil-fuel related energy R&D received 27 percent of federal subsidies ($3.1 billion) while the entire class of renewable power technologies received a miserly 12 percent ($1.4 billion). Again, these numbers underestimate the amount awarded to conventional sources because they exclude subsidies such as limited nuclear liability provided under the Price Anderson Act and low-cost financing given to federal power entities that operate nuclear power plants. Looking at tax credits and incentives, a review of U.S. Department of Treasury data indicates that the energy sector, excluding nuclear power, received $18.2 billion in lost tax expenditures from 2002 to 2007. Here, fossil fuels received 75 percent of all energy related tax credits ($13.7 billion) while clean power systems received a meager 15 percent ($2.8 billion).[31] Shockingly, the DOE intends to worsen this bias even further in their request for FY 2009 appropriations by calling for a 34 percent increase in R&D funding for fossil energy and a 44 percent increase for nuclear power but a *decline* in funding for renewables and clean power sources.

These subsidies artificially lower the costs of innovation in mature industries, increase barriers to entry for newer, cleaner, and emerging technologies, and obscure costs and risks of conventional fuel cycles.[32] They thus make it *impossible* for clean power technologies to compete. Consider the modern case of nuclear power. The Energy Policy Act of 2005 is "festooned with lavish subsidies" for nuclear energy, including $13 billion worth of loan guarantees covering up to 80 percent of project costs, $3 billion in R&D, $2 billion of public insurance against delays, $1.3 billion in tax breaks for decommissioning, an extra 1.8 ¢/kWh in operating subsidies for the first 8 years and 6 GW (equivalent to about $842 per installed kW), government funds for licensing, compensation for project delays for the first six reactors to be developed, and limited liability for accidents (capped at $10.9 billion).[33] These subsidies are in addition to numerous "other" benefits the nuclear industry enjoys: free off-site security, no substantive public participation or judicial review of licensing, and payments to operators to store waste.[34] The existing subsidy established by the Price Anderson Act for nuclear power is estimated to be worth more than the *entire* DOE R&D budget for most of the 1990s.[35] Forcing clean power sources to compete with nuclear sources (but without equal subsidization) is like trying to race a bicycle against a Ferrari.

Third, abolishing energy subsidies would free up approximately $30 billion of government revenue to reduce the national deficit or fund other programs. The Congressional Budget Office calculated that the national debt for the U.S. government was more than $9.2 trillion and that the deficit grew about $800 billion in 2007 (almost 7 percent of national GDP). Every incremental reduction of the national debt strengthens the dollar, decreases inflation, increases employment, and decreases the amount of interest on foreign loans (the federal government spent $406 billion alone on such *interest* in 2006). The elimination of energy subsidies could also untangle much-needed revenue for underfunded programs such as social security, education, and health care. The $30 billion from energy subsidies, for instance, could completely fund two more National Aeronautics and Space Administrations or increase the Department of Education's budget by 50 percent.

Fourth and finally, the removal of American energy subsidies would benefit the global market. Subsidies become self-replicating because, once enacted, they continue to shape energy choices through the long-lived infrastructure and capital stock they create. This justifies further expenditures to operate, maintain, and improve existing technologies. Coal and nuclear plants built 40 years ago, for example, still receive subsidies for coal mining and uranium enrichment.[36] The federal government subsidizes fossil fuels so much that among the 30 industrialized nations forming the Organization for Economic Cooperation and Development (OECD)—including the EU, Japan, Australia, and Korea—the United States was responsible for 70 percent of *all* subsidies for coal worldwide in 2001.[37] This subsidization creates higher demand for fossil fuel imports globally, forcing other counties to subsidize their own energy sectors.[38] Removing subsidies, however, can reverse this trend. One group of economists calculated that by merely cutting fuel subsidies for gasoline by 80 percent, global demand for oil would immediately drop by 5 percent—the equivalent of removing 2.5 million barrels of oil a day from the market.[39] In this way, repealing American subsidies could start reversing the global commitment to conventional energy sources.

No. 3: Get the Price Right

The fundamental disconnect between cost and price in the electricity industry has its roots in the historical design of franchise rates for wholesale power. Public utility commissions and regulators typically set retail electricity rates based on a combination of utility revenue (including administrative, financial, and marketing costs), operating expenses (including personnel, fuel, purchased power, and maintenance charges), and recovery of capital investment (plant and equipment committed to public service less depreciation).[40] A percentage profit, or rate of return, on all investments is often included in electricity rates. Typically, because fuel prices can

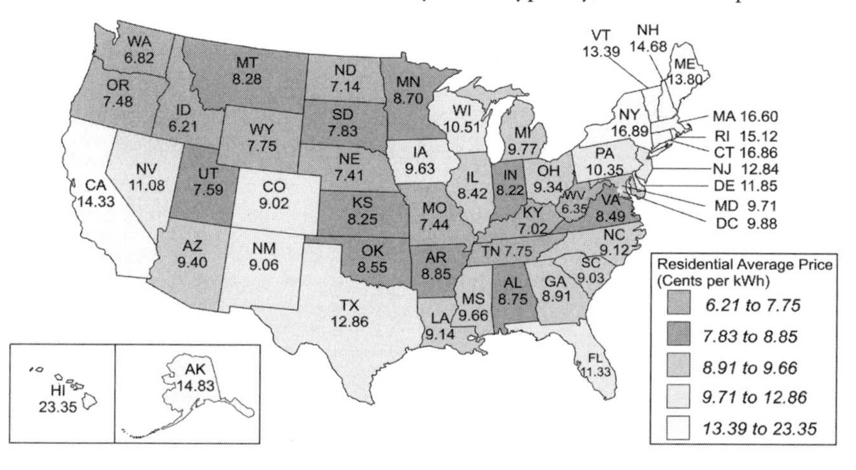

Map 16
Average Residential Electricity Prices by State, 2006

Given the different fuel mixes and resources within individual states, residential electricity prices already differ immensely in the various regions of the country. The highest rates are in Hawaii, Alaska, California, Texas, and the Northeast, while the cheapest rates are in the Midwest and parts of the South.

Source: U.S. EIA *Residential Electricity Prices: A Consumer's Guide* (2007).

vary considerably in response to market conditions, most states have a separate fuel adjustment clause, a mechanism intended to protect utilities from price swings. Consequently, retail rates for electricity vary greatly across the country, with power costing about 7 ¢/kWh in Washington and North Dakota but more than twice as much in New York and three times as much in Hawaii.

Consumers, however, are generally unaware of daily, weekly, and seasonal changes in price, and instead see only a monthly electricity bill. To avoid billing spikes in high usage months, some companies even allow customers to average costs over entire years, so that no price variations are seen. To insulate utility shareholders (and ratepayers) from the vagaries of fluctuating fuel prices, nearly all states allow utilities to adjust prices to customers using fuel adjustment clauses to transfer the costs of fuel from the utility directly to the ratepayer so that changing fuel costs do not affect profits.[41] Put another way, a utility that spends more on fuel to meet increased demand can recover every cent by raising the price and spreading the extra cost among all consumers. Because increases in the costs of fuel are passed directly onto customers, little incentive remains for utilities to improve efficiency or pursue alternative sources.

Ralph Loomis, an executive vice president for the largest utility in the country—Exelon—admits that the current system of electricity pricing encourages waste and overconsumption. He argues that:

> In the marketplace you promote [efficient energy use] by allowing prices to go up, customers get the signal, and they think to themselves, "Aha, this is a valuable commodity that needs to be carefully used and conserved. I need to use it wisely." However, if, as a matter of public policy you have a long history going back one hundred years of promoting economic development by having a very regulated utility industry that insures that prices don't go up, then residential customers never get that price signal. And, indeed, over time, customers become un-attuned to making intelligent decisions.[42]

Undersecretary David Garman agrees and notes that, unfortunately, only "accurate market signals would incentivize consumers to do the right things" and make "smarter choices" about their consumption of energy. Instead, electricity markets in the United States continue to shelter consumers and distort electricity prices.[43]

Policymakers should implement four changes concerning how we price electricity in this country by:

- Abolishing price caps;
- Eliminating declining block rate pricing;
- Reflecting time of use in electricity rates and bills;
- Internalizing the cost of externalities.

Abolishing price caps would enable electricity rates to reflect current market prices and volatility. At least 22 states and the District of Columbia have some type of price cap on residential or industrial electricity rates.[44] By keeping prices artificially low, price caps fuel excessive consumption, inhibit exploration and investment, undervalue technologies such as conservation, and distort the ability for consumers to make rational decisions based on the actual cost of electricity.[45] They can also force utilities to go bankrupt when unexpected outages and price increases cannot be passed on to consumers. During the 2001 electricity crisis in California, for instance, price caps of $250/MWh prevented PG&E and Southern California Edison from recovering on imported power and fuel costs (up to five times greater than electricity rates) to its industrial and residential customers.

The widespread adoptions of low, fixed-price rate caps in default service plans across the United States suppress the value of clean power. Price caps blind individual customers to what is really going on in wholesale markets, blunting the effectiveness of energy efficiency measures that would moderate wholesale price volatility. Low default service rates reduce incentives for creative load management, and when costs are passed onto consumers, they create a "safe harbor" for conventional practices and make it nearly impossible for innovative firms to enter the market.[46]

Price caps can also adversely affect the development of renewable energy systems. In Colorado, regulators unintentionally created a "catch-22" situation for renewable energy developers by creating a price cap on the amount renewable generators could charge customers. In designing the "safety valve," regulators pegged the value of renewable electricity not to its full market value, but to the avoided cost of natural gas generation.[47] In other words, the regulations limited the difference in the cost of renewable electricity relative to the cost of the same amount of electricity if it had been generated using natural gas. The problem is that the more renewable energy is deployed, the more it depresses the cost of natural gas. As renewable resources reach certain levels in the market, they offset natural gas consumption and decrease gas prices—lowering the value of those same renewable resources throughout the state.[48]

Eliminating declining block rate pricing would create an incentive for industries to promote energy efficiency and consume less electricity. Economic theory holds that the more one consumes a precious commodity, the more its price should go up. The opposite occurs with declining block rate pricing, where consumers of small amounts of electricity pay higher prices for electricity and larger consumers (often industrial or commercial customers) pay lower prices. Since big consumers of electricity pay the least, they thus consume more than they normally would. Declining block rate pricing is so prevalent among industrial customers that the average industrial price of electricity is about 41 percent less than the residential rate.[49] The EIA reports that utilities in Virginia, Indiana, Kentucky, Ohio, Nebraska, Montana, North Dakota, West Virginia, and Arkansas still rely on declining block rates for either residential or industrial customers (or both).[50] The promotion of inverse block rate pricing, where customers are charged higher rates for electricity the more they consume, encourages more rational use.

At least three utility companies have experimented with this type of an approach for residential customers. In California, PG&E customers pay about 11.6 ¢/kWh for normal electricity use but those consuming 30 percent more than the average pay 13.3 ¢/kWh, those consuming twice as much pay 22.8 ¢/kWh, and those consuming three times as much pay about 24.8 ¢/kWh. The Arizona Public Service Company also structures their rates this way: for 0 to 400 kWh, customers pay about 7.6 ¢/kWh, but have to pay 10.6 ¢/kWh for 401 to 800 kWh and 12.4 ¢/kWh for more than 800 kWh. The Idaho Power Company charges their customers 6.2 ¢/kWh for up to 800 kWh but 8.4 ¢/kWh for customers consuming more than 2,000 kWh. These programs, while all relatively new, have decreased rate volatility and thus rate charges for large customers by as much as 9.5 percent.[51]

Reflecting time of use through "real-time," "interval metering," "time-of-use," or "seasonal" rates would show customers how electricity production and consumption varies according to the time of day, week, and month. The regulation of utilities has been based historically on average prices, set in rates revealed to customers in monthly bills. There are some markets where hourly electricity prices can vary by factors of 100 or more, but these prices are all always averaged, so consumers never see them. Most electricity bills combine charges for several appliances, lighting, water heating, space heating, and cooling all into a lump sum. It is impossible for consumers to tell, without careful monitoring and experimentation, how much of the bill results from the individual use of appliances or technologies, or how much the bill could be decreased by using more efficient models.[52]

Yet, as David Garman explains:

> Consumers have to get market signals . . . A kWh of electricity at 8am is a very different thing than at 3am. It is almost a completely different commodity, generated in a different fashion with different consequences. Thus far we have shielded consumers from market signals, rather than recognizing those differences.[53]

If other commodities are priced the way electricity is, telephone bills might give a single dollar figure for long-distance or roaming service without having calls itemized. Groceries would have no price markings, and customers would instead be billed via a monthly statement that would say something like "Amount Due: $1,056.55 for 2371 food units in January."[54] Economist Alfred Kahn noted that charging a flat rate for electricity regardless of when it was used seemed like "charging a flat price per pound for all items in a grocery department store. What would happen if everything that came out of the cow—steak, hamburger, suet, bones, and hide—were priced at average cost per pound?"[55] The result is that everyone would always eat steak (possibly even vegetarians!).

The Energy Policy Act of 2005 implicitly recognized this flaw in electricity pricing and encouraged utilities to provide time-based rate schedules reflecting variations during the day to all individual customers requesting it. However, the Act also said, ambiguously, that each state regulatory authority can decide whether to implement that provision.[56] Correspondingly, LBNL estimates that only about 100 utilities (less than 2 percent) offered some sort of time-of-use rate for electricity customers in 2007.[57] Time differentiated or de-averaged rates would not necessarily mean instant minute-by-minute prices, but would at least reflect meaningful differences between peak and off-peak consumption. Studies have shown that three forms of relatively simple information would be most valuable and effective: energy use histories, such as month-to-month comparisons of current year consumption with past years, adjusted for weather and price fluctuations; comparisons to the usage of other customers in the same neighborhood or with similar sized residences; and calculations of change in energy use before and after investment in a new appliance or home improvement.[58]

New York has experimented with some of these "alternative rate designs" by offering time-of-use rates, day-ahead real time pricing, critical peak pricing, and pricing at real-time market rates. Researchers at LBNL surveyed 149 commercial and industrial customers in the Niagara Mohawk Power Corporation service area where the utility offered time-of-use tariffs for large customers with peak demand needs. They found that more than 30 percent of industrial customers responded by foregoing discretionary electricity usage and 15 percent shifted usage from peak periods to off-peak periods; 45 percent of respondents installed demand reduction enabling technologies on site; and peak load for the utility was reduced by an astounding 15 percent.[59] Further south, Georgia Power introduced time-of-use meters for large industrial customers and, from 1992 to 2002, enrolled 1,650 customers to reduce peak demand by 17 percent.[60] In the West, a recent pilot program in California also had very promising results with time-of-use rates at the residential level. The program installed more than 2,400 time-of-use meters for residential customers and, after combining them with time-of-use tariffs, found that participations shifted more than 20 percent of their peak consumption to off-peak hours.[61]

Internalizing external costs would drastically raise electricity prices but would also ensure that electricity is accurately priced. Chapter 1 shows that utilities sold approximately $360 billion of electricity last year, but the social cost of power outages and air pollution were *more* than $407.9 billion.

While some states, such as New York, Massachusetts, Nevada, and California, internalize some externalities in retail prices, none of them internalize all externalities, and 18 states do not require explicit consideration of environmental externalities at all.[62] Some of this sheltering is intentional to keep electricity prices low. Marilyn Brown from ORNL explains that because "the current externalities associated with fossil fuel consumption are hidden," the result is "an artificially low price of fossil fueled electricity."[63] Energy trader Michael Pomorski states that electricity prices "have not risen to a level that forces people to make conscious decisions about their energy use. Electricity prices are just too cheap, and costumers don't see the true environmental costs of their energy decisions reflected in electricity prices."[64] Zia Haq from the EIA elaborates that "consumers have no financial incentive to consider alternatives that might have a more attractive price once externalities are included."[65]

A preponderance of evidence suggests that pricing electricity more accurately will greatly improve the efficiency of the electricity industry, provide customers with proper price signals, and reduce wasteful energy use. One study provided residents with daily electricity prices for a month and found a 10.5 percent reduction in electricity use.[66] Another analysis of residential electricity use from 1973 to 1980 found that "feedback" in the form of information detailing daily and weekly electricity prices reduced consumption between 6 and 20 percent.[67] When Princeton University researchers gave residents of Twin Rivers, New Jersey, information about their level of electricity and natural gas use on a daily basis, consumption dropped 10 to 15 percent.[68] Another study reviewed 19 sets of data from experimental studies where households were informed frequently (often daily) about how much electricity they were using and found a 20 percent reduction in consumption.[69] A random sample of 414 Delaware residents, matched with utility company records, also found that merely telling consumers that peak consumption was more expensive reduced electricity use all year round.[70] Another study involved eight experiments tracking electricity use at 602 households over the course of many years. In some experiments, feedback was given three to four times a week, and in one experiment it was given continuously and informed households of the cost of their consumption every half hour. The researchers found that frequent, credible feedback about electricity prices resulted in 10 to 13 percent less electricity use than control groups.[71]

No. 4: Inform the Public and Protect the Poor

Finally, a national systems benefit charge (SBC) should be created to distribute public information, protect poor households, and promote energy efficiency. SBCs (also called public benefit funds, system benefit funds, and clean energy funds) originated in the 1990s at a time when state policymakers were considering electric utility restructuring. Afraid that gains made in pursuing research, development, and implementation of environmentally preferable technologies would end after markets were deregulated, advocates won concessions in some states for a new funding mechanism for high-risk or long-term projects. A SBC places a very small tax, often a tenth of a cent, on every kWh of electricity generated and utilizes those funds to pursue socially beneficial energy projects.[72]

SBCs were first implemented in Washington State in 1994 and were endorsed by FERC in 1995 as a way to fund services that had previously been included in customers' bills from regulated utility companies. As part of the negotiations for California's restructuring law, environmental advocates won a provision for a public benefit fund that expended $872 million on energy-efficiency work from 1998 to the end of 2001 and allocated $540 million for renewable energy projects. To develop renewable energy technologies and other programs expected to struggle after deregulation, the CEC created its Public Interest Energy Research program, which initially drew about $62 million annually from the state's SBC.[73] As of 2006, 15 states have SBCs estimated to bring $4 billion in revenues by 2017. These funds, managed through a nonprofit organization called the Clean Energy States Alliance, are used to sponsor original research, collect information and analyses, expand the use of clean power technologies, and promote energy efficiency programs.

At the federal level, a national SBC of 0.1 ¢/kWh would raise about $3.7 billion per year to then be split evenly to inform the public, protect the poor, and manage energy efficiency and DSM programs.

Using one-third of the funds to *educate the public* would go a long way toward minimizing the aesthetic and environmental objections some people have toward clean power technologies. Thomas Jefferson used to say that "a democratic society depends upon an informed and educated citizenry," but that in order for education to occur, people had to be informed "even against their will." A national electricity information and education campaign could include grade-school classes on energy and the environment; public demonstrations and tours of clean power facilities; mandatory disclosure of electricity usage for the construction of new buildings and the renting and leasing of existing ones; free energy audits and training sessions for industrial, commercial, and residential electricity customers; improved labeling, rating, and certification programs for appliances and electricity-using devices; and a federal information "clearing house" consisting of Web sites, free books, indexing services, and libraries to help consumers gather and process information in order to make more informed choices about their electricity use.

The federal government already manages about 20 information programs and assessment centers related to electricity, but most programs focus on industrial customers and are very narrow. The DOE's State Energy Program, for instance, provides technical expertise to help industrial plant managers identify opportunities for energy-efficiency improvements. The DOE's "Golden Carrot" program offers targeted information for school lunch programs and refrigerator manufacturers. The DOE's "Climate Vision" program and the EPA's "National Pollution Prevention Vendor Database" have more than 1,200 listings of pollution prevention products and services. EPA's "VendInfo" database helps industrial clients find energy service companies willing to install energy-efficient products. The Department of Commerce runs a "Manufacturing Extension Partnership" to help train plant managers and provide specialized energy-efficiency knowledge.

Such information programs, however, must be carefully tailored and significantly ramped up. Information is less likely to be used if it requires effort or arrives when a household owner or business manager is busy with other things. Households and industries consume different fuels, in different kinds of buildings, and vary in their income, housing and facility tenure, and individual needs. To avoid creating information that is merely a distraction, "general" or "generic" distribution strategies must be avoided. Because of the invisibility of energy, most people have little experience with energy consuming devices, and may find information about them difficult to comprehend. Moreover, government agencies may be convinced its information is accurate and reliable, but energy users may be skeptical over past performance.[74]

Using one-third of the funds to *protect poorer households* would minimize environmental racism and classism. Since less affluent families spend a larger proportion of their income on electricity (some as much as 40 percent), more accurately pricing it by removing price caps, eliminating block rate pricing, creating time-of-use rates, and internalizing externalities will likely raise electricity prices and hurt them the most. States such as California and New York, along with countries such as Australia and Denmark, have already recognized this and offer a range of concessions (including a percentage reduction in energy bills, special loans, and extra rebates) to the unemployed, elderly, disabled, and low-income households to offset rising electricity prices. Nationally, low-income assistance and weatherization can serve as a hedge against these rising prices.

The DOE currently runs a Weatherization Assistance Program that provides households below the poverty line with free energy efficiency improvements. These improvements are not just traditional "weatherizing" (such as caulking, leak plugging, and adding weather stripping to doors and windows to save energy) but also a wide variety of energy efficiency measures that encompass the building envelope, home heating and cooling systems, electricity, and electrical appliances. The basic premise is to make multifamily and low-income homes the most energy efficient to permanently reduce energy bills. From 1976 to 2006, the program has provided weatherization services to more than 5.5 million families.

Despite having weatherized millions of homes, the DOE estimates that they have served only 16 percent of eligible households, and project that more than 27 million homes are currently eligible for assistance (if the DOE program had the funds). By reducing the energy bills of low-income families instead of distributing one-time dispensations of aid, weatherization helps minimize dependency by lowering heating bills an average of 31 percent and overall energy bills by $358 per year. Because weatherization enhances the infrastructure of homes and buildings, it also increases the value of housing stock. Local industry is stimulated as well, and the DOE estimates that the national weatherization program already supports 8,000 technical jobs in low-income communities.

For those especially hard hit, SBC funds could provide further low income assistance, grants, no interest loans, and other targeted subsidies to offset rising electricity prices.

Using one-third of the SBC funds—$1.22 billion—to *promote energy efficiency* could produce $8.5 billion in economic savings and displace the need to invest in expensive T&D systems, and it is already well-known which energy efficiency measures are most effective. The DOE recently surveyed 18 energy efficiency project areas and found that the five most effective mechanisms in terms of energy savings were the following:

- Workshops and training, accounting for 22.1 percent of all savings;
- Building codes and standards, accounting for 19.8 percent of all savings;
- Energy audits, accounting for 15.9 percent;
- Building retrofits, accounting for 10.9 percent;
- Technical assistance, accounting for 7.0 percent.[75]

The survey found that the least effective energy efficiency measures were carpools and vanpools, interest reduction programs, procurement, and Home Energy Rating systems, each responsible for less than 1 percent of savings.

The OTA also found, after reviewing 58 DSM programs in the early 1990s, that the biggest areas of energy efficiency opportunity remain improvements in thermal integrity of building shells and envelopes, electric equipment, and lighting, along with substituting energy fuels and installing energy management controls that shift time of electricity use.[76] The OTA concluded that the most effective energy efficiency and DSM programs had the following five elements:

- Marketing strategies that used multiple approaches, such as direct mail and media, combined with personal contacts with the target audience. Particularly successful were those DSM programs that developed regular, person-to-person contacts and follow-ups after installation to ensure measures are working properly, and those that offered assistance with further projects.
- Approaches and programs that targeted specific audiences, such as customers, architects, equipment suppliers, and engineers, and different types of investment decisions, such as new construction, remodeling, retrofitting, or replacement.
- Technical assistance that helped targeted customers assess energy efficiency opportunities and implement DSM measures, such as energy audits, advice on equipment, recommendation of contractors, computer modeling of possible savings, and information on new technologies.
- Simple program procedures and materials that made it easier for customers to understand program potential.
- Financial incentives that attracted customer attention and reduced first costs. Those offering free measures produced the highest participation rates.

A third assessment of utility DSM programs found that the most successful programs among the states had consistent definitions of energy efficiency resources, did not rely on voluntary action, provided incentives to utilities, and established strict penalties for noncompliance or poor performance.[77]

More recently, the DOE and ORNL surveyed the "most effective" energy efficiency policies in 2007 and found that the best programs were designed to address multiple barriers at once. They typically combined forms of influence (such as information, persuasion, and financial incentives), and attempted to understand behavior from the household's perspective rather than presuming customer motives. Successful programs also recognized that household behavior faced constraints beyond the owner's control (such as the practices of repair personnel or manufacturers) and required continual monitoring so they could be adjusted as needed.[78]

All Together Now

Unfortunately, pursuing each of these "big four policy mechanisms" will not work in isolation. Making clean power mandatory through a national FIT, for example, but not removing subsidies for clean power and continuing to price electricity inaccurately decreases the economic viability of clean power projects and interferes with the ability of users to sell power back to the grid (or conserve it). Making clean power mandatory without promoting public information and education will ensure that consumers remain uninformed about energy-efficient technologies and practices. Promoting a national FIT without funding energy efficiency and DSM would force utilities to procure significantly more electricity supply. Relying solely on changes in pricing also becomes risky when prices unexpectedly change.

For example, removing energy subsidies without fundamentally changing the way electricity is priced would be futile. Many large apartment complexes from New York to Nebraska, for example, do not measure electricity consumption according to individual use, but instead rely on metering individual buildings (that divide energy costs by the number of tenants). Several landlords and property managers prefer to have units metered collectively to take advantage of declining block rate pricing structures. Removing subsidies without reforming electricity prices still distorts the true value of clean power technologies.

Moreover, economic studies looking at the consequences of removing energy subsidies have found that it would be inadequate to correct market barriers alone and could even be counterproductive. The OECD modeled the effect of eliminating coal subsidies in five European countries and concluded that such actions would have a minimal effect on reducing the use of coal, since those countries would merely switch to cheaper international imports if domestic prices rose.[79] In a follow-up study on the feasibility of removing all fossil fuel subsidies internationally, researchers found that a slight increase in global CO_2 emissions would occur as those fossil fuels were merely exported to the developing world.[80] Similarly, the decline of coal production in the United Kingdom after the elimination of domestic coal purchase obligations on large power producers only led to a switch to other fossil fuels, such as oil and natural gas.[81] The evidence suggests that changes in behavior and significant greenhouse gas reductions will happen only if policy reforms include at least the removal of subsidies *and* more accurate electricity pricing. One study projected that if done properly, removing subsidies and accurately pricing electricity in all OECD countries would reduce CO_2 emissions by 16 percent and increase economic growth by 1 percent.[82]

Simply changing the price of electricity is insufficient as well. For most people, the only visible sign of electricity use is at payment time, when utility bills periodically reach the household.[83] Extensive interviews with residential electricity consumers have found more than half of electricity customers (55 percent) pay all of their bills the same time each month. This "processing and batch" treatment of electricity bills suggests that for the majority of consumers, electricity prices will be ignored because they are injected into an activity primarily concerned with verifying dollar amounts and writing checks. Only 40 percent of those surveyed, for instance, looked at their actual usage of electricity when paying the bill.[84]

Furthermore, decisions about energy efficiency are often made by people who are not paying the energy bills, such as landlords or developers of commercial office space. Many buildings, moreover, are occupied for their entire lives by temporary owners and renters, each unwilling to make long-term investments in efficiency.[85]

Price-based mechanisms face additional barriers when used individually. People question whether energy efficiency will actually lower prices, since its benefits are usually spread out over all consumers. Some energy services fulfill social functions independent of cost, so that people will ignore price changes for as long as possible until it becomes completely prohibitive and a threshold is passed. Without compensatory modification, rising prices are inherently inequitable to low-income families.[86] Pricing electricity accurately but not coupling it with information programs also does nothing to eliminate unrealistic payback rates among property owners and investors.

History confirms the inadequacy of relying solely on price to influence behavior. Between 1974 and 1979, home electricity costs increased as much as 108 percent in some parts of the country but did not reduce consumption more than 20 percent.[87] Only small proportions of the population adopted energy consuming practices during these years, as less than 10 percent of residents changed their home heating systems, installed additional insulation in their homes, tried carpooling, or attempted to ride the bus.[88] A 17 percent change in the price of gasoline in 1974 produced no change in sales among a survey of consumers, and the researchers found that a price increase of at least 170 percent would be needed to reduce consumption by 20 percent.[89] A similar study found that the use of monetary payments, provision of information, and daily feedback on electricity consumption did nothing by themselves to change behavior, and were effective only when used together.[90]

More recently, psychologist Robert Cialdini found in 2005 that relying on social norms (in this case, a card asking hotel guests to reuse their towels) increased reuse rates from 35 percent to 58 percent without any changes in price.[91] Another 2007 study delivered notices to household doorsteps informing homeowners how their energy consumption compared to the neighborhood average and reduced consumption without any additional financial incentive.[92] During 1991 and 1996, utilities in Seattle, Washington, saved peak electric load 12 times faster than in Chicago, Illinois, even though power in Chicago was twice as expensive, implying that price alone cannot always explain or induce energy savings.[93]

Without forcing utilities and system operators to use clean power, information and pricing can be manipulated. During the energy crisis of the 1970s, many utilities, afraid of losing revenue, increased electricity prices in direct proportion to the level that consumers invested in energy efficiency. When customers invested in efficiency but saw that their bill appeared the same, they concluded that conservation efforts simply did not work, and that participation had no correlation with saving money.[94] All of these reasons suggest that more accurately pricing electricity is an important part of promoting clean power. But it is not a panacea.

Improved information and education, by itself, are also inadequate. Surveys of customers have found that less than 38 percent read bill inserts, a problem compounded when some inserts are mailed out at periodic intervals and information is repetitive, contributing to an "information glut" that reduces attentiveness.[95] An assessment of energy efficiency programs involving refrigerators, natural gas ranges, washers and dryers, dishwashers, and room air conditioners found that energy labels and Energy Star programs were inadequate alone to cause substantial changes in consumer preferences.[96] One study investigated the effectiveness of informational material designed to increase knowledge about electricity distributed to apartment complexes in Wisconsin and found that neither information nor short-term increases in prices reduced consumption.[97]

To their credit, utilities offered free or low-cost home energy audits in the early 1980s, but typical response rates were less than 5 percent during the duration of programs. In California, electric and gas utilities publicized their audits in media releases and bill inserts, and publicly discussed the benefits of audits in terms of savings and the environment at public meetings, but by 1982 just 2 percent of eligible Californian customers took advantage of the program. Follow-up interviews found that consumers did not trust utilities as a source of information, did not believe that one could get "something" for "nothing," were unable to arrange a convenient time for the energy audit to take place, did not see the announcement or publicity, or were unwilling or unable to act on information about energy efficiency investments because they could not find a reliable contractor.[98]

The Obstacles Are Surmountable

That said, when comprehensive policy action is undertaken, efforts to promote clean power systems have been successful. Texas currently features the largest amount of installed renewable power capacity in the country when large geothermal and hydroelectric sources of electricity are excluded, more than twice that of any other state.[99] To achieve such an abundance of renewable energy technologies, policymakers radically reconfigured the state electricity market, including mandates and legislation that unbundled transmission, power generation, and retail sales (so that any one category of utility cannot own or operate any of the other two); eliminated restrictions on market electricity transactions (so that renewable producers could enter the market at no charge); promoted rapid permitting so that projects could be implemented in less than 3 years; abolished the ability for utilities to levy stranded costs and other discriminatory practices against renewable technologies; outlawed "extensive pre-interconnection studies"; mandated the use of net metering; placed the burden of proof on utilities to show that renewable projects would be unsafe; promoted precertification and equipment testing so that renewable generators could interconnect more easily with the grid; and adopted a strict renewable portfolio standard (RPS) mandating that Texan utilities provide more than 2,000 MW of renewable energy by 2009.

While arguably further behind than Texas regarding the deployment of small-scale renewable energy, California continues to lead the country in deployment of clean power systems. California had to remove excessive utility tariffs, increase tax credits for renewable energy systems, and institute a large consumer awareness program before clean power sources were widely used. Regulators offered streamlined permitting for clean power projects and published a handbook describing city, county, and state air district permitting processes for DG units, and established 14 permit assistance centers to help applicants. Policymakers required net metering for all customer classes. Solar income tax credits were available to cover up to 15 percent of new system costs for residences and businesses. And the state government created a rigorous buy down program that offered cash rebates of $4,500/kW or 50 percent of the system price (whichever was less) for small scale wind turbines and solar PV systems.[100] Consequently, even the DOE has argued that California remains at "the forefront of using DG technologies."[101] DG use jumped from a few hundred MW in the 1990s to more than 2,000 MW in 2002. California also passed one of the country's first and most ambitious RPS policies, which requires that an additional 1 percent of electricity sales must come from renewable resources each year (so that 20 percent of statewide generation becomes renewable by 2010).

In Europe, where production of wind capacity grew 40 percent between 1990 and 2000, regulators implemented aggressive incentives and rebates for consumers, FITs, and environmental taxes on carbon and other pollutants to promote renewable energy technologies.[102]

The experience with Texas, California, and Europe proves that most people will not adopt renewable energy technologies in the face of their barriers without sustained, consistent, and aggressive legislative action. But they also imply that, given the right configuration of political support, utility acceptance, and social awareness, clean power technologies can achieve significant market penetration.

Bring in the Feds

While the benefits of these big four policy mechanisms are numerous, five strong reasons suggest that federal action is essential.[103]

First, there is the matter of *distributive justice*. Unless the dirtiest polluters and utilities can be persuaded to join those promoting clean power, state and regional attempts to improve the environment, particularly when they are voluntary, will achieve little.

Second, local actors face *constitutional challenges* to dealing with the problem individually. Innovative state programs dealing with interstate problems always face challenges based on the contention that they will interfere with interstate commerce in contravention to the Constitution's Commerce Clause. Moreover, attempts to forge interstate cooperation face challenges based upon the Compacts Clause and the Supremacy Clause of the Constitution. For these reasons, federal policy is needed to remove the underlying tensions between state-by-state energy policy and the U.S. Constitution.

Third, local and state regulations can impose *additional costs* on businesses and consumers. Differing state statutes can complicate efforts to conduct business in multiple states. They risk duplicating costly R&D. They promote complexity and force companies to grapple with inconsistencies among regulations. And they can significantly increase transaction costs associated with enforcing and monitoring a plethora of distinct individual programs. Federal action provides investors with a degree of comparative simplicity and clarity.

Fourth, the *matching principle* in environmental law suggests that the level of jurisdictional authority should best "match" the geographic scale of environmental problems. In the case of clean power, this principle calls for national action, not local or regional intervention. The current state-by-state approach ensures that the distribution of the costs and benefits of providing public goods remains uneven. When interstate spillovers or public goods are involved, federal intervention is needed to equalize disparities between upstream and downstream communities.

Fifth, there is *historical precedence*. The creation of a federal role in the regulation of interstate electric power began in 1927, after the Supreme Court ruled in *Rhode Island Public Utilities Commission v. Attleboro Steam and Electric Co.* that state regulatory agencies were constitutionally prohibited from setting the prices of electricity sold across state lines. This facilitated the passage of the Public Utility Act of 1935 and the Federal Power Act of 1935. These laws, as amended, endow the federal government with jurisdiction over the prices, terms, and conditions of wholesale power sales involving privately owned power companies and the transmission of electricity.[104] The federal government already approves rates for public power sold and transported by five federal power marketing administrations and TVA, and oversees and licenses nonfederal hydroelectric projects on navigable waters. The federal government offers initiatives to expand national DSM programs, endorses financial and technical assistance to state energy offices, funds R&D on energy technologies, sets standards for energy efficiency labeling, provides loans to electric cooperatives, and is the country's largest energy consumer. For these reasons, clean power should be seen as a federal, rather than local or state, issue, and should be promoted through federal action.

CHAPTER 8

Conclusions

Historian Lynn White once remarked, after ruminating on the nature of technological change, that "a new device merely opens a door; it does not compel one to enter." [1] His statement infers that technologies, by themselves, achieve nothing. They must be embraced socially before they will become widely used. In order for clean power technologies to achieve social acceptance, they must function technically as well as economically, politically, socially, and culturally.

The subtle, yet powerful, impediments facing clean power technologies are more about *culture* and *institutions* than engineering and science. American consumers believe that they are entitled to cheap and abundant forms of electricity. Utilities wish to retain the means to distribute electricity, and businesses and investors seek to ensure that any investments in energy have rapid payback schedules. More than just an interesting sociological insight, the cultural and social attitudes of consumers, utilities, business leaders, and systems operators have multiple implications for the extensive deployment of clean power technologies. This chapter offers six conclusions concerning clean power and the modern American electric utility system.

No. 1: The Country Faces an Impending Electricity Crisis

If the American electricity industry remains configured the way it is today, engineers and architects would need to construct as many as 310,000 miles of new natural gas pipelines, 10,000 natural gas plants, 4,950 coal plants, 190 nuclear reactors, 4 large uranium enrichment plants, 5 fuel fabrication plants, and 3 waste disposal sites the size of Yucca Mountain by 2040. The costs of power outages would exceed $412 billion, the industry would consume and withdraw more water than the agricultural sector (threatening widespread shortages), more than 65,000 Americans would die prematurely from power plant pollution, and the country's electricity generators would dump 4.5 billion tons of CO_2 into the atmosphere every 12 months.

The dominant electricity strategy in the United States can be called "strength through exhaustion." Such a strategy depends primarily on mammoth, centralized fossil fuel and nuclear power plants to meet the nation's electricity needs. Over time, more plants are built than are immediately needed in the hopes that capacity will always remain just one step ahead of demand. Despite massive amounts of government subsidization over almost a century, however, this approach suffers from numerous problems. Large generators cannot be mass produced. They take much longer to build, and are therefore exposed to escalating interest rates, inaccurate demand forecasts, and unforeseen labor conflicts. Their centralization requires costly and expansive T&D systems. They are inefficient, often wasting as much as two-thirds of the energy contained in the fuel they burn. They are also unreliable, requiring expensive reserve capacity and risking catastrophic system failures that cost consumers billions of dollars. The current system is thus subject to highly uncertain projections about fuel availability, centrally administered by a technocratic elite, and vulnerable to the ebb and flow of international politics, and it requires garrison-like security measures at multiple points in the supply chain.

Continued reliance on this inefficient, expensive, brittle, elitist, and polluting system could eventually unleash a full-blown crisis. The country's lakes, rivers, streams, air, and land face irreversible and rising levels of pollution, and the country's power grid remains incredibly CO_2-intensive. One recent assessment compared the American energy landscape in 1970 to 2005 and found that in almost every indicator, the nation has *backslid*.[2]

Energy problems almost always flow directly from poor energy policies, and major legislative mandates of the past 30 years designed to spur work on clean power have resulted in only limited success. The depressingly low number of clean power technologies in the United States implies that the $6 billion worth of subsidies, energy R&D programs, and tax preferences spent on such systems has not catalyzed their wider use, and defies international trends. Nonhydroelectric renewable resources represent about 5 percent of global power capacity, more than twice the percentage achieved in the United States, and in 2007 the construction and installation of renewable generators was a more than $100 billion industry.[3] In Europe, Asia, Africa, and the Middle East, renewable energy makes up a substantial portion of electricity generation capacity, and that amount continues to grow. Austria, Canada, Denmark, Finland, New Zealand, Portugal, Sweden, and Switzerland all produce more than 20 percent of their power from renewables. Denmark plans to produce 60 percent of their electricity from renewables by 2025, and Germany is seeking to generate 100 percent of its electricity from renewables by 2030.[4] Developing countries, such as China (17 percent), Egypt (15 percent), and Morocco (10 percent) even have more installed capacity (as a percentage) than the United States. Globally, the production of power from DG and renewables surpasses all of the power produced by all nuclear plants, and more than 100 GW of wind capacity were installed globally at the close of 2007. In contrast, the federal government has failed to set a national target for renewable energy among regulated utilities, disapproved of a national FIT, underinvested in energy efficiency, and refused to create a nationwide systems benefit charge.

Within the conventional system, we already know that the marketplace is highly risk-averse and will not endorse new power technologies on its own. Consumers cannot understand life-cycle costing and discount rates for electricity investments. Electricity is such a low fraction of a typical firm's input costs that it will waste fuel regardless of the price. Trading on the world market for electricity fuels undermines national security, increases inflation and unemployment, and erodes American leadership.[5] Solving the problems inherent with conventional electricity generation by deploying more conventional units is much like addressing the problems of drug addiction by doing more drugs.

Up until this point experts have primarily focused on the "energy problem" as a purely technical one, and policymakers continue to rely extensively on technology to provide solutions to what are, in essence, deeper social, political, and cultural problems. In his 2006 State of the Union Address, for example, President George W. Bush unveiled his idea for the Advanced Energy Initiative, which he promised would help fight American dependence on foreign sources of oil. "The best way to beat this oil addiction," insisted President Bush, "is through technology." The intellectual program of energy policy making thus seems guided by some deeply acknowledged but unseen prioritization of technical prescriptions to energy troubles. Beyond the "hard" analysis, modeling, and conjecture about the particular characteristics of conventional and alternative clean electricity technologies, however, the important issues are really moral, ethical, social, and political. Contrary to the commentary in newspapers, television shows, blogs, public discussions, classrooms, and coffee shops, there are still unresolved and even unasked questions about acceptable levels of social and environmental costs, as well as how the costs and benefits of the existing system are distributed within and between generations.[6]

No. 2: Clean Power Offers an Optimal Solution

The solution to these challenges is clear. Four clean power technologies—energy efficiency, renewables, DG, and CHP units—are commercially available and, when all costs and benefits are included, are 2 to 15 times cheaper than conventional fossil fuel and nuclear systems. This is because clean power technologies greatly improve energy security, strengthen industrial competitiveness, enhance local economic growth, and reduce environmental burdens associated with electricity production.

Clean power technologies reduce dependence on foreign sources of fuel, thereby creating a more secure form of supply that minimizes exposure to economic and political changes abroad. They decentralize electricity supply so that an accidental or intentional outage affects a smaller amount of capacity than one at a larger gas, coal, or nuclear facility. Clean power systems improve the reliability of power generation by conserving or producing power close to the end user, and minimize the need to produce, transport, and store hazardous fuels.

The use of clean power reduces the price of fossil fuels and improves the stability of electricity prices. Wind and solar generators lower demand for natural gas and thereby decrease natural gas rates for both electricity users and natural gas consumers. Clean power technologies diversify the energy base, thereby providing more stable energy prices and insulating the industry from price spikes, interruptions, shortages, accidents, delays, and international conflicts. Unlike generators relying on oil, natural gas, uranium, and coal, renewable generators are not subject to the rise and fall of fuel costs. They thus provide a hedge against future environmental regulations (such as a carbon tax) that could make the price of conventional power unexpectedly rise, and improve the competitiveness of the domestic energy sector by offering lucrative export opportunities for American companies to sell clean power technologies overseas. They can also respond more rapidly to supply and demand fluctuations, improving the efficiency of the market because of their modularity.

The construction, installation, and operation of clean power technologies produce economic benefits such as local growth and reduced electricity rates. They improve property tax revenues, provide landowner revenues and local sales dollars, and increase domestic employment.

Most significantly, clean power technologies have environmental benefits since their use tends to avoid air pollution and the dangers and risks of extracting fossil fuels and uranium. They displace or generate electricity without releasing significant quantities of CO_2 and other greenhouse gases that contribute to climate change as well as life-endangering nitrogen oxides, sulfur dioxides, particulate matter, and mercury. They also create power without relying on the extraction of fossil fuels and its associated digging, drilling, mining, transporting, storing, combusting, and reclaiming of land.

No. 3: Clean Power Is Impeded by a Seamless Web of Social Obstacles

Americans want contradictory things from their power technologies. People desire technologies that can take advantage of abundant sources of energy, delivered through reliable distribution systems, but they also yearn for inexpensive prices and minimal harm to the environment. Additionally, people want energy systems to be unobtrusive and invisible. Because the values Americans espouse toward energy and the environment are contradictory, it should come as no surprise that the technologies people support (and oppose) remain inconsistent as well.

Decisions about technologies have always been about more than just technical feasibility. Large-scale technological systems possess significant political, social, economic, and cultural capital, and the reasons for their growth (or decline) are always both social *and* technical. From an intellectual standpoint, the clash between traditional centralized, fossil-fueled electrical generators and novel, dispersed, clean power resources is more than a conflict over technology. It represents a battle about new technology versus unconscious and subtle social impediments; a contest over how best to manage power systems and control a vast and expanding electricity industry; and a competition about conceptions of centralization versus decentralization and profiteering versus social welfare.

In his work on the sociology of technology, Wiebe Bijker once wrote that:

> Purely social relations are to be found only in the imaginations of sociologists, among baboons, or possibly on nudist beaches; and purely technical relations are to be found only in the sophisticated reaches of science fiction. The technical is socially constructed, and the social is technically constructed—all stable ensembles are bound together as much by the technical as by the social. Where there was purity, there is now heterogeneity. Social classes, occupational groups, firms, professions, machines—all are held in place by intimately linked social and technical means . . . Society is not determined by technology, nor is technology determined by society. Both emerge as two sides of the socio-technical coin.[7]

Technologies and social structures, in other words, are mutually interdependent. Technology reflects and influences values and interests but also vice versa. In the electricity sector, lingering utility monopoly rules continue to favor the design, financing, construction, and use of large-scale generators. Managerial practices and methods within the industry continue to rely on fossil-fueled technologies. Even more pervasive, deeply held social attitudes regarding a technological society, material standards of living, and consumption motivate people to use more electricity, but also impede the acceptance of new generators near population centers where they are needed most.

The growth of a technological system as complicated as the electric power grid, encompassing hundreds of thousands of miles of transmission and distribution lines, thousands of large generators, and hundreds of coal mines and LNG terminals, builds constituencies from all reaches of society in order to function. Engineers are needed to repair broken technologies, local utility commissioners to set electricity prices, politicians to enact favorable legislation, banking managers and investors to try to turn a profit, and customers to create demand for electricity. Almost a century of utility preference for large, centralized plants has influenced the way that all of these actors conceive of power generation. Government policy for clean power has helped advance its technical and economic feasibility, but legislative action has occurred inconsistently. Affected by a host of political and ideological factors, parochial and intermittent support has often taken the form of constantly expiring tax benefits. Business managers and political leaders who still retain much of the control of the electric utility system have erected barriers discouraging new participants from entering the industry. The public's lack of knowledge about electricity production and its use often leads to greater opposition to a large number of small, distributed power units (even those that create fewer environmental problems) than to a smaller number of conventional plants located far from load centers, disassociating the more direct problems of fossil fuel combustion and nuclear generation from society.

Clean power systems therefore suffer assault from both sides: they are too radical for some conservative utility managers and investors, but too "ugly," intrusive, and unfamiliar for some environmental activists and property owners. Put differently, the decentralized and modular nature of clean power technologies, which provide huge potential benefits to the utility system, make them especially visible and objectionable. As Americans (and others around the globe) become increasingly aware that all energy technologies, including cleaner ones, contain environmental

Table 13
The Impediments, Challenges, Obstacles, and Barriers Facing Clean Power Technologies in the United States

A jumble of financial, market-based, political, regulatory, cultural, behavioral, aesthetic, and environmental barriers prevent consumers, utilities, politicians, and businesses from embracing clean power technologies.

Category	Barrier	Explanation
Financial and market impediments	*Information failure*	Producers do not distribute accurate or readily available information about clean power projects
		Consumers lack information about clean power technologies, a trend exacerbated by transaction costs and "bounded rationality"
		Real time electricity costs are masked through customer aggregation, average billing, and regulated rate plans
	Returns on investment	Homeowners lack available capital or access to it to purchase clean power technologies
		A large gap exists between private and social discount rates energy investments
		Consumers, businesses, and utilities are more concerned with "first costs" than "lifetime costs"
	Split incentives / Principal Agent Problem	Builders make energy decisions for homeowners
		Landlords make energy decisions for tenants
		Businesses remained focused on core missions and maximizing profit
		Fiscal or regulatory policies discourage energy efficiency
		A limited supply and availability of energy-efficient technologies exists
	Predatory market power	Strenuous interconnection requirements and stranded costs prevent access to the grid
		The intermittent nature of some renewable resources convinces utilities that they are ill-suited to provide base-load and peaking power
		Intellectual property rights, patent blocking, and patent suppression are used to prevent entry into the industry
Political and regulatory obstacles	*Flawed expectations*	Early clean power advocates had inflated hopes and expectations
	Variable and inconsistent incentives	Clean power programs and subsidies, such as the production tax credit, were allowed to expire or were never fully implemented

	Varying state standards	State programs have differing and sometimes contradictory definitions, standards, goals, and requirements for clean power
	Underfunded R&D	Public funding of R&D has declined precipitously since the 1980s Private funding of R&D has been reduced as utilities and energy companies consolidate and restructure for competitive electricity markets
	Hubris	A "top-down" approach to energy R&D plagues DOE programs on wind and solar systems
Cultural and behavioral barriers	*Public apathy and misunder-standing*	People remain uninformed and apathetic about electricity technologies and express preference for familiar energy systems
	Consumption and abundance	Historical antagonism toward nature, industriousness, industrialization, and the promotion of leisure have resulted in values predisposed toward excess consumption and waste
	Psychological resistance	Comfort, freedom, control, and trust are prioritized more than energy conservation and clean power use
Aesthetic and environ-mental challenges	*Environmental costs not included in the price of power*	Consumers cannot make rational comparisons between conventional and clean power sources
	Environmental objections to clean power technologies	Clean power technologies are believed to be aesthetically unpleasing and to harm the environment and degrade land and property
	Symbolism	Clean power technologies can symbolize distrust in government or a clash between rural/urban and rich/poor constituents
	Internal fighting	Clean power technology firms fight amongst themselves instead of coordinating policy

downsides, the new distributed systems bring painfully to the foreground what previously seemed "invisible."

Such social factors have erected an immense cultural and technological commitment to the existing electricity industry, honed and embedded through decades of operation, convincing utilities that the present system was the *best* model for generating, distributing, and selling electricity. And so, Americans prioritize abundant supply and cheap prices as important energy needs. To meet these needs, the electric utility sector has been able to promote abundance and cheapness only with the assistance of regulators and politicians, who subsidize all forms of energy to shield consumers from the true costs of energy extraction, generation, distribution, and use. Thus, the dominant electric utility system has become a naturalized, taken-for-granted component of our modern landscape, barely recognized by most customers.[8]

No. 4: History Tells Us This Is Nothing New

It is often mistakenly assumed that technological evolution occurs in a Darwinian world in which all technological possibilities begin on equal footing and advance or stagnate according to relative efficiency or social merit.[9] Such an idyllic notion obscures the fact that each of the obstacles, barriers, impediments, and challenges facing clean power technologies existed, in some form, for conventional technologies. The diffusion of energy technologies, historically, has been a slow process, ranging from a decade or two to more than a century. In an analysis of the market penetration of 20 different energy technologies, Peter Lund found that market penetration can take anywhere from less than 10 years to more than 70 years.[10] The diesel engine, for example, needed about 60 years before it became fully embraced as a commercial technology. The transition from wood to coal challenged an energy system based on local public resources such as forests and replaced it with one based on commercial enterprises. The shift from a regionally based coal system dependent on public rail transport to an oil system based on international production and trade necessitated the successful enlistment of state and national governments to enforce low wages through anti-union labor policies and government assistance in the organization of markets and the development of infrastructure.

Even the existing, naturalized electric utility industry departed from the past norms and practices to a new set of rules based on monopolization of the market, public financing of rural electrification programs and large hydroelectric projects, redistribution of capital risk so that private ventures were publicly subsidized, and the emergence of the American military as a major force in energy policy making. Over the years, the electric utility industry has been augmented and reinforced by the seemingly natural roles and responsibilities fulfilled by electricity producers and consumers. It has supported the distribution and alignment of political resources in a manner favorable to conventional technologies, and erected financial and capital infrastructure geared to serve a centralized system. It has resulted in a body of law and public policy that legitimizes and incentivizes current patterns of production and use.[11]

The ability for policymakers in business and government to overcome the immense infrastructural challenges facing the coal, oil, and electricity industries suggests that, given the right mix of incentives, similar obstacles could be overcome for clean power technologies. The "energy problem," then, is not so much a failure of technical capacity, hard work, or imagination, but one of policy and political will.[12]

No. 5: Comprehensive Electricity Policy Changes Are Needed

The tendency for the impediments to clean power technologies to be social rather than technical means that policymakers continue to promote alternative

Table 14
The Big Four Policy Mechanisms

Concerted policy action consisting of a national FIT, repeal of conventional energy subsidies, accurate electricity pricing, and a national systems benefit charge would do much to eliminate the barriers facing clean power systems.

Impediment	Policy	Details
Utilities and businesses will not invest in clean power	Make clean power mandatory	Create a national feed-in tariff (FIT) and guarantee clean power access to the grid
Political support for clean power has been inconsistent and unfair	Eliminate subsidies	Immediately repeal federal government subsidies for nuclear and fossil fuel technologies
Consumers do not receive accurate price signals for electricity	Get the price right	Abolish electricity rate caps, eliminate declining block rate pricing, reflect time of use in electricity rates, and internalize external costs
Property owners and environmentalists are uninformed about clean power technologies	Inform the public and protect the poor	Establish a national systems benefit charge (SBC) to generate revenue to distribute information and educate the public, provide low-income assistance and weatherization, and promote energy efficiency and DSM programs

technologies in the wrong way. Many renewable energy technologies, notably wind turbines and biomass generators, have already caught up to fossil-fueled generators in terms of their levelized costs (despite years of unfair subsidies and inaccurate electricity rates). Many more would clearly compete if externalities were included in the price of electricity. Instead of creating government incentives that aim to further increase the efficiency and technical capacity of such systems, policymakers should shift their focus away from the technical to focus on efforts aimed at increasing public understanding of energy systems and challenging entrenched utility practices.

Because the barriers facing clean power are diffuse, a multitude of policies must be *comprehensively* implemented to eliminate them. As Lee Schipper put it more than 30 years ago, "the non-technical barriers to efficient energy utilization are many, and removing these barriers entails political action as well as straightening out of the economic system in nearly every phase of social activity, because all activity today uses energy." [13] Carl Blumstein and his colleagues from the University of California add that the barriers to clean power are "deeply embedded in the social and institutional fabric; they are not only tenacious but resilient, and when broken down, they tend to reappear in altered form." [14] Chapter 7 details that federal policymakers should implement four mechanisms to properly encourage clean power sources: they must make clean power mandatory through a national FIT, eliminate subsidies for conventional energy technologies, price electricity accurately, and establish a national SBC to distribute public information, provide low-income assistance, and promote energy efficiency.

Historical experience, numerous quantitative and qualitative studies, and perhaps common sense suggest that implementing just one or two of these mechanisms independently will be insufficient. Making clean power mandatory without changing electricity prices will not send proper price signals to consumers. Removing subsidies does little to eliminate the market power already afforded to conventional systems. More accurate electricity prices without low-income assistance will hurt poorer families the most. Distributing information about clean power without internalizing the costs of negative externalities erodes the incentive to follow through with energy efficiency investments.

No. 6: Recognize the Politicized Nature of Electricity

Because of the vast and entrenched interests associated with the existing system, electricity is highly politicized, and reaching agreement about electricity policy is unlikely to occur. Greenpeace estimated that American oil, natural gas, and coal companies spend approximately $31 million *every year* on lobbying and campaign contributions.[15] During important elections, the numbers jump significantly: oil and gas companies contributed about $255 million to political campaigns and electric utilities an additional $20 million for the 2004 presidential election cycle. From 2003 to 2006, fossil fuel lobbyists contributed about $58 million to state-level campaigns alone. Over the same period, renewable energy lobbyists spent just $500,000.[16]

Even if clean power advocates matched the expenditures from the oil, gas, coal, and nuclear lobbies, electricity and energy issues would remain polemical. Historian John G. Clark put it best when he noted that:

> There is no end to the energy problem. The only reasonable basis for predicting the future course of energy developments and the repercussions of such development is the historical record. But this record speaks with a forked tongue . . . there is something in it for everyone.[17]

The energy crisis of 1973 itself demonstrates the inability to forge consensus. The American public tended to see the oil crisis primarily as a maneuver of oil companies to reap the benefits of higher prices. Anti-environmentalists saw it as a blessing of their war against environmental organizations (such as the Sierra Club) and their allies. Environmentalists saw high prices as a boon for discouraging oil consumption. Independent oil companies saw it as a way to attack the "major" oil companies such as Exxon and Shell. Tax reformers saw it as a chance to end certain provisions that benefited the major oil companies abroad. In essence, there was something in the energy crisis for everyone, but none of "it" was consistent.[18]

A similar survey of 130 "energy advocates" found that many had diametrically opposed views. When asked about the perceived causes of the nation's energy problem, some strongly believed that we waste too much energy while others firmly argued that the restrictions in production were to blame. When asked what technologies the United States should focus on developing, some said exclusively conservation and solar power while others said exclusively fossil fuels and nuclear power. Asked if economic growth required increased energy consumption, 90 percent of "solar" activists said economic growth was possible with reduced consumption, but 80 percent of "conventional" activists felt that economic growth demanded increased consumption.

Surprisingly, all "advocates" had exceptionally high levels of education, had similar income, and represented a broad spectrum of interests, implying that their views could not be explained by education, class, or political affiliation. The researchers concluded that two virtually dichotomous networks of energy researchers were emerging, separated by deep epistemic gulfs.[19]

Another study also found that "education" did little to ameliorate differences between advocates of conventional and alternative energy technologies. Researchers discovered that the two groups that had the most in common, petrochemical industry executives and conservationists, were on both extremes of the energy policy gamut. Both groups were constituted with educated, fairly affluent, strongly committed individuals who believed the nation was confronting a serious and lasting crisis; they just concluded the *opposite* about what should be done about it.[20]

Two more recent events demonstrate the politicized nature of electricity reform. The first is the rise of faux "grass roots" consumer organizations that turn out to be funded entirely by utilities and energy companies wishing to shape public opinion. About half the states get more than 50 percent of their electricity from coal, and some corporations have proven quite devious about trying to keep demand for coal-fired electricity surging (see Map 17). When state environmental officials rejected two coal-fired power plants in Kansas because of the millions of tons of carbon dioxide they would produce in 2008, a pro-coal lobbying group, Kansans for Affordable Energy, was soon formed. The group, funded mostly from Peabody Coal (the supplier of coal to Kansas from its mines in Wyoming) and Sunflower Electric Corporation (the local utility), placed newspaper advertisements with pictures of the smiling faces of Presidents Mahmoud Ahmadinejad of Iran, Vladimir Putin from Russia, and Hugo Chavez from Venezuela, suggesting that if the coal plants were nixed, these natural gas exporting countries would benefit. Even though the ads were completely false (not one of those countries exports natural gas to Kansas), the campaign convinced the state legislature to approve the coal plants, a move that almost succeeded until it was vetoed by Governor Kathleen Sebelius.[21] In Illinois, when Commonwealth Edison, a subsidiary of Exelon, was confronted with the possibility of an extension of electricity rate caps to protect consumers in 2007, the utility spent $10 million to form Consumers Organized for Reliable Electricity. The group ran a comprehensive array of television and newspaper ads taking a stance against the rate freeze "on behalf of the public," but never once mentioned their ties

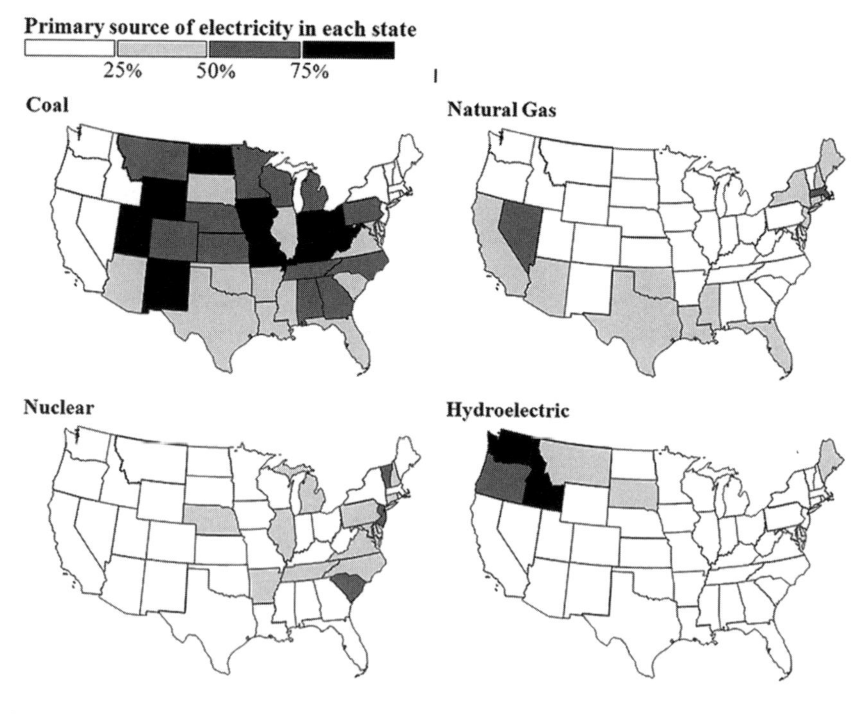

Map 17
Coal, Natural Gas, Nuclear, and Hydroelectric States, 2007

This map shows how each of the lower 48 states produces its electricity. More than half of all states receive more than half of their electricity from coal-fired power plants. The map also gives a rough idea of the political power afforded to each fuel; coal and natural gas, for instance, have strong political lobbies because they are used in most states, while nuclear and hydroelectric sources have less support because their use is more contained.

Source: Felicity Barringer, "States' Battles over Energy Grow Fiercer with U.S. in a Policy Gridlock," *New York Times,* March 20, 2008, 3.

to the utility.[22] And, nationally, the American coal lobby has responded to all of the negative publicity connecting coal use to climate change by creating two sleek non-profit groups termed "Americans for Balanced Energy Choices" and the "American Coalition for Clean Coal Electricity." These organizations remain dedicated to "correcting misinformation about coal" and reminding the public that "coal keeps America's economy running strong, protects our natural resources, and ensures our nation's energy security." (This claim about "protecting" natural resources may strike readers as exceptionally peculiar, since coal mining is the essence of extracting coal from the earth, not preserving it there.)[23] Almost every aspect of the Kansan power plant permitting process, the proposed electricity rate freeze in Illinois, and the American coal lobby's reaction to climate change, in other words, was highly political.

The second relates to efforts to repeal energy subsidies. The U.S. House of Representatives tried to repeal $22 billion worth of subsidies for the oil and gas industry in 2007 during preliminary discussions of the Energy Independence and Security Act. The Senate, however, squashed the idea under direct threat of a veto from the White House when finalizing the bill.[24] Even the president, it appears, gets personally involved in protecting the interests of existing energy companies.

These studies and recent examples suggest that the true contest between conventional and clean power paths has little to do with the technologies themselves, and much more with the way that people *think* about electricity and energy. In a theoretical sense, fossil fuels, nuclear, renewables, DG, CHP, and energy efficiency can be pursued simultaneously and coexist harmoniously. One could conceivably integrate a collection of 1 kW solar panels, for instance, on the cooling tower of 1,000 MW nuclear facility. Yet while both paths may not be theoretically exclusive, they are still "culturally and institutionally antagonistic." [25] The tools required for each inhibit the availability of the other.

The philosophical ideas behind clean power attempt to recognize that people are more important than goods. Electricity, energy, and technology are seen as a means, not an ends. Clean power technologies endorse a philosophy of "nature knows best" and recognize that because we know little of the natural world, we should take care to preserve it until we know the true consequences of our actions. Clean power promotes the idea that the national interests of the country lie less in traditional geopolitical balancing acts and big technology than in striving to attain a just and equitable world, even at the expense of commercial advantage.

These views are incommensurable with the dominant philosophy underpinning conventional technologies. The current approach sees the energy problem consisting of how to expand available domestic resources to meet anticipated demands. It assumes that electricity consumption is essential to economic growth, demands larger and more centralized facilities, and places faith in the ability of technological ingenuity to overcome the vexing problem of resource depletion.

One path values centrism, autarchy, vulnerability, and technocracy, where the other is antithetical to each. The two paths are as follows:

- Culturally incompatible, each entailing an evolution process that makes the other kind of world hard to imagine;
- Institutionally incompatible, each requiring organizations and policy actions that inhibit each other;
- Logistically incompatible, requiring money, materials, fuels, work, skills, political attention, time, and commitment in a world of finite and limited resources;
- Physically incompatible, as many investments mark the landscape with capital-intensive projects that narrow future choice through their physical irreversibility.

As a pernicious example of this last form of incapability—physical—consider the recent experience in Ontario, Canada, where the Ontario Power Authority (OPA) has purposely stalled renewable energy deployment so that nuclear power plants can come online. The OPA, which is planning to construct 14,000 MW of new nuclear capacity over the next two decades, decided to halt key aspects of its FIT in 2008, placing a moratorium on all new renewable power contracts over 10 kW for many parts of Ontario and limiting developers to one 10 MW project per transformer station across the entire service area. Ostensibly, OPA officials are reserving all remaining T&D capacity for the expansion of nuclear power in the provinces at the direct expense of solar and wind development.[26]

Where we are now illustrates the path dependence and burgeoning exclusivity associated with the conventional system: just the past few decades of commitments to traditional technologies have already given society many analysts who cannot imagine the industry any other way, an environment populated by rigid institutions and utilities that resist pressure to reform. A utility investing $10 billion dollars in two nuclear reactors does not *want* such exclusivity to wane. They prefer to be certain that they can get an operating license in 2030 and run to 2050 without unexpected and onerous future changes.[27]

Consequently, the choice between conventional and clean power systems requires completely distinct orders of thinking and values. A policy based on depleting fossil fuels and polluting the environment is more than just "different" from one utilizing nondepleting and nonpolluting ones. Living off clean power differs from depleting fossil fuels as chess differs from checkers. The very rules of the game are different, though the board on which the two games are played looks the same. One game recognizes permanence and ecological discipline as rules restricting legitimate moves. The other game has no such rules. A good move in the checkers game of digging, damming, and drilling to extract energy resources is not usually a good move in the chess game of pursuing more benign and efficient clean power sources. In checkers, all pieces are of equal power; in chess, essential and qualitative differences exist among the capacities of each piece. A definite decision must be made about the direction we move, clean or dirty, before we decide about the rate at which we move in that direction.[28] One strategy is based on depleting natural resources, the other minimizing their depletion; both strategies have their own tactics, but those tactics make sense *only* if one game is being played—not both.

This tendency is illustrated most clearly with an analogy from mass media and communication studies.[29] Some *forms* of communication essentially limit the *type* of communication possible. Neil Postman, the late professor of media ecology at New York University, often stated that smoke signals used by Native Americans, for example, would not enable the discussion of philosophical argument or narrative. Puffs of smoke are insufficiently complex to express ideas on the nature of existence, and Cherokee philosophers would run short of either wood or blankets before they reached their second axiom. The form of smoke signals thus excludes the content of philosophy.

In much the same way, the "form" of conventional energy technologies excludes the "content" of clean power (or meaningful attempts to promote clean alternatives). As an example, consider the mutual exclusivity of building new nuclear power stations or promoting DG.[30] Building new nuclear power stations locks countries into a centralized and inflexible paradigm that necessitates transmission-based infrastructure, passive distribution networks delivering power one way, and relatively few units. DG, in contrast, has exactly the opposite characteristics and utilizes transmission networks that balance power flows around the system, support active T&D with two way flows of power, and endorse a wide range of generating technologies diversified by location and source. New nuclear plants reduce pressure for a network of DG units, consume sparse R&D funds, and minimize chances for behavioral change; DG does the same for nuclear power.

As a result, consensus on electricity issues should be neither expected nor sought. The physicist Max Planck often said that "a new truth does not triumph by convincing its opponents and making them see the light, but rather because its opponents eventually die, and a new generation grows up that is familiar with it." [31] Electricity policy is a site of conflict, not cooperation, and encompasses matrices of different interests and ideologies. Organized consumer groups contend with organized producer groups to question energy rates and prices. Business and labor groups in the different industries, such as nuclear, oil, coal, and gas, compete for preferential treatment by governments. Federal and state governments differ over the division of authority. Geographic regions seek to influence policy in ways favoring their situations as energy producers or consumers.[32]

The former Committee on Nuclear and Alternative Energy Systems, a panel set up by the National Academies in the 1970s to look at the energy crisis, vividly demonstrates the difficulties faced by any group trying to arrive at consensus on electricity issues. "It simply can't be done," its Chair, Harvey Brooks, concluded, "at least not within any group that honestly represents the spectrum of defensible views in today's academic, intellectual, and industrial community." [33] Physicist Amory Lovins reached a similar conclusion when he remarked that:

> Underlying much of the energy debate is a tacit, implicit divergence on what the energy problem "really" is. Public discourse suffers because our society has mechanisms only for resolving conflicting interests, not conflicting views of reality . . . Most countries, especially the U.S., are full of diverse people unable to agree about the nature of the energy problem, the role of governance and of price, the future of big oil companies, and the desirable direction of social evolution. If resolving these conflicts is a prerequisite to addressing the energy problem, hell will freeze over first.[34]

Put in slightly different terms, there is a deep conceptual split between different stakeholders that makes harmony and unanimity in electricity policy functionally impossible without unforeseen changes in industry interests and values.

Concluding Thoughts

The long-standing and self-evident vision of the American electric utility sector has been to achieve the lowest possible unit price of electricity through the fullest exploitation of resources and maximum consumption.[35] Cheap electricity, following this paradigm, has attracted a stronger constituency than social welfare.[36] Adherence to such a singular goal, however, has downplayed and sometimes ignored far more important issues relating to the *quality* of energy and electricity as well as the *quality* of human life. We need to move beyond such a quantitative approach that prioritizes cheapness, exploitation, and consumption to a new qualitative framework appreciating people and the natural world. This type of approach would afford value to power technologies based on their full social cost instead of their market power or compatibility with convention.

It is perhaps here that the fundamental strength of "the electricity problem," the capital intensity and long-lived nature of electrical generators, can be turned to an advantage. Every plant built now will be with us for at least 30 years. During most of our lifetimes, we will witness a complete turnover of building stock or power plant capacity just one or two times. Thankfully, our prospect is not determined by destiny, the future is not fixed, and much of this capacity has yet to be built. But any transition to clean power technologies will largely be determined by the choices made in the next few decades. If certain decisions are not made today, then the path toward a clean power future slowly becomes locked out.

Basically, the issue boils down to two simple questions: Do we want a nuclear or fossil fuel economy, centrally administered by technical specialists completely dependent on research subsidies and future technological breakthroughs, sure to promote international proliferation and worsen inequity and vulnerability, which requires draconian security measures, wastefully generates and distributes electricity, must uselessly degrade energy to fit the majority of end uses, remains based on highly uncertain projections about fuel availability, and trashes the planet for many future generations? Or do we want a small to medium scale, decentralized electricity system that is more efficient, virtually free from subsidization, subject to local control by the same people who want the energy, operates with minimal disruption of ecological services, remains resilient to disruptions and terrorist assaults, is equally available to all future generations, and is highly beneficial to all income groups?[37]

If we want the second, a system based on elegance rather than brute force, there are many things that each of us can do. As homeowners we can install geothermal heat pumps, integrate solar panels into our roof, and improve the energy efficiency of our residences with CFLs, appliances, insulation, and a variety of other technologies and devices. As electricity customers, we can demand that utilities offer green power programs, purchase renewable energy credits, or operate clean power plants. As taxpayers and voters, we can write letters, march, and cast a local, state, or national ballot for those candidates that we believe can best promote clean power. As employees, we can request changes at the workplace. As employers or managers, we can start focusing more on energy savings. As shareholders and investors, we can support clean power companies. As citizens, we can participate in meetings, permitting and siting discussions, and state hearings. As parents, we can educate our children, and as children, we can educate our parents.

Hubris takes many forms, not just arrogantly overreaching, but also stubbornly holding back. In the end, the answer to the question "what is blocking clean power?" is simple: "we all are."

NOTES

Preface

1. For an outstanding introduction to renewable energy generally, see Godfrey Boyle, *Energy Systems and Sustainability* (Oxford: Oxford University Press, 2003), 3–6.
2. The most important of these was my dissertation, which was the starting point for the book, and is also available online. Readers are encouraged to see *The Power Production Paradox: Revealing the Socio-technical Impediments to Distributed Generation Technologies,* available at http://scholar.lib.vt.edu/theses/available/etd-04202006-172936/. Readers are also invited to sample *Energy and American Society: Thirteen Myths,* ed. with Marilyn A. Brown (New York: Springer Press, 2007), xi and 340pp; *Carbon Lock-In: Barriers to the Deployment of Climate Change Mitigation Technologies,* with Marilyn A. Brown, Jess Chandler, and Melissa V. Lapsa (Oak Ridge, TN: Oak Ridge National Laboratory, ORNL/TM-2007/124, November 2007), 174pp; *Renewing America: The Case for a National Renewable Portfolio Standard,* with Christopher Cooper (New York: Network for New Energy Choices, 2007), ix and 158pp; *A Study of Increased Use of Renewable Energy Resources in Virginia,* with Mike Karmis, Jason Abiecunas, Jeffrey Alwang, Stephen Aultman, Lori Bird, Paul Denholm, Donna Heimiller, Richard F. Hirsh, Anelia Milbrandt, Ryan Pletka, and Gian Porro (Blacksburg, VA: Virginia Center for Coal and Energy Research, Report for the Virginia Commission on Electrical Utility Restructuring November 11, 2005), 107pp; "Promoting a Level Playing Field for Energy Options: Electricity Alternatives and the Case of the Indian Point Energy Center," with Marilyn Brown, *Energy Efficiency* 1, no. 1 (2008): 35–48; "Valuing the Greenhouse Gas Emissions from Nuclear Power: A Critical Survey," *Energy Policy* 36, no. 8 (August 2008): 2940–2953; "Island Wind-Hydrogen Energy: A Significant Potential U.S. Resource," with Richard Hirsh, *Renewable Energy* 33, no. 8 (August 2008): 1928–1935; "The Best of Both Worlds: Environmental Federalism and the Need for Federal Action on Renewable Energy and Climate Change," *Stanford Environmental Law Journal* 27, no. 2 (June 2008): 397–476; "The Costs of Failure: A Preliminary Assessment of Major Energy Accidents, 1907 to 2007," *Energy Policy* 36, no. 5 (May 2008): 1802–1820; "Is the Danish Wind Energy Model Replicable for Other

Countries?" with Hans Henrik Lindboe and Ole Odgaard, *Electricity Journal* 21, no. 2 (March 2008): 27–38; "Distributed Generation (DG) and the American Electric Utility System: What Is Stopping It?" *Journal of Energy Resources Technology* 130, no. 1 (March 2008): 16–25; "Replacing Tedium with Transformation: Why the U.S. Department of Energy Needs to Change the Way it Conducts Long-term R&D," *Energy Policy* 36, no. 3 (March 2008): 923–928; "A Matter of Stability and Equity: The Case for Federal Action on Renewable Portfolio Standards in the U.S.," *Energy & Environment* 19, no. 2 (February 2008): 241–261; "Distributed Generation in the US—Three Lessons," *Cogeneration and Onsite Power Production* 10, no. 1 (January/February 2008): 69–72; "Developing an 'Energy Sustainability Index' to Evaluate Energy Policy," with Marilyn Brown, *Interdisciplinary Science Reviews* 32, no. 4 (December 2007): 335–349; "Solving the Oil Independence Problem: Is it Possible?" *Energy Policy* 35, no. 11 (November 2007): 5505–5514; "State Efforts to Promote Renewable Energy: Tripping the Horse with the Cart?" with Christopher Cooper, *Sustainable Development Law & Policy* 8, no. 1 (Fall 2007): 5–9, 78; "Necessary but Insufficient: State Renewable Portfolio Standards and Climate Change Policies," with Jack Barkenbus, *Environment* 49, no. 6 (July/August 2007): 20–30; "Coal and Nuclear Technologies: Creating a False Dichotomy for American Energy Policy," *Policy Sciences* 40, no. 2 (June 2007): 101–122; "Big *Is* Beautiful: The Case for Federal Leadership on a National Renewable Portfolio Standard," with Christopher Cooper, *Electricity Journal* 20, no. 4 (May 2007): 48–61; "Abandoned Treaties, Environmental Damage, Fossil Fuel Dependence: The Coming Costs of Oil and Gas Exploration in the Arctic National Wildlife Refuge," *Environment, Development, and Sustainability* 9, no. 2 (May 2007): 187–201; "Green Means 'Go?' A Colorful Approach to a National Renewable Portfolio Standard," with Christopher Cooper, *Electricity Journal* 19, no. 7 (August/September 2006): 19–32; "Assessing U.S. Energy Policy," with Marilyn Brown and Richard Hirsh, *Daedalus: Journal of the American Academy of Arts and Sciences* 135, no. 3 (Summer 2006): 5–11; "Eroding Wilderness: The Ecological, Legal, Political, and Social Consequences to Oil and Natural Gas Exploration in the Arctic National Wildlife Refuge (ANWR)," *Energy & Environment* 17, no. 4 (April 2006): 549–567; "Using Distributed Generation and Renewable Energy Systems to Empower Developing Countries," *The International Journal of Environmental, Cultural, Economic, and Social Sustainability* 2, no. 3 (2006): 77–86; "PUHCA Repeal: Higher Prices, Less R&D, and More Market Abuses?" *Electricity Journal* 19, no. 1 (January/February 2006): 85–89; "Technological Systems and Momentum Change: American Electric Utilities, Restructuring, and Distributed Generation," with Richard F. Hirsh, *Journal of Technology Studies* 32, no. 2 (Spring 2006): 72–85; "Why Has There Been a Relative Failure of Renewable Energy Systems in the USA?" in *Renewable Energy 2006*, ed. David Flin (London: World Renewable Energy Network and United Nations Educational, Scientific, and Cultural Organization, 2006), 24–32; "Think Again: Nuclear Power," *Foreign Policy* 150 (September/October 2005), available at http://www.foreignpolicy.com/story/cms.php?story_id=3250. Some, though not all, of these publications can be found at the Virginia Tech Consortium on Energy Restructuring's Web site (http://www.history.vt.edu/Hirsh/CERdocuments.html) and the Centre on Asia and Globalisation's Web site (http://www.spp.nus.edu.sg/cag/Books_Articles.aspx). All of these articles (and more!) are on file with the author, and interested readers are invited to email him directly to request copies.

3. Walt Patterson, *Keeping the Lights On: Towards Sustainable Electricity* (London: Earthscan, 2007).
4. Sovacool, *Power Production Paradox.*

Introduction

1. To reach the 913 tons per hour mark, I worked backwards from Lawrence Livermore National Laboratory's analysis of 5 to 8 million tons per year. Eight million tons per year means 21,918 tons per day, or 913.24 tons per hour. Data taken from S. Julio Friedman, "Technical Feasibility of Rapid Deployment of Geological Carbon Sequestration," *Hearing Before the House Energy and Commerce Committee* (March 6, 2007).
2. Using data from 2006, the U.S. Energy Information Administration reports that each residential customer in the United States uses approximately 920 kilowatt-hours of electricity per month, or 31 kWh per day. The National Energy Technology Laboratory calculates that every single kWh of electricity requires about 25 gallons of water to produce. Data taken from the 2006 *Electric Power Annual* and Jeff Hoffmann, Tom Feeley, and Barbara Carney, "DOE/NETL's Power Plant Water Management R&D Program—Responding to Emerging Issues" (presentation at the 8th Electric Utilities Environmental Conference, Tucson, AZ, January 24–26, 2005).
3. In 2003, the U.S. Department of Energy estimated that power outages and power quality disturbances cost the customers $180 billion annually in $2003. One gets $206 billion when adjusting the amount for inflation, whereas the Department of Education's budget was only $56 billion. See U.S. Department of Energy, Office of Electric Transmission and Distribution, *Transforming the Grid to Revolutionize Electric Power in North America* (Washington, DC: U.S. Department of Energy, July 2003), 3–5.
4. John A. "Skip" Laitner, *Testimony Before the House Committee on Science and Technology* (September 25, 2007), 4.
5. Harvard School of Public Health, *Impact of Pollution on Public Health* (January 3, 2001), quoting a study conducted by John Spengler and Jonathan Levy entitled *Estimated Public Health Impacts of Criteria Pollutant Air Emissions from Nine Fossil-Fueled Power Plants in Illinois.*
6. According to the EIA, utilities spent $58.34 billion on natural gas, $38.76 billion on coal, and $7.27 billion on oil in 2006. Since about two-thirds of this fuel is "wasted," one can argue that consumers pay $68.9 billion too much for their power every year, or $689 billion every decade. Data taken from Table 4.5, "Receipts, Average Cost, and Quality of Fossil Fuels for the Electric Power Industry, 1995 to 2006," from the U.S. EIA's *Electric Power Annual.* For similar estimates, see Richard Munson, *From Edison to Enron: The Business of Power and What It Means for the Future of Electricity* (New York: Praeger, 2005), 4; and Thomas R. Casten, *Turning Off the Heat: Why Americans Must Double Energy Efficiency to Save Money and Reduce Global Warming* (New York: Prometheus, 1998), 3–5.
7. Chapter 1 fully explains this calculation when talking about externalities.
8. Chapter 2 explains this estimate when talking about renewable resource potential.
9. According to the EIA, the country's residences consumed 1,351,520,036 megawatt-hours (MWh) in 2006 and had 122,471,071 customers. This equates to 11.03542269 MWh/res customer per year, or 919.6 kWh/res customer per month. Each home thus consumes about 30 kWh per day, and renewables would provide just

2.7 percent of this electricity, or 0.81 kWh. Two 60-watt light bulbs running for 10 hours would consume around 1.2 kWh, more than all the daily energy provided by wind, geothermal, solar, and biomass. See U.S. EIA, *Summary Statistics for the U.S.* (Washington, DC: U.S. Department of Energy, 2006), Table ES1 and Table 7.1.

10. Running the hot-water faucet for 5 minutes uses as much energy as burning a 60 W light bulb for 14 hours. See Peter H. Gleick, *Water and Energy (and Climate): Critical Links* (Oakland, CA: Pacific Institute, February 2008), p. 6.
11. Quoted in Benjamin K. Sovacool, *The Power Production Paradox: Revealing the Socio-technical Impediments to Distributed Generation Technologies* (doctoral dissertation, Virginia Polytechnic Institute & State University, Blacksburg, VA, April 2006).
12. Ibid.
13. Ibid.
14. Lee Raymond, "The Global Outlook for Petroleum" (speech at the World Petroleum Congress, Beijing, October 13, 1997).
15. Benjamin K. Sovacool, "Distributed Generation (DG) and the American Electric Utility System: What is Stopping It?" *Journal of Energy Resources Technology* 130, no. 1 (March 2008).
16. Ibid.
17. Ibid.
18. See David Moskovitz, "Why Regulatory Reform for DSM?" in *Regulatory Incentives for Demand-Side Management,* ed. Steven Nadel, Michael W. Reid, and David R. Wolcott (Washington, DC: American Council for an Energy-Efficient Economy, 1992), 1; and Alan H. Sanstad and Richard B. Howarth, " 'Normal' Markets, Market Imperfections and Energy Efficiency," *Energy Policy* 22, no. 10 (1994): 811–818.
19. Amory B. Lovins, L. Hunter Lovins, and Paul Hawken, "A Road Map for Natural Capitalism," *Harvard Business Review* (May/June 1999): 148–149.
20. Ralph Cavanagh, *Energy Efficiency in Buildings and Equipment: Remedies for Pervasive Market Failures* (Washington, DC: National Commission on Energy Policy, 2003).
21. Steven Nadel, R. Neal Elliott, and Therese Langer, *A Choice of Two Paths: Energy Savings Pending Federal Energy Legislation* (Washington, DC: ACEEE, 2005).
22. John Dernbach, "Stabilizing and Then Reducing U.S. Energy Consumption: Legal and Policy Tools for Efficiency and Conservation," *Environmental Law Review* 37 (2007): 10003–100031.
23. Karen Ehrhardt-Martinez and John A. "Skip" Laitner, *The Size of the U.S. Energy Efficiency Market: Generating a More Complete Picture* (Washington, DC: ACEEE, May 2008).
24. EIA, *2003 Electric Power Annual,* Table ES, 5.
25. Richard F. Hirsh, *Technology and Transformation in the American Electric Utility Industry* (Cambridge: Cambridge University Press, 1989), 9. Prices have been adjusted to $2007.
26. For an excellent introduction to conventional vs. alternative conceptions of producing power, see Amory Lovins et al., *Small Is Profitable: The Hidden Benefits of Making Electrical Resources the Right Size* (Snowmass, CO: Rocky Mountain Institute, 2002).
27. Brian O'Shaugnessy, "Power Generation Resource Incentives and Diversity," *Hearing Before the Committee on Energy and Natural Resources of the U.S. Senate* (May 8, 2005) (Washington, DC: Government Printing Office, National Energy Policy Development Group, 2005), 68.

28. Electric Power Research Institute, *Electricity Technology Roadmap: Meeting the Critical Challenges of the 21st Century* (New York: EPRI, 2003), 3.
29. Daniel Gross, "Solving 'Fission Impossible,'" *Newsweek,* October 29, 2007, E24.
30. David K. Garman, "Electricity Generation Portfolio Standards," *Testimony before the Senate Committee on Energy and Natural Resources* (March 8, 2005).
31. Amory Lovins, "Energy Strategy: The Road Not Taken," *Foreign Affairs* 55 (1976/1977): 93.
32. For outstanding explorations of the history of the electric utility system and how it has resisted change, see Thomas P. Hughes, *Networks of Power: Electrification in Western Society, 1880–1930* (Baltimore: Johns Hopkins University Press, 1993) (arguing that the existing system developed "momentum" with social and political actors as it expanded); and Richard F. Hirsh, *Power Loss: The Origins of Deregulation and Restructuring in the American Electric Utility System* (Cambridge, MA: MIT Press, 1999) (arguing that utility executives often take a conservative approach toward managing their operations).
33. F. Paul Bland, "Problems of Price and Transportation: Two Proposals to Encourage Competition from Alternative Energy Sources," *Harvard Environmental Law Review* 10 (1986): 345–386.
34. Nancy Rader and Richard Norgaard, "Efficiency and Sustainability in Restructured Electricity Markets," *Electricity Journal* 9, no. 6 (July 1996): 40.
35. Loren Lutzenhiser, "Innovation and Organizational Networks: Barriers to Energy Efficiency in the US Housing Industry," *Energy Policy* 22 (1994): 867–876.
36. Hughes, *Networks of Power;* David E. Nye, *Electrifying America: Social Meanings of a New Technology, 1880–1940* (London: MIT Press, 1990).
37. David E. Nye, *Consuming Power: A Social History of American Energies* (London: MIT Press, 1999).
38. Vaclav Smil, *Energy in World History* (New York: Westview, 1994); Martin V. Melosi, *Coping with Abundance: Energy and Environment in Industrial America* (New York: Knopf, 1985).
39. U.S. Energy Information Administration, *International Energy Outlook* (May 2007), Tables H1 and H7.
40. U.S. Department of Transportation, *Number of U.S. Aircraft, Vehicles, Vessels, and Other Conveyances* (Washington, DC: Bureau of Transportation Statistics, 2007), Table 1.11.
41. U.S. Department of Energy Office of Policy and International Affairs and Oak Ridge National Laboratory, *Behavioral Research Workshop: Residential Buildings Energy Efficiency Draft Summary Report* (February 26, 2008), 8.
42. Marilyn A. Brown, "Ending the Energy Stalemate: Facts and Figures from the National Commission on Energy Policy" (presentation to the Oak Ridge Center for Advanced Studies Seminar, June 30, 2005), 15.

Chapter 1

1. Donald L. Klass, "A Critical Assessment of Renewable Energy Usage in the USA," *Energy Policy* 31 (2003): 353–367; T. L. Shaw, "Renewable Energy Sources in Perspective," in *Policy and Development of Energy Resources,* ed. T. L. Shaw, D. E. Lennard, and P. M. S. Jones (New York: Wiley & Sons, 1984), 171–192.

2. Jan Berry and Steve Fischer, "More Than 200 Hospitals Nationwide Are Recycling Energy for Peak Performance," *Distributed Energy* (January/February 2004): 32.
3. This excellent example has been borrowed from Patricia Nelson Limerick, Claudia Puska, Andrew Hildner, and Eric Skovsted, *What Every Westerner Should Know About Energy* (Boulder, CO: Center of the American West, 2003).
4. Amory B. Lovins, "Energy Myth Nine—Energy Efficiency Improvements Have Already Reached Their Potential," in *Energy and American Society—Thirteen Myths,* ed. Benjamin K. Sovacool and Marilyn A. Brown (New York: Springer, 2007), 239.
5. My thinking on this matter has been greatly influenced by the work of Amory Lovins, Jane Summerton, and Ted Bradshaw. See Jane Summerton and Ted K. Bradshaw, "Toward a Dispersed Electrical System: Challenges to the Grid," *Energy Policy* (January/February 1991): 24–34; Amory B. Lovins and L. Hunter Lovins, "Soft Energy Paths: Enjoying the Inevitable," in *Technology and Energy Choice,* ed. John Byrne, Mary Callahan, and Daniel Rich (Newark, DE: University of Delaware, 1983), 1–29; Denton E. Morrison and Dora G. Lodwick, "The Social Impacts of Soft and Hard Energy Systems: The Lovins' Claims as a Social Science Challenge," *Annual Review of Energy* 6 (1981): 357–378; Amory B. Lovins, *Soft Energy Paths: Towards a Durable Peace* (New York: Harper Collins, 1979); Amory Lovins, "Soft Energy Technologies," *Annual Review of Energy* 3 (1978): 466–517; Amory Lovins, "Energy Strategy: The Road Not Taken," *Foreign Affairs* 55 (1976/1977): 65–96; Amory Lovins, "Scale, Centralization, and Electrification in Energy Systems," in *Future Strategies for Energy Development: A Question of Scale* (Knoxville, TN: U.S. Department of Energy and Oak Ridge National Laboratory, 1976), 88–171.
6. Amory Lovins, "A Target Critics Can't Seem to Get in Their Sights," in *The Energy Controversy: Soft Path Questions and Answers,* ed. Hugh Nash (San Francisco: Friends of the Earth, 1979), 26.
7. U.S. Energy Information Administration, *Existing Capacity by Producer Type* (Washington, DC: U.S. Department of Energy, 2006), Table 2.3.
8. The average energy conversion for a human being is 60 to 90 W for maintenance and 40–70 W for work and other activities, or the equivalent of the electric power used by a typical incandescent light bulb. The maximum rate at which a human being delivers work is 330 W for periods of a few hours and 2,000 W for durations of around one minute. See Bent Sorensen, "A History of Renewable Energy Technology," *Energy Policy* (January/February 1991): 8–12.
9. The two important terms here are "frequency" and "voltage." In terms of frequency, the design of consumer equipment such as motors, clocks, and electronics often assume a relatively constant power frequency of 60 Hz for proper operation in the United States. Actual frequencies rarely deviate beyond 59.9 and 60.1 Hz, well within the tolerance of equipment. Frequency fluctuations result from an imbalance between supply and demand for power in a system. In any instant, if the total demand for power exceeds total supply, such as when a generator fails or demand increases unexpectedly, the rotation of all generators slows down, causing power frequency to decrease. A similar process occurs in reverse when generation exceeds load. The usefulness of a particular generator in regulating frequency varies from unit to unit because of differences in the vamp rate, or the rate at which generation output can increase or decrease. Large steam generating units such as nuclear power plants and large coal plants change output levels slowly, meaning they have difficulty reacting to changes in frequency. In terms of

voltage, many types of customer equipment require voltage to fall within a narrow range in order to function properly. If delivered voltage is too low, electric lights dim and electric motors function poorly and can overheat. Overly high voltages shorten lives of lamps substantially and increase motor power, which can damage attached equipment. Unlike frequency, which is the same at all locations in the power system, voltage varies from point to point. The voltages throughout a power system depend on the voltage output of individual generators and voltage control devices, as well as the flow of power through the system. Maintaining voltage involves balancing the supply and demand of reactive power in the system. Reactive power is created when current and voltage in an alternating current system are not in phase due to interactions with electric and magnetic fields around circuit components. Reactive power is often referred to as VARS (for volt amperes reactive), and is regulated by adjusting magnetic fields within the generators. See U.S. Office of Technology Assessment, *Energy Efficiency: Challenges and Opportunities for Electric Utilities* (Washington, DC: U.S. Government Printing Office, September 1993).

10. David J. Bjornstad and Marilyn A. Brown, *A Market Failures Framework for Defining the Government's Role in Energy Efficiency* (Knoxville, TN: Joint Institute for Energy and Environment, June 2004, JIEE 2004-02).
11. According to the EIA's *Electric Power Annual*, utility revenue for residential, commercial, industrial, and transportation customers was $326.5 billion in 2006 and an estimated $360 billion in 2007. The pornography industry had approximately $13.3 billion in revenues in 2006, and the Associated Press reported that Americans gave $96.82 billion to religious organizations the same year.
12. Figures taken from the EIA's *Financial Statistics of Major U.S. Publicly Owned Electric Utilities* (Washington, DC: U.S. Department of Energy, 2000) and *Electric Power Annual 2003* (Washington, DC: U.S. Department of Energy, 2006), and adjusted upwards according to projections from the 2008 Annual Energy Outlook to get values for 2008.
13. According to the National Restaurant Association, 7,482 Burger King Restaurants operate in the United States and Canada while the United States has more than 8,900 power providers.
14. U.S. EIA, *Annual Energy Outlook 2008 with Projections to 2030* (Early Release) (February 2008), 6.
15. U.S. EIA, *Annual Energy Outlook 2007 with Projections to 2030* (February 2007), 82–88.
16. Paula Berinstein, *Alternative Energy: Facts, Statistics, and Issues* (New York: Oryx Press, 2001), 2–3.
17. Benjamin K. Sovacool, *The Power Production Paradox: Revealing the Socio-technical Impediments to Distributed Generation Technologies* (doctoral dissertation, Virginia Polytechnic Institute & State University, Blacksburg, VA, April 2006).
18. Ibid.
19. Hans Blix, "Nuclear Energy in the 21st Century," *Nuclear News* (September 1997), 34–48.
20. See the EIA's *International Energy Outlook* (July 2005).
21. "Tide May Be Shifting Versus Coal," *Electricity Journal* 21, no. 2 (March 2008): 3–7.
22. Sherri K. Stuewer, "Addressing the Risk of Climate Change" (seminar at the Lee Kuan Yew School of Public Policy, Singapore, August 17, 2007).

23. John Hoffmeister, "How the U.S. Can Ensure Energy Supply for the Future" (remarks at the World Affairs Council of Greater Richmond in Richmond, Virginia, May 9, 2007). Hoffmeister commented that "The nature of our business is to constantly look for more fossil fuels, more fossil fuels, and more fossil fuels to try to supply the demand."
24. NEMS tracks the geographical differences in regional energy markets at substate levels, including specific census divisions and NERC subregions. NEMS is so rigorous it is used as a benchmark for models employed by the Union of Concerned Scientists and the Tellus Institute in their own projections of renewable energy production.
25. Data based on Table 2.5 from the EIA's *Electric Power Annual 2006,* which projects that 78,558 MW out of a total of 86,944 MW of capacity additions from 2007 to 2011 will be fueled by coal, petroleum, natural gas, and other gases.
26. National Energy Technology Laboratory, *Tipping Point or Opportunity for Clean Coal Technologies?* (February 10, 2004).
27. See the EIA's *2007 Annual Energy Outlook,* 3–10.
28. Berinstein, *Alternative Energy,* 141.
29. Robert Zubrin, "The Hydrogen Hoax," *The New Atlantis* 15 (Winter 2007): 11.
30. S. Julio Friedman and Thomas Homer-Dixon, "Out of the Energy Box," *Foreign Affairs* 83, no. 6 (2004), 78.
31. Joint Research Center, *Well-to-Wheels Analysis of Future Automotive Fuels and Powertrains in the European Context* (Brussels: Joint Research Centre of the EU Commission, 2004).
32. Richard Heinberg, *Power Down: Options and Actions for a Post-Carbon World* (New York: New Society Publications, 2004), 129.
33. American Physical Society, *The Hydrogen Initiative* (Washington, DC: American Physical Society, 2004).
34. Robert Zubrin, *The Hydrogen Hoax,* 14.
35. National Research Council, *The Hydrogen Economy* (Washington, DC: National Academy Press, 2004), ix.
36. Brent D. Yacobucci and Aimee E. Curtright, "A Hydrogen Economy and Fuel Cells: An Overview," *CRS Report For Congress* (January 14, 2004), 6.
37. C. Geffen, J. Edmonds, and S. Kim, "Transportation and Climate Change: The Potential for Hydrogen Systems," *Environmental Sustainability in the Mobility Industry* (Detroit, MI: Society of Automotive Engineers, 2004), 13–20.
38. See Brian Tulloh, "Letters from Readers," *EnergyBiz Insider* (February 8, 2007).
39. The four most dominant types of "capture" processes are postcombustion, precombustion, oxy-fuel combustion, and purification. Postcombustion capture requires separation of carbon dioxide from flue gas after it is burned. Precombustion decarbonization separates carbon dioxide and captures it before it is burned (IGCC uses this approach). Oxy-fuel combustion, using oxygen instead of air for combustion, concentrates the carbon dioxide exhaust stream. Purification is the rarest and is used to capture CO_2 in limited quantities from industrial practices that do not rely on combustion. For a well-done introduction to clean coal, see Jeffrey Logan, John Venezia, and Kate Larsen, "Opportunities and Challenges for Carbon Capture and Sequestration," *WRI Issue Brief 1* (October 2007), 1–8.
40. Ibid.

41. Mark de Figueiredo, "The Liability of Carbon Dioxide Storage" (Ph.D. thesis, MIT University, February 2007).
42. James Katzer et al., *The Future of Coal: Options for a Carbon-Constrained World* (Cambridge, MA: An Interdisciplinary MIT Study, 2007).
43. Eileen Claussen, "Global Climate Change and Coal's Future" (remarks before the American Coal Council, May 18, 2004).
44. William L. Sigmon Jr., "Fossil and Hydro Generation in the Energy Industry," *William & Mary Law and Policy Review Annual Symposium* (February 2, 2008).
45. Sovacool, *Power Production Paradox.*
46. Historically, the capital intensity of worldwide nuclear projects has made them very expensive outside of the United States as well. In Argentina, the 698 MW Attucha II reactor cost $6,017 per installed kilowatt, the Brazilian 626 MW Angra I reactor, $2,874. Reactor projects started, but never finished, in Egypt and Iran have cost around $4,000 per installed kilowatt. See Bill Keepin and Gregory Kats, "Greenhouse Warming: Comparative Analysis of Nuclear and Efficiency Abatement Strategies," *Energy Policy* (December 1988): 538–561.
47. Lance E. Echavarri, "Is Nuclear Energy at a Turning Point?" *Electricity Journal* 20, no. 9 (November 2007), 89–97.
48. Each of these four estimates are taken from Pam Radtke Russell, "Prices Are Rising: Nuclear Cost Estimates Under Pressure," *EnergyBiz Insider* (May-June, 2008).
49. E. S. Beckjord et al., *The Future of Nuclear Power: An Interdisciplinary MIT Study* (Cambridge, MA: MIT Press, 2003).
50. Friedman and Homer-Dixon, "Out of the Energy Box," 74.
51. For every 700 new nuclear plants that are constructed, 11 to 22 large enrichment plants, 18 fuel fabrication plants, and 10 waste disposal sites the size of Yucca Mountain need to be built. Data taken from The Keystone Center, *Nuclear Power Joint Fact-Finding* (July 2007).
52. Nathan E. Hultman, Jonathan G. Koomey, Daniel M. Kammen, "What History Can Teach Us About the Future Costs of U.S. Nuclear Power," *Environmental Science & Technology* (April 1, 2007): 2088–2099.
53. Richard Weitz, "Global Nuclear Energy Partnership: Progress, Problems, and Prospects," *Weapons of Mass Destruction Insights* (March 2008).
54. National Research Council, *Review of DOE's Nuclear Energy Research and Development Program* (Washington, DC: National Academies Press, 2007).
55. Peter R. Orszag, *Costs of Reprocessing Versus Directly Disposing of Spent Nuclear Fuel* (Washington, DC: Congressional Budget Office, November 14, 2007).
56. Timothy Abram, *Generation IV Nuclear Power: The State of the Science* (London: Office of Science and Innovation, 2006).
57. K. L. Murty and I. Charit, "Structural Materials for Gen-IV Nuclear Reactors: Challenges and Opportunities," *Journal of Nuclear Materials* (2007).
58. D. Haas and D. J. Hamilton, "Fuel Cycle Strategies and Plutonium Management in Europe," *Progress in Nuclear Energy* 49 (2007): 582.
59. A doubling of demand would mean about 998 GW of capacity would need to be built. If each coal unit had a capacity of 200 MW, almost 5,000 facilities would be needed.
60. This figure presumes that the 998 GW of capacity are met by natural gas units of 100 MW each.

61. The study estimated that by 2020 the country would need to construct 393,000 MW of new generating capacity, or an average of 60 to 90 new plants per year. See T. Randall Curlee and Michael J. Sale, "Water and Energy Security," *Water in the 21st Century* (Oak Ridge, TN: UT-Battelle, 2003), 8.
62. Wendell H. Wiser, *Energy Resources: Occurrence, Production, Conversion, Use* (New York: Springer, 2000), 121.
63. University of Wyoming, *The Unit Train* (May 19, 2003), 1.
64. See Steve Clemmer et al., *Clean Power Blueprint: A Smarter National Energy Policy for Today and the Future* (Washington, DC: Union of Concerned Scientists, October 2001), 31; and N. C. Parker, *Using Natural Gas Transmission Pipeline Costs to Estimate Hydrogen Pipeline Costs* (University of California, Davis: Institute of Transportation Studies Report No. UCD-ITS-RR-04-35, 2004).
65. See EIA's 2006 *Electric Power Annual,* Table 2.6.
66. Scott Sklar and Todd R. Burns, *Hydroelectric Energy: An Overview* (Washington, DC: The Stella Group, 2003), 5. Adjusted to $2007.
67. Sovacool, *Power Production Paradox.*
68. Ibid.
69. Ibid.
70. The Public Service Commission states that "Utah Power residential ratepayers currently pay about 6.3 ¢/kWh. About 3 cents of this are attributed to the actual generation of electricity. The transmission, distribution, and general administration and overhead costs to deliver power to households and businesses make up the rest." See Public Service Commission of Utah, *Electric Energy Crisis from Utah's Perspective* (March 2004).
71. Sovacool, *Power Production Paradox.*
72. Ibid.
73. Edison Electric Institute, *U.S. Transmission Capacity: Present Status and Future Prospects* (Washington, DC: EEI, August 2004), v.
74. Sovacool, *Power Production Paradox.*
75. M. Reutter, "Transmission Congestion Threatens to Clog Nation's Power Grid," *News Bureau University of Illinois at Champagne-Urbana* (July 26, 2006).
76. Eric Hirst, "Transmission Investment: All Talk and Little Action," *Public Utilities Reports* (July 2004): 2.
77. Eric Hirst and Brendan Kirby, "Expanding Transmission Capacity: A Proposed Planning Process," *Electricity Journal* (October 2002), 54. Adjusted to $2007.
78. E. K. Datta and D. Gabaldon, "Energy Technology: Winner Take All," Public Utilities Report (October 15, 2003).
79. Massoud Amin, "Energy Infrastructure Defense Systems," *Proceedings of the IEEE* 93, no. 5 (May 2005): 863.
80. Kristina Hamachi LaCommare and Joseph H. Eto, *Cost of Power Interruptions to Electricity Consumers in the U.S.* (Lawrence Berkeley National Laboratory LBNL-58164, February 2006), 19.
81. Scott Sklar, "New Dawn for Distributed Energy," *Cogeneration Onsite Power Production Magazine* (July/August 2005), 115–123.
82. International Energy Agency, *Distributed Generation in Liberalized Electricity Markets* (Paris: International Energy Agency, 2002), 44. Prices have been adjusted to $2007.
83. Advanced Technology Solutions/Nielson, *Industrial Survey 2005*. Prices have been adjusted to $2007.

84. Jackie Lin, "Power Outage Hits Industrial Park Hard," *Taipei Times,* April 11, 2004, 10. Prices have been adjusted to $2007.
85. M. Freeman, "NERC Forecast: 22 Actions Required to Save U.S. Electric Grid," *Executive Intelligence Review* (October 27, 2007).
86. U.S. DOE, *National Transmission Grid Study* (Washington, DC: U.S. DOE, May 2002), 17.
87. Hirst, "Transmission Investment," 2.
88. Steve Huntoon and Alexandra Metzner, "The Myth of the Transmission Deficit," *Public Utilities Fortnightly* 141 (November 1, 2003): 31.
89. Seth Blumsack, Lester B. Lave, and Marija Ilic, "The Real Problem with Merchant Transmission," *Electricity Journal* 21, no. 2 (March 2008): 9–19.
90. Ibid.
91. See FERC, *Proposed Pricing Policy for Efficient Operation and Expansion of the Transmission Grid,* Notice of Proposed Policy Statement, Docket Number PL03-1-000, 102 FERC 61.032 (2003), 15.
92. R. Deb and K. White, "Valuing Transmission Investments: The Big Picture *and* the Details Matter—and Benefits Might Exceed Expectations," *Electricity Journal* 18, no. 7 (August/September 2005), 42.
93. National Commission on Energy Policy, *Siting Critical Energy Infrastructure: An Overview of Needs and Challenges* (Washington, DC: NCEP, June 2006), 6.
94. J. Fuquay, "TXU Facing Power Fight," *Fort-Worth Star Telegram,* April 25, 2007.
95. National Council on Electric Policy, *Electricity Transmission: A Primer* (Washington, DC: U.S. DOE, June 2004), 24.
96. Leonard S. Hyman, "The Next Big Crunch: T&D Capital Expenditures," *Energy Industry Commentary* (January 2004), 92.
97. Paul Joskow, "Transmission Policy in the U.S.," *Utilities Policy* 13 (2005): 114.
98. Amory Lovins and Hunter Lovins, *Brittle Power: Energy Strategy for National Security* (Andover, MA: Brick House Publishing Company, 1982).
99. Lovins, "A Target," 19.
100. International Energy Agency, *Energy Security in a Dangerous World* (Paris: International Energy Agency, 2004).
101. Thomas Homer-Dixon, "The Rise of Complex Terrorism," *Foreign Policy* (January/February 2002): 34–41.
102. Sovacool, *Power Production Paradox.*
103. Alexander Farrell, Hisham Zerriffi, and Hadi Dowlatabadi, "Energy Infrastructure and Security," *Annual Review of Environment and Resources* 29 (2004): 421–422.
104. Ibid.
105. John Robb, "Security: Power to the People," *Fast Company Magazine* 103 (March 2006), 120.
106. Benjamin Sovacool and Saul Halfon, "Reconstructing Iraq: Merging Discourses of Security and Development," *Review of International Studies* 33 (2007): 239–240.
107. National Research Council, *Making the Nation Safer: The Role of Science and Technology in Countering Terrorism* (Washington, DC: National Academies Press, 2002), 178.
108. Scott Avedisian, "Liquefied Natural Gas," *Hearing Before the Senate Committee on Energy and Natural Resources* (February 15, 2005), 69–70.
109. Paul W. Parformak, "Liquefied Natural Gas Infrastructure Security," CRS Report for Congress, September 9, 2003, 9.

110. Quoted in Mark Clayton, "A Prized Source of Energy, or Potent Terror Target?" *Christian Science Monitor,* April 6, 2004, 1.
111. Cindy Hurst, *The Terrorist Threat to Liquefied Natural Gas: Fact or Fiction?* (Washington, DC: Institute for the Analysis of Global Security, February 2008), 13.
112. Ibid., 1.
113. David Greene and Sanjana Ahmad, *Costs of U.S. Oil Dependence: 2005 Update* (January 2005), Report to the U.S. DOE, ORNL/TM-2005/45. Numbers have been adjusted to $2007.
114. Mark A. Delucchi and James J. Murphy, "US Military Expenditures to Protect the Use of Persian Gulf Oil for Motor Vehicles," *Energy Policy* 36 (2008): 2253–2264.
115. Sovacool, *Power Production Paradox.*
116. Benjamin K. Sovacool, "The Costs of Failure: A Preliminary Assessment of Major Energy Accidents, 1907 to 2007," *Energy Policy* 36, no. 5 (May 2008), 1802–1820.
117. Data taken from International Atomic Energy Agency, *In Focus: Chernobyl Twenty Years Later* (Paris, France: IAEA, 2006); and T. S. Gopi Rethinaraj, "Nuclear Safety Issues: Review" (presentation to the ISEAS Sustainable Development and Energy Security Seminar, April 22, 2008).
118. Sovacool, "The Costs of Failure."
119. Vladimir Kirillovich Savchenko, *The Ecology of the Chernobyl Catastrophe: Scientific Outlines of an International Programme of Collaborative Research* (Kiev, Ukraine: Informa Health Care, 1995), 70–72.
120. Sergey I. Dusha-Gudym, "Transport of Radioactive Materials by Wildland Fires in the Chernobyl Accident Zone: How to Address the Problem," *International Forest Fire News* 32 (January/June 2005): 119–125; Ryszard Szczygieł and Barbara Ubysz, *Chernobyl Forests. Two Decades After the Contamination* (Kiev: Forest Research Institute, 2006).
121. Beckjord et al., *Future of Nuclear Power,* 22 and 48.
122. Bernard Papin and Patrick Quellien, "The Operational Complexity Index: A New Method for the Global Assessment of the Human Factor Impact on the Safety of Advanced Reactor Concepts," *Nuclear Engineering and Design* 236 (2006): 1113–1121.
123. David Lochbaum, *U.S. Nuclear Plants in the 21st Century: The Risk of a Lifetime* (Washington, DC: Union of Concerned Scientists, 2004).
124. U.S. Department of Energy, *A Roadmap to Deploy New Nuclear Power Plants in the U.S. by 2010* (Washington, DC: DOE, 2001).
125. Lovins and Lovins, *Brittle Power,* 158.
126. Public Citizen, *The Big Blackout and Amnesia in Congress* (Washington, DC: Public Citizen, 2004).
127. P. Gaukler, S. Barnett, and D. Rosinski, "Putting Nuclear Terrorism in Perspective," *Natural Resources & Environment* 16, no. 3 (2002): 141–187.
128. David Lochbaum, "Beware of Aging Reactors, a Weak Regulator, and Vulnerability to terrorists," *Testimony Before the Clean Air, Wetlands, Private Property, and Nuclear Safety Subcommittee of the U.S. Senate Committee on Environment and Public Works* (May 8, 2001).
129. Robert F. Kennedy Jr., "Nuclear Plants Vulnerable to Attack," *Seattle Post-Intelligencer,* August 5, 2005, 12.
130. Arjun Makhijani and Scott Saleska, "The Nuclear Power Deception," *Report to the Institute of Energy and Environmental Research* (Takoma Park, MD: Institute for Energy and Environmental Research, 1996).

131. Frank Barnaby and James Kemp, *Secure Energy? Civil Nuclear Power, Security, and Global Warming* (Oxford: Oxford Research Group, March 2007).
132. Foster Electric Report, "PG&E Unable to Locate Missing Nuclear Waste" (August 25, 2004), 16.
133. Rensselaer Lee, "Nuclear Smuggling and International Terrorism: Implications for U.S. Policy," *CRS Report for Congress* (October 22, 2002): 8; U.S. Government Accountability Office, "Combating Nuclear Smuggling," *GAO Report to the Department of Homeland Security* (June 21, 2005).
134. Matthew E. Chen, "Oil Companies and Human Rights," *Orbis* 51, no. 1 (Winter, 2007): 41–54; Tarek F. Maassarani, Margo T. Drakos, and Joanna Pajkowska, "Extracting Corporate Responsibility: Towards a Human Rights Impact Assessment," *Cornell International Law Journal* 40 (Winter 2007): 135–165.
135. Sovacool, *Power Production Paradox.*
136. Matthias Ruth, Steven Gabriel, Karen Palmer, Dallas Burtraw, Anthony Paul, Yihsu Chen, Benjamin Hobbs, Daraius Irani, Jeffrey Michael, Kim Ross, Russel Conklin, and Julia Miller, "Economic and Energy Impacts from Participation in the Regional Greenhouse Gas Initiative: A Case Study of the State of Maryland," *Energy Policy* 36 (2008): 2279–2289.
137. Randall Baker, "Energy Policy: The Problem of Public Perception," in *Energy: Science, Policy, and the Pursuit of Sustainability,* ed. Robert Bent, Lloyd Orr, and Randall Baker (Washington, DC: Island Press, 2002), 133–134.
138. Alan Hathaway, "The Impact of Renewable/ee Portfolio Standard on Future Rate Hikes in Virginia" (presentation to the Energy Virginia Conference, October 17, 2006).
139. Tom Gray, "Trans-Praire and Interior West Wind 'Pipelines,'" *Wind Energy Weekly,* July 31 2002, 12.
140. Comments at the EUCI 3rd Annual Renewable Portfolio Standards (RPS) Conference in Westminster, Colorado, April 23–24, 2007.
141. K. Silverstein, "Rockies Project: Laying the Groundwork," *EnergyBiz Insider,* April 30, 2007.
142. The year 2007 was no exception. In its report on short-term energy and summer 2007 fuels outlook, the DOE said it expected natural gas prices over the summer season to be 18 percent above its predictions a year earlier. See Alan Nogee, Jeff Deyette, and Steve Clemmer, "The Projected Impacts of a National Renewable Portfolio Standard," *Electricity Journal* 20, no. 4 (May 2007): 33–47.
143. A. Greenspan, "Natural Gas Supply and Demand Issues," *Testimony before the House Committee on Energy and Commerce* (June 10, 2003).
144. R. Gupta, "Enhancing Energy Security," *Hearing Before the House Committee on Natural Resources* (March 19, 2003).
145. See M. A. Adelman and G. C. Watkins, "U.S. Oil and Natural Gas Reserve Prices, 1982–2003," *Energy Economics* 27 (2005): 553–571; Robert S. Pindyck, "Volatility in Natural Gas and Oil Markets," *The Journal of Energy and Development* 30, no. 1 (2004): 1–19; and U.S. Energy Information Administration, *Annual Energy Outlook 2006.*
146. Paul J. Hibbard, *US Energy Infrastructure Vulnerability: Lessons from the Gulf Coast Hurricanes* (Washington, DC: National Commission on Energy Policy, March 2006), 5.
147. Ed Vine, Marty Kushler, and Dan York, "Energy Myth Ten—Energy Efficiency Measures Are Unreliable, Unpredictable, and Unenforceable," in *Energy and American*

Society—Thirteen Myths, ed. B. K. Sovacool and M. A. Brown (New York: Springer, 2007).

148. M. Davidson, "Natural Gas Spikes Keep Market Guessing," *Platt's Insights* (December, 2006).
149. R. Smith, "Emboldened States Take Charge of Energy Issues," *Wall Street Journal,* October 12, 2006, A6.
150. D. Kopecky, "The Trend Toward On-Site Power Generation," *Energy Pulse* (March 19, 2007).
151. National Economic Research Associates, *The Role of Energy Costs in Industry Products* (January 1996).
152. Alan Nogee, "Renewable Energy and Electricity: Creating Jobs, Saving Consumers Money, and Increasing Our Energy Security," *Testimony before the House Committee on Energy and Commerce, Subcommittee on Energy and Air Quality* (February 16, 2005).
153. Gupta, "Enhancing Energy Security."
154. Dianne Feinstein, "Wholesale Electricity Prices in California and the Western U.S.," *Hearing Before the Senate Committee on Energy and Natural Resources* (May 3, 2001), 6–8.
155. Joseph Barton, "FY 2005 Budget Priorities for the Department of Energy," *Hearing Before the House Committee on Energy and Commerce* (April 1, 2004), 3.
156. William Stanley Jevons, *The Coal Question: An Inquiry Concerning the Progress of the Nation, and the Probable Exhaustion of Our Coal Mines* (London: Macmillan and Company, 1866).
157. Joseph Barton, "Future Options for Generation of Electricity from Coal," *Hearing Before the Subcommittee on Energy and Air Quality of the House Committee on Energy and Commerce* (June 24, 2003), 1; Steven F. Leer, "Energy Production on Federal Lands," *Hearing Before the Senate Committee on Energy and Natural Resources* (February 27, 2003), 37–41.
158. Jerry M. Eyster and Trygve Gaalaas, *Coal Price Volatility Is Here to Stay* (Washington, DC: PA Consulting Group, May 2004).
159. John Walsh, "Problems of Expanding Coal Production," *Science* 184, no. 4134 (April 19, 1974): 336–339.
160. Mel Horwitch, "Coal: Constrained Abundance," in *Energy Future: Report of the Energy Project at the Harvard Business School,* ed. Robert Stobach and Daniel Yergin (New York: Random House, 1979), 79–107.
161. Vaclav Smil, *Energy at the Crossroads: Global Perspectives and Uncertainties* (Cambridge, MA: MIT Press, 2003), 232–233.
162. Frank A. Verrasto, "EIA 2005 Annual Energy Outlook," *Hearing Before the Senate Committee on Energy and Natural Resources* (February 3, 2005), 33–37.
163. World Wildlife Foundation, *Coming Clean: The Truth and Future of Coal in the Asia Pacific* (Washington, DC: WWF, 2007).
164. "Volatile Coal Prices Reflect Supply, Demand Uncertainties," *Platt's Insights Magazine,* March 30, 2005.
165. Ibid.
166. Ibid.
167. International Energy Agency, "Investment in the Coal Industry," *Background Paper on the Meeting with the IEA Governing Board* (November 2003).
168. "Uranium: Glowing," *The Economist,* 380, no. 8491 (2006): 53.

169. U.S. Department of Energy, *Report to Congress on the Maintenance of Viable Domestic Uranium, Conversion, and Enrichment Industries* (2000).
170. International Atomic Energy Agency, *Analysis of Uranium Supply to 2050* (Geneva: IAEA, 2001), 11.
171. National Energy Technology Laboratory, *Brownfield IGCCs as an Option in the National Energy Modeling System (NEMS)* (DOE/NETL-2008/1311).
172. National Environmental Trust, *Toxic Power: What the Toxics Release Inventory Tells Us About Power Plant Pollution* (Washington, DC: National Campaign Against Dirty Power, August 2000).
173. Smith, *Emboldened States.* Prices adjusted to $2007.
174. U.S. Department of Energy, *Energy Demands on Water Resources: Report to Congress on the Interdependency of Energy and Water* (Washington, DC: U.S. DOE, December 2006).
175. Ibid.
176. Thomas J. Feeley, "Tutorial on Electric Utility Water Issues" (presentation to the 28th International Technical Conference on Coal Utilization and Fuel Systems, National Energy Technology Laboratory, March 10–13, 2003), 4.
177. U.S. Department of Energy, *The Wind/Water Nexus* (Washington, DC: U.S. Department of Energy, April 2006, DOE/GO-102006-2218).
178. See Southern Alliance for Clean Energy, *Water* (Asheville, NC: SACE, 2007).
179. DOE, *Energy Demands.*
180. Bill Smith, Bob Goldstein, and Keith Carns, *Water and Sustainability—The EPRI Research Plan* (Washington, DC: EPRI, July 25, 2002).
181. U.S. Global Climate Change Research Program, *U.S. National Assessment of Potential Consequences of Climate Variability and Change: Regional Paper, Pacific Northwest* (Washington, DC: GCCRP, 2004); R. Palmer and M. Hahn, *The Impacts of Climate Change on Portland's Water Supply* (Portland, OR: Portland Water Bureau, 2002).
182. B. Foss, "Natural Gas Jumps 14%; Oil Drifts Higher," *Arkansas Democrat Gazette,* August 1, 2006.
183. John Norton, "Water, Taxes in Pueblo, Colo., Hamper Xcel Energy Power Plant Expansion," *The Pueblo Chieftain,* November 20, 2003, 12.
184. John Veil, *Impacts of Electric Power and Coal Industries on Water Resources* (Argonne, IL: Argonne National Laboratory, 2006).
185. J. Bartram et al., *Toxic Cyanobacteria in Water: A Guide to Their Public Health Consequences, Monitoring and Management* (Geneva: World Health Organization, 2001).
186. Benjamin K. Sovacool, "Coal and Nuclear Technologies: Creating a False Dichotomy for American Energy Policy," *Policy Sciences* 40, no. 2 (June 2007): 101–122.
187. Roe-Han Yoon, "Future Options for Generation of Electricity from Coal," *Hearing Before the House Committee on Energy and Commerce* (June 24, 2003), 84.
188. U.S. Nuclear Regulatory Commission, *Regulating Nuclear Fuel* (Washington, DC: NRC, September 2001).
189. Environment News Service, "Illinois Sues Exelon for Radioactive Tritium Releases Since 1996," March 21, 2006.
190. Abby Luby, "Leaks at Indian Point Created Underwater Lakes," *Riverkeeper* (February 28, 2008).
191. Rita J. King, "Entergy Facing Lawsuit Over Radiation Leak," *North County News,* July 12, 2007, 6.

192. T. Lackson et al., "The 2003 North American Electrical Blackout: An Accidental Experiment in Atmospheric Chemistry," *Geophysical Research Letters* 31 (2004): 3106.
193. According to the Centers for Disease Control, the average weight for adults (adjusting for both men and women) is 74.5 kilograms, while the average amount of air pollution from power plants per capita is 7,926 kg/year per person.
194. According to the American Medical Association, the average person farts around 5 grams per day. Eight tons of pollution per person per year, however, works out to around 219,000 grams per day. It would take one approximately 120 years to fart that much.
195. American Lung Association, *Summary of the American Lung Association's Annual Clean Air Test* (Washington, DC: ALA, 2005).
196. Joseph J. Romm and Christine A. Ervin, *How Energy Policies Affect Public Health* (Washington, DC: Solstice, 2005). Adjusted to $2007.
197. Debra A. Jacobson, "Increasing the Value and Expanding the Market for Renewable Energy and Energy Efficiency with Clean Air Policies," *Environmental Law Review* 37 (2007): 10135–10137.
198. See Mark A. Delucchi, James J. Murphy, and Donald R. McGubbin, "The Health and Visibility Cost of Air Pollution: A Comparison of Estimation Methods," *Journal of Environmental Management* 64 (2002): 139–152; Isabelle Romieu, Jonatham M. Samet, Kirk R. Smith, and Nigel Bruce, "Outdoor Air Pollution and Acute Respiratory Infections Among Children in Developing Countries," *Journal of Occupational and Environmental Medicine* 44 (2002): 640–649; K. Katsouyanni and G. Pershagen, "Ambient Air Pollution Exposure to Cancer," *Cancer Causes and Control* 8 (1997): 284–291.
199. Harvard School of Public Health, *Impact of Pollution on Public Health* (Cambridge, MA: Harvard University, January 3, 2001).
200. Deanne M. Ottaviano, *Environmental Justice: New Clean Air Act Regulations and the Anticipated Impact on Minority Communities* (New York: Lawyer's Committee for Civil Rights Under Law, 2003).
201. Conrad G. Schneider, *Death, Disease, and Dirty Power: Mortality and Human Health Damage due to Air Pollution from Power Plants* (Boston: Clean Air Task Force, 2000); Maria T. Padian, *New York's Dirty Street: The Power Plant Pollution Loophole* (New York: Pace Energy Project, 1998); Michael T. Kleinman, *The Health Effects of Air Pollution on Children* (Los Angeles: University of California, 2000).
202. U.S. Environmental Protection Agency, *Asthma Facts* (Washington, DC: EPA, 2004).
203. U.S. Environmental Protection Agency, *Air Pollution Facts* (Washington, DC: EPA, 2003).
204. David R. Wooley, *A Guide to the Clean Air Act for the Renewable Energy Community* (Washington, DC: Renewable Energy Policy Project, 2000), 7–14.
205. Rodney Sobin, "Energy Myth Seven: Renewable Energy Systems Could Never Meet Growing Electricity Demand in America," in *Energy and American Society—Thirteen Myths,* ed. B. K. Sovacool and M. A. Brown (New York: Springer, 2007), 171–199.
206. EPA, *Air Pollution.*
207. R. A. Kowalik, D. M. Cooper, C. M. Evans, and S. J. Ormerod, "Acid Episodes Retard the Biological Recovery of Upland British Streams from Acidification," *Global Change Biology* 13 (2007): 2439–2452.
208. EPA, *Air Pollution.*

209. Armond Cohen, "National Energy Policy: Coal," *Hearing Before the House Committee on Energy and Commerce* (March 14, 2001), 65–71.
210. Romm and Ervin, *How Energy Policies Affect Public Health.*
211. Alan J. Krupnick and Winston Harrington, "Ambient Ozone and Health Effects: Evidence from Daily Data," *Journal of Environmental Economics and Management* 18 (1990): 1–18.
212. Sobin, "Energy Myth Seven."
213. U.S. EPA, *Mercury Study—Report to Congress* (Washington, DC: EPA, 1997); U.S. EPA, *Fact Sheet on Mercury* (Washington, DC: EPA, 2001).
214. U.S. EPA, *2003 Mercury Advisory Listing* (Washington, DC: EPA, 2004).
215. Quoted in Carl Pope, "The State of Nature: Our Roof Is Caving In," *Foreign Policy* (July/August 2005): 67–71.
216. Bruce Hannon, "Energy Use and Moral Restraint," *Journal of Social and Biological Structures* 1 (1978): 357–375.
217. David R. Wooley, *A Guide to the Clean Air Act for the Renewable Energy Community* (Washington, DC: Renewable Energy Policy Project, 2000), 7–14.
218. U.S. EPA, *Particulate Matter: Health and Environment* (Washington, DC: EPA, 2006).
219. Angela Ledford, *The Dirty Secret Behind Dirty Air* (Boston, MA: Clean Air Taskforce, June 2004).
220. Douglas W. Dockery, C. Arden Pope, X. Xu, J. D. Spengler, J. H. Ware, M. E. Fay, B. G. Ferris, and F. E. Speizer, "Mortality Risks of Air Pollution: A Prospective Cohort Study," *New England Journal of Medicine* 329 (1993): 1752–1759.
221. C. Arden Pope, Michael J. Thun, Mohan M. Namboodiri, Douglas W. Dockery, John S. Evans, Frank E. Speizer, and Clark W. Heath Jr., "Particulate Air Pollution as a Predictor of Mortality in a Prospective Study of U.S. Adults," *American Journal of Respiratory Critical Care Medicine* 151 (1995): 669–674.
222. Deborah Sheiman Shprentz, *Breath-Taking: Premature Mortality Due to Particulate Air Pollution in 239 Cities* (Washington, DC: Natural Resources Defense Council, 1996).
223. National Center for Environmental Assessment, *Provisional Assessment of Recent Studies of Health Effects of Particulate Matter Exposure* (Washington, DC: U.S. Environmental Protection Agency, July 2006, EPA/600/R-06/063).
224. EPA, *Air Pollution.*
225. C. Arden Pope, David W. Bates, and Mark E. Raizenne, "Health Effects of Particulate Air Pollution: Time for Reassessment?" *Environmental Health Perspectives* 103, no. 5 (May 1995): 478.
226. Ledford, *Dirty Secret.*
227. Deanne M. Ottaviano, *Environmental Justice: New Clean Air Act Regulations and the Anticipated Impact on Minority Communities* (New York: Lawyer's Committee for Civil Rights Under Law, 2003), 19.
228. Sobin, *Energy Myth Seven.*
229. According to the U.S. EPA, coal accounted for 29 percent of greenhouse gas emissions and natural gas 16 percent. See U.S. Environmental Protection Agency, *U.S. Inventory of Greenhouse Gas Emissions and Sinks* (Washington, DC: EPA, April 2007, EPA 430-F-07-004).
230. Katzer et al., *The Future of Coal.*
231. David G. Hawkins, Daniel A. Lashof, and Robert H. Williams, "What to Do About Coal?" *Scientific American* (September 2006): 69–72.

232. Peter Fontaine, "The Gathering Storm," *Public Utilities Fortnightly* 8, no. 142 (August 2004): 50–61.
233. Quoted in Elaine Robbins, "The Backyard Drill: A Surge in Natural Gas Drilling," *Planning Magazine* 8, no. 70 (August 1, 2004): 16–22.
234. Foster Natural Gas Report, "Greenpeace Issues Report on California LNG" (September 23, 2004), 10–14.
235. Arjun Makhijani, Lois Chalmers, and Brice Smith, *Uranium Enrichment: Just Plain Facts to Fuel an Informed Debate on Nuclear Proliferation and Nuclear Power* (Tacoma Park, MD: Institute for Energy and Environmental Research, 2005).
236. Makhijani et al., *Uranium Enrichment.* The calculation works like this: it takes approximately 55 kWh of electricity to enrich one separative work unit (SWU) of uranium; it also takes 100,000 SWU to produce 1,000 MW of electricity. Therefore, it means that 5,500 MWh are needed to generate 1,000 MW of electricity.
237. Peter Asmus, "Nuclear Dinosaur," *The Washington Post,* July 6, 2005, A7.
238. Benjamin K. Sovacool, "Valuing the Greenhouse Gas Emissions from Nuclear Power: A Critical Survey," *Energy Policy* 36, no. 8 (August 2008): 2940–2953.
239. The calculation works like this: In 2005, 435 nuclear plants supplied 16 percent of the world's power, constituting 368 GW of installed capacity generating 2,768 TWh of electricity. With every TWh of nuclear electricity having carbon-equivalent life-cycle emissions of 66,000 tons of CO_2, these plants emitted a total of some 182.7 million tons. If each ton cost $24, the grand total would be about $4.4 billion every year.
240. Barnaby and Kemp, *Secure Energy?*
241. IPCC, "Summary for Policymakers," *Climate Change: 2007* (Washington, DC: Government Printing Office, 2007).
242. Klaus S. Lackner and Jeffrey D. Sachs, "A Robust Strategy for Sustainable Energy," *Brookings Papers on Economic Activity* 2 (2004): 215–248.
243. United Nations Development Programme, *Energy After Rio: Prospects and Challenges* (Geneva: United Nations, 1997).
244. Cynthia Rosenzweig, David Karoly, Marta Vicarelli, Peter Neofotis, Qigang Wu, Gino Casassa, Annette Menzel, Terry L. Root, Nicole Estrella, Bernard Seguin, Piotr Tryjanowski, Chunzhen Liu, Samuel Rawlins, Anton Imeson, "Attributing Physical and Biological Impacts to Anthropogenic Climate Change," *Nature* 453 (May 15, 2008): 353–357.
245. Eileen Claussen and Janet Peace, "Energy Myth Twelve—Climate Policy Will Bankrupt the U.S. Economy," in *Energy and American Society—Thirteen Myths,* ed. B. K. Sovacool and M. A. Brown (New York: Springer, 2007), 311–340.
246. Alan Carlin, "Why a New Approach Is Required if Global Climate Change Is to Be Controlled Efficiently, Or Even at All," *William & Mary Law and Policy Review Annual Symposium* (February 2, 2008).
247. These examples are taken respectfully from John Allen Paulos, "How Iraq Trillion Could Have Been Spent," *ABC News,* February 4, 2007.
248. Volkan S. Ediger, Enes Hosgor, A. Nesen Surmeli, and Huseyin Tatlidil, "Fossil Fuel Sustainability Index: An Application of Resource Management," *Energy Policy* 35 (2007): 2969–2977.
249. David Waskow and Carol Welch, "The Environmental, Social, and Human Rights Impacts of Oil Development," in *Covering Oil: A Reporter's Guide to Energy and*

Development, ed. Svetlana Tsalik and Anya Schiffrin (New York: Open Society Institute, 2005), 101–123
250. Ibid.
251. Ibid.
252. Ibid.
253. Charles H. Peterson et al., "Ecological Consequences of Environmental Perturbations Associated with Offshore Hydrocarbon Production: A Perspective on the Long-Term Exposures in the Gulf of Mexico," *Canadian Journal of Fisheries and Aquaculture Science* 53 (1996): 2637–2654.
254. Patricio Silva, "National Energy Policy," *Hearing Before the House Subcommittee on Energy and Air Quality* (Washington, DC: Government Printing Office, February 18, 2001), 113–116.
255. Waskow and Welch, "Impacts of Oil Development," 107.
256. Silva, "National Energy Policy."
257. Benjamin K. Sovacool, "Abandoned Treaties, Environmental Damage, Fossil Fuel Dependence: The Coming Costs of Oil and Gas Exploration in the Arctic National Wildlife Refuge," *Environment, Development, and Sustainability* 9, no. 2 (May 2007): 187–201; Benjamin K. Sovacool, "Eroding Wilderness: The Ecological, Legal, Political, and Social Consequences to Oil and Natural Gas Exploration in the Arctic National Wildlife Refuge (ANWR)," *Energy & Environment* 17, no. 4 (April 2006): 549–567.
258. U.S. Energy Information Administration, "Why Do Natural Gas Prices Fluctuate So Much?" *Natural Gas 1998 Issues and Trends* (Washington, DC: U.S. DOE, 1998).
259. U.S. Energy Information Administration, *U.S. Underground Natural Gas Storage Developments: 1998–2005* (Washington, DC: U.S. Department of Energy, 2006).
260. U.S. EIA, *U.S. LNG Markets and Uses: June 2004 Update* (Washington, DC: EIA, 2004), 16.
261. Marcello J. Lippman and Sally M. Benson, *Relevance of Underground Natural Gas Storage to Geologic Sequestration of Carbon Dioxide* (Berkeley, CA: Lawrence Berkeley National Laboratory, 2004).
262. Susan E. Nissen et al., *Geologic Factors Controlling Natural Gas Distribution Related to the January 2001 Gas Explosions in Hutchinson, Kansas* (Lawrence, KS: Kansas Geological Survey, 2001).
263. EIA, *Underground Natural Gas.*
264. Sovacool, "The Costs of Failure."
265. Waskow and Welch, "Impacts of Oil Development," p. 108.
266. U.S. General Accounting Office, *Pipeline Safety: The Office of Pipeline Safety is Changing How it Oversees the Pipeline Industry* (Washington, DC: GAO/RCED-00-128, May 2000).
267. Blacksmith Institute, *World's Worst Polluted Places* (New York: Author, 2008).
268. Sobin, "Energy Myth Seven."
269. Richard F. Bonskowski, *U.S. Metallurgical Coal and Coke Supplies—Prices, Availability, and the Emerging Futures Markets* (Washington, DC: EIA, 2002).
270. Brandon Ortiz, "Advocates Blame Industry for Accidents; Industry Points at the Public," *The Lexington Herald-Ledger* (January 16, 2005), A2.
271. Dallas Burtraw, Ken Harrison, and JoAnne Pawlowski, "Coal Transportation and Road Damage," in *Fuel Cycle Externalities: Analytical Methods and Issues* (Knoxville, TN: Oak Ridge National Laboratory, July 1994), 15-1 to 15-6. Updated to $2007.

272. David Fleming, *The Lean Guide to Nuclear Energy: A Lifecycle in Trouble* (London: The Lean Economy Connection, 2007).
273. U.S. Department of Energy, *Plutonium Recovery from Spent Fuel Reprocessing by Nuclear Fuel Services at West Valley, New York from 1966 to 1972* (Washington, DC: DOE, February 1996).
274. Dan Watkiss, "The Middle Ages of Our Energy Policy—Will the Renaissance Be Nuclear?" *Electric Light & Power* (May/June 2008).
275. William H. Hannum, Gerald E. Marsh, and George S. Stanford, "Smarter Use of Nuclear Waste," *Scientific American* (December 2005): 84–87.
276. Watkiss, "The Middle Ages."
277. Ibid.
278. Sovacool, "Think Again."
279. Lawrence Flint, "Shaping Nuclear Waste Policy at the Juncture of Federal and State Law," *Boston College Environmental Affairs Law Review* 28 (2000): 163–191.
280. Mark Holt, "Civilian Nuclear Waste Disposal," *CRS Issue Brief for Congress* (Washington, DC: Congressional Research Service, August 31, 2004).
281. Watkiss, "The Middle Ages."
282. Jason Hardin, "Tipping the Scales: Why Congress and the President Should Create a Federal Interim Storage Facility for High-Level Radioactive Waste," *Journal of Land, Resources, and Environmental Law* 19 (1999): 293–338.
283. Watkiss, "The Middle Ages."
284. As of 2008, according to the NRC 13 U.S. plants are in the process of decommissioning: Dresden Nuclear Power Station, Unit 1; GE VBWR (Vallecitos); Humboldt Bay Power Plant, Unit 3; Fermi 1 Power Plant; Indian Point Unit 1; LaCrosse Boiling Water Reactor; Millstone Nuclear Power Station, Unit 1; N.S. Savannah; Peach Bottom Unit 1; Rancho Seco Nuclear Generating Station; San Onofre Nuclear Generating Station, Unit 1; Three Mile Island Nuclear Station, Unit 2; Zion Nuclear Power Station, Units 1 and 2. See United States Nuclear Regulatory Commission, *Decommissioning Nuclear Power Plants* (Washington, DC: NRC, 2008).
285. National Research Council, *Affordable Cleanup? Opportunities for Cost Reduction in the Decontamination and Decommissioning of the Nation's Uranium Enrichment Facilities* (Washington, DC: National Academy Press, 1996). All figures have been updated to $2007.
286. U.S. General Accounting Office, *Uranium Enrichment: Decontamination and Decommissioning Fund Is Insufficient to Cover Cleanup Costs* (Washington, DC: GAO-04-692, July 2004). All figures have been updated to $2007.
287. Alvin M. Weinberg, "Nuclear Energy: A Faustian Bargain?" in *Energy and the Way We Live: Article Booklet* (San Francisco: Boyd & Fraser Publishing Company, 1980), 31–33.
288. David W. Orr, "Problems, Dilemmas, and the Energy Crisis," in *Social and Political Perspectives on Energy Policy*, ed. Karen M. Gentemann (New York: Praeger, 1981), 1–17.

Chapter 2

1. See David Kathan, *Policy and Technical Issues Associated with ISO Demand Response Programs* (Washington, DC: National Association of Regulatory Utility Commissioners,

July 2002); Ed Moscovitch, "DSM in the Broader Economy: The Economic Impacts of Utility Efficiency Programs," *The Electricity Journal* (May 1994): 14–28; Eric Hirst and Marilyn Brown, "Closing the Efficiency Gap: Barriers to the Efficient Use of Energy," *Resources, Conservation and Recycling* 3 (1990): 267–281.

2. Amory B. Lovins, *Energy End-Use Efficiency* (Snowmass, CO: Rocky Mountain Institute, September 19, 2005), 1.
3. Amory B. Lovins, "Negawatts: Twelve Transitions, Eight Improvements, and One Distraction," *Energy Policy* 24, no. 4 (1996): 331–343; John Byrne and Daniel Rich, "Energy Markets and Energy Myths: The Political Economy of Energy Transitions," in *Technology and Energy Choice,* ed. John Byrne, Mary Callahan, and Daniel Rich (Newark, DE: University of Delaware, 1983), 124–160; Lee Schipper and Joel Darmstadter, "The Logic of Energy Conservation," *Technology Review* 80 (1978): 41–48.
4. M. Rufo and F. Coito, *California's Secret Energy Surplus: The Potential for Energy Efficiency* (San Francisco: The Energy Foundation, 2002).
5. Amory B. Lovins, "Energy Myth Nine—Energy Efficiency Improvements Have Already Reached Their Potential," in *Energy and American Society—Thirteen Myths,* ed. Benjamin K. Sovacool and Marilyn A. Brown (New York: Springer, 2007), 250.
6. See Richard F. Hirsh, *Power Loss: The Origins of Deregulation and Restructuring in the American Electric Utility System* (Cambridge, MA: MIT Press, 1999), for history of mainstreaming of conservation and DSM in 1980s. See also U.S. Department of Energy, *Benefits of Demand Response in Electricity Markets and Recommendations for Achieving Them* (Washington, DC: U.S. Department of Energy, February 2006); U.S. Office of Technology Assessment, *Energy Efficiency: Challenges and Opportunities for Electric Utilities* (Washington, DC: U.S. Government Printing Office, September 1993); Andrew Warren, "Saving Megabucks by Saving Megawatts," *Energy Policy* (December 1987): 522-528; Clark W. Gellings and John H. Chamberlin, *Demand-Side Management: Concepts and Methods* (Liburn, GA: Fairmont Press, 1988), 2.
7. Marilyn A. Brown and Benjamin K. Sovacool, "Promoting a Level Playing Field for Energy Options: Electricity Alternatives and the Case of the Indian Point Energy Center," *Energy Efficiency* 1, no. 1 (2008), 35–48.
8. Benjamin K. Sovacool, *The Power Production Paradox: Revealing the Socio-technical Impediments to Distributed Generation Technologies* (doctoral dissertation, Virginia Polytechnic Institute & State University, Blacksburg, VA, April 2006).
9. John Carlin, "Renewable Energy in the U.S.," *Encyclopedia of Energy* 5 (2004): 347–363.
10. R. Madlener and N. Wohlgemuth, "Small Is Sometimes Beautiful: The Case of Distributed Generation in Competitive Energy Markets," *Proceedings of the 1st Austrian-Czech-German Conference on Energy Market Liberalization in the Central and Eastern Europe, Prague, Czech Republic, September 6–8, 1999* (Prague: CZAEE, 1999), 95–96.
11. A. Goett and R. Farmer, "Prospects for Distributed Electricity Generation: A CBO Paper," Congressional Budget Office, 2003, 6.
12. P. B. Meherwan, *Handbook for Cogeneration and Combined Cycle Power Plants* (New York: ASME Press, 2002), 25.
13. Daniel Yergin, "Conservation: The Key Energy Source," in *Energy Future: Report of the Energy Project at the Harvard Business School,* ed. Robert Stobaugh and Daniel Yergin (New York: Random House, 1979), 137.
14. Lovins, *End-Use Efficiency.*

15. DOE, *Benefits of Demand Response.*
16. Eric Hirst, "Price and Cost Impacts of Utility DSM Programs," *The Energy Journal* 13, no. 4 (October 1992): 75–91.
17. New York State Energy Research and Development Authority, *New York Energy Smart Program Evaluation and Status Report* (Albany, NY: NYSERDA, 2004).
18. Ralph Cavanagh, "Restructuring for Sustainability: Toward New Electric Service Industries," *Electricity Journal* (July 1996): 72.
19. Marilyn A. Brown and Phillip E. Mihlmester, "What Has DSM Achieved in California?" *Proceedings of the 7th National Demand-Side Management Conference* (EPRI TR-104196 Project RP3084 Proceedings, June 1995), 229–235.
20. Mark D. Levine and Richard Sonnenblick, "On The Assessment of Utility Demand-Side Management Programs," *Energy Policy* 22 (1994): 848–856; Albert L. Nichols, "Demand-Side Management: Overcoming Market Barriers or Obscuring Real Costs?" *Energy Policy* 22, no. 10 (1994): 840–847.
21. The original figure was $438 billion in 2000, adjusted to $2007.
22. Steven Nadel, "National Energy Policy: Conservation and Energy Efficiency," *Hearing before the House Committee on Energy and Commerce* (June 22, 2001), 46–51.
23. Arnold P. Fickett, Clark W. Gellings, and Amory B. Lovins, "Efficient Use of Electricity," *Scientific American* 263, no. 3 (September 1990): 65–74.
24. Fereidoon P. Sioshansi, "The Myths and Facts of Energy Efficiency: Survey of Implementation Issues," *Energy Policy* 19, no. 3 (April 1991): 231–243.
25. Cavanagh, "Restructuring for Sustainability."
26. Lee Schipper, "Energy and Society: An American View," *Energy in Transition: A Report on Energy Policy and Future Options* (Los Angeles: University of California Press, 1980), xi–xxvi; Lee Schipper and Allan J. Lichtenberg, "Efficient Energy Use and Well-Being: The Swedish Example," *Science* 194, no. 4269 (December 3, 1976), 1001–1013.
27. See Lee Schipper, "Energy Conservation Policies in the OECD," *Energy Policy* (1987): 538–548; Jane Carter, "Energy Conservation: The Need for a Sharper Institutional Focus," *Energy Policy* (April 1985): 118–119; Joel Darmstadter, Joy Dunkerley, Jack Alterman, *How Industrial Societies Use Energy: A Comparative Analysis* (London: Johns Hopkins University Press, 1977); Joel Darmstadter, "Intercountry Comparisons of Energy Use," in *Energy and the Way We Live,* ed. Melvin Kranzberg, Timothy A. Hall, and Jane L. Scheiber (San Francisco: Boyd & Fraser Publishing, 1980), 174–180; Lee Schipper, "Energy Use in Sweden and the U.S.," in *Energy and the Way We Live,* ed. Kranzberg, Hall, and Scheiber, 181–184.
28. Marilyn Brown, Benjamin Sovacool, and Richard Hirsh, "Assessing U.S. Energy Policy," *Daedalus: Journal of the American Academy of Arts and Sciences* 135, no. 3 (Summer 2006): 5–11.
29. Quoted in Arnold P. Fickett, Clark W. Gellings, and Amory B. Lovins, "Efficient Use of Electricity," *Scientific American* 263, no. 3 (September 1990), 65–74.
30. Allan Mazur and Eugene Rosa, "Energy and Life-Style," *Science* 186, no. 4164 (November 15, 1974): 607–610.
31. Richard Cowart, *Efficient Reliability: The Critical Role of Demand-Side Resources in Power Systems and Markets* (Washington, DC: National Association of Regulatory Utility Commissioners, June 2001).
32. Antonio Herzog, Timothy Lipman, and Jennifer Edwards, "Renewable Energy: A Viable Choice," *Environment* 43, no. 10 (December 2001), 8–20.

33. David M. Nemtzow, "National Energy Policy: Conservation and Energy Efficiency," *Hearing before the House Committee on Energy and Commerce* (June 22, 2001), 76–81.
34. Ahmad Faruqui, Ryan Hledik, Sam Newell, and Hannes Pfeifenberger, "The Power of 5 Percent," *Electricity Journal* 20, no. 8 (October 2007), 68–77.
35. Chris Neme, John Proctor, and Steven Nadel, *Energy Savings Potential from Addressing Residential Air Conditioner and Heat Pump Installation Problems* (Washington, DC: ACEEE, 1999).
36. Joseph Eto, S. Kito, L. Shown, and R. Sonnenblick, "Where Did the Money Go? The Cost and Performance of the Largest Commercial Sector DSM Programs," *The Energy Journal* 21, no. 2 (2000): 23–49.
37. International Energy Agency, *The Experience With Energy Efficiency Policies and Programs in IEA Countries: Learning from the Critics* (Paris, France: International Energy Agency, August 2005).
38. See Table 2 of Kenneth Gillingham, Richard Newell, and Karen Palmer, "Energy Efficiency Policies: A Retrospective Examination," *Annual Review of Environmental Resources* 31 (2006): 161–192.
39. U.S. EIA, *Annual Energy Outlook 2007 with Projections to 2030* (February 2007), 82–88.
40. Cowart, *Efficient Reliability.*
41. Charles Komanoff, "Securing Power Through Energy Conservation and Efficiency in New York: Profiting from California's Experience," *Report for the Pace Law School Energy Project and the Natural Resources Defense Council* (May 2002), 1–22.
42. Ibid.
43. Lovins, "Energy Myth Nine."
44. Ibid.
45. Ibid.
46. Amory Lovins et al., *Small Is Profitable: The Hidden Benefits of Making Electrical Resources the Right Size* (Snowmass, CO: Rocky Mountain Institute, 2002).
47. International Energy Agency, *Distributed Generation in Liberalized Electricity Markets* (Paris: International Energy Agency, 2002).
48. Howard J. Wenger, Thomas E. Hoff, and Brian K. Farmer, "Measuring the Value of Distributed Photovoltaic Generation: Final Results of the Kerman-Grid Support Project" (presentation at the First World Conference on Photovoltaic Energy Conversion Conference Proceeding, Waikaloa, Hawaii, December 1994) (Washington, DE: IEEE, 1994), 792–796.
49. Charles D. Feinstein, Ren Orans, and Stephen W. Chapel, "The Distributed Utility: A New Electric Utility Planning and Pricing Paradigm," *Annual Review of Energy and the Environment* 22 (1997): 95.
50. T. Hoff and D. S. Shugar, "The Value of Grid-Support Photovoltaics in Reducing Distribution System Losses," *IEEE Transactions on Energy Conversion* 10 (September 1995): 569–576.
51. Goett and Farmer, "Prospects."
52. Richard Perez, Marek Kmiecik, Tom Hoff, John G. Williams, Christy Herig, Steve Letendre, and Robert M. Margolis, *Availability of Dispersed Photovoltaic Resource During the August 14th 2003 Northeast Power Outage* (Albany, NY: University of Albany, 2007).

53. Stephen Salsbury, "Facing the Collapse of the Washington Public Power Supply System," in *Social Responses to Large Technical Systems: Control or Anticipation,* ed. Todd R. La Porte (London: Kluwer, 1991), 61–97.
54. Ed Vine, Marty Kushler, and Dan York, "Energy Myth Ten—Energy Efficiency Measures Are Unreliable, Unpredictable, and Unenforceable," in *Energy and American Society—Thirteen Myths,* ed. B. K. Sovacool and M. A. Brown (New York: Springer, 2007).
55. Ralph Cavanagh, "Least-Cost Planning Imperatives for Electric Utilities and Their Regulators," *Harvard Environmental Law Review* 10 (1986): 299–344.
56. J. Murawski, "Cost of Power Plant Jumps," *The News & Observer,* November 17, 2006.
57. Lovins, "Energy Myth Nine."
58. Amory B. Lovins, L. Hunger Lovins, and Leonard Ross, "Nuclear Power and Nuclear Bombs," *Foreign Affairs* 58 (1979/1980): 1171.
59. Christopher Flavin, Janet L. Sawin, John Podesta, Ana Unruh Cohen, and Bracken Hendricks, *American Energy: The Renewable Path to Energy Security* (Washington, DC: Worldwatch Institute & the Center for American Progress, September 2006), 16.
60. According to PSE's Director of Resource Acquisition, Roger Garratt, PSE poured the first foundation on May 18, 2005 and the Hopkins Ridge Wind Project began commercial operations on November 27, 2005.
61. Comments at the EUCI 3rd Annual Renewable Portfolio Standards (RPS) Conference, April 23–24, 2007, Westminster, CO.
62. Lovins et al., *Small Is Profitable.*
63. Ibid.
64. Alfred Marshall, *Principles of Economics: An Introductory Volume* (New York: The Macmillan Company, 1890 [1920]).
65. Donald MacKenzie, *Knowing Machines: Essays on Technical Change* (London: MIT Press, 1996).
66. Sam Schoofs, *A Federal Renewable Portfolio Standard: Policy Analysis and Proposal* (New York: IEEE, August 6, 2004).
67. Appendix E of Office of Energy Efficiency and Renewable Energy, *Projected Benefits of Federal Energy Efficiency and Renewable Energy Programs (FY2007–FY2050)* (Washington, DC: DOE, 2007).
68. Karen Palmer and Dallas Burtraw, "Cost-Effectiveness of Renewable Energy Policies," *Energy Economics* 27 (2005): 873–894.
69. Janet Sawin, "The Role of Government in the Development and Diffusion of Renewable Energy Technologies: Wind Power in the U.S., California, Denmark, and Germany, 1970–2000" (Ph.D. dissertation, Tufts University, Boston, MA, 2001).
70. U.S. Department of Energy, Energy Efficiency and Renewable Energy Program, *Annual Report on U.S. Wind Power Installation, Cost, and Performance Trends: 2007* (Washington, DC: U.S. DOE, May 2008), 16.
71. U.S. EIA, "Capacity Factor by Energy Source" (Monthly Nonutility Power Plant Report, 2000).
72. J. Alan Beamon and Thomas J. Leckey, *Trends in Power Plant Operating Costs* (Washington, DC: U.S. EIA, 1999).
73. World Nuclear Association, *Early Orders for New Reactors* (London: World Nuclear Association, 2003).
74. Paul Gipe, *California Updates Wind Stats* (Tehachapi, CA: Wind-Works, 2002).

75. Chris Namovicz, "Issues in Wind Resource Supply Data and Modeling" (presentation at the ASA Committee on Energy Statistics, October 5, 2006).
76. U.S. Department of Energy, *Annual Report on U.S. Wind,* 23–24.
77. Daniel M. Kammen, "Renewable Energy: Taxonomic Overview," *Encyclopedia of Energy* 5 (2004): 385–412.
78. Philippe Menanteau, "Learning from Variety and Competition Between Technological Options for Generating Photovoltaic Electricity," *Technological Forecasting and Social Change* 63 (2000): 63–80.
79. Gregory F. Nemet, "Beyond the Learning Curve: Factors Influencing the Cost Reductions in Photovoltaics," *Energy Policy* 34 (2006): 3218–3232.
80. M. King Hubbert, "Energy Resources of the Earth," *Scientific American* (September 1971): 61.
81. Asia Pacific Energy Research Centre, *A Quest for Energy Security in the 21st Century: Resources and Constraints* (Tokyo, Japan: Institute of Energy Economics, 2007).
82. Tim Jackson, "Renewable Energy: Great Hope or False Promise?" *Energy Policy* (January/February 1991): 7.
83. Quoted in Kate Connolly, "Endless Possibility," *The Guardian* (April 16, 2008).
84. See U.S. Department of Energy, *Characterization of U.S. Energy Resources and Reserves* (Washington, DC: DOE/CE-0279, 1989).
85. Electricity generation taken from U.S. Energy Information Administration, *Electricity Net Generation from Renewable Energy by Energy Use Sector and Energy Source, 2002–2006* (Washington, DC: U.S. DOE, 2007). Achievable onshore wind potential assumes class 1–7 wind regimes in all 50 states (and is based on the DOE estimate that onshore wind could supply "more than one and a half times the current electricity consumption of the U.S."). See Energy Efficiency and Renewable Energy Program at the U.S. Department of Energy, *Wind Energy Resource Potential* (Washington, DC: DOE, 2007). Achievable offshore wind potential assumes water depths from zero to 900 meters. The estimate excludes 266,200 MW of offshore potential for waters currently deeper than 900 meters because such technology is not commercially available. Data taken from Walt Musial, *Offshore Wind Energy Potential for the U.S.* (NREL, May 19, 2005), 9. Achievable solar photovoltaic potential assumes prices of $2 to $2.50 per installed watt. Data were taken from Maya Chaudhari, Lisa Frantzis, and Tom E. Hoff, *PV Grid Connected Market Potential* (The Energy Foundation, September, 2004). Achievable solar thermal potential includes parabolic troughs and power towers, and is taken from National Renewable Energy Laboratory, *Concentrating Solar Power Resource Maps* (Golden, CO: NREL, December 2007). NREL states that "realistically, the potential of concentrating solar power in the Southwest could reach hundreds of gigawatts or greater than 10% of U.S. electric supply." Achievable geothermal potential taken from Bruce D. Green and R. Gerald Nix, *Geothermal—The Energy Under Our Feet* (Golden, CO: National Renewable Energy Technology, November, 2006, NREL/TP-840-40665). Achievable biomass potential (combustion) converted from estimates provided in Oak Ridge National Laboratory and U.S. Department of Energy, *Biomass as Feedstock for a Bioenergy and Bioproducts Industry: The Technical Feasibility of a Billion-Ton Annual Supply* (Washington, DC: U.S. Department of Energy, 2005, DOE/GO-102995-2135). Achievable biomass potential (landfill gas) taken from U.S. Environmental Protection Agency *An Overview of Landfill Gas Energy in the U.S.* (Washington, DC: Landfill Methane Outreach Program, May

2007). Achievable hydroelectric potential excludes all nationally protected lands and areas, and is taken from U.S. Department of Energy, *Water Resources of the U.S. with Emphasis on Low Head/Low Power Resources* (Washington, DC: DOE/ID-11111, April 2004).

86. U.S. Government Accountability Office, "Renewable Energy: Wind Power's Contribution to Electric Power Generation and Impact on Farms and Rural Communities." *GAO Report to the Ranking Democratic Member, Committee on Agriculture, Nutrition, and Forestry, U.S. Senate* (Washington, DC: GAO, 2004), 15–16.
87. Scott Kennedy, "Wind Power Planning: Assessing Long-Term Costs and Benefits," *Energy Policy* 33 (2005): 1661–1675.
88. Willett Kempton, Jeremy Firestone, Jonathan Lilley, Tracy Rouleau, and Phillip Whitaker, "The Offshore Wind Power Debate: Views from Cape Cod," *Coastal Management* 33 (2005): 119–149.
89. U.S. Department of Energy, "Wind and Hydropower Technologies Program: Wind Energy Multi Year Program Plan for 2005–2010" (November 2004), 6.
90. In the United States, researchers characterize wind resources by the power density of flowing air. Class 7 winds provide power densities greater than 800 Watt/meter^2 at heights of 50 meters or greater than 400 Watt/meter^2 at heights of 10 meters.
91. Lackner and Sachs, "A Robust Strategy."
92. U.S. Department of Energy, *Parabolic Trough Solar Thermal Electric Plants* (Washington, DC: DOE/GO-102033-1740, June 2003), 2.
93. Lovins, "Energy Myth Nine."
94. U.S. Department of Energy, *Solar Energy Technologies Program: Applications* (Washington, DC: DOE, February 8, 2007), 3.
95. P. Denholm and R. Margolis, "Very Large-Scale Deployment of Grid-Connected Solar Photovoltaics in the United States: Challenges and Opportunities" (presentation at the Solar 2006 Conference, NREL/CP-620-39683, April 2006).
96. Lovins, "Energy Myth Nine."
97. Richard Perez, *Photovoltaics Can Add Capacity to the Utility Grid: Mapping the Effective Load-Carrying Capacity of PV to Highlight Service Territories that Can Benefit From Photovoltaics* (New York: Department of Atmospheric Science at the State University of New York, 1996).
98. G. Silcker, "Peak Power Requirements," *Proceedings of the Solar Power 2004 Conference* (Washington, DC: Solar Electric Power Association, 2004).
99. Chris Robertson and Jill K. Cliburn, "Utility-Driven Solar Energy as a Least Cost Strategy to Meet RPS Policy Goals and Open New Markets" (presentation at the ASES Solar 2006 Conference).
100. R. Perez, R. Margolis, M. Kmiecik, M. Schwab, and M. Perez, *Effective-Load Carrying Capability of Photovoltaics in the U.S.* (Golden, CO: National Renewable Energy Laboratory, June 2006, NREL/CP-620-400GB).
101. F. Beck, "Powering the South: A Clean & Affordable Energy Plan for the Southern U.S." (Renewable Energy Policy Project, 2001).
102. Bruce D. Green and R. Gerald Nix, *Geothermal—The Energy Under Our Feet* (Golden, CO: National Renewable Energy Technology, November 2006, NREL/TP-840-40665), 223–224.
103. Wendell A. Duffield and John H. Sass, *Geothermal Energy: Clean Power from the Earth's Heat* (Washington, DC: U.S. Department of Interior/U.S. Geological Survey, 2003).

104. Green and Nix, *Geothermal.*
105. Geothermal Energy Association, *Geothermal Energy: Estimates of Potential Use* (Washington, DC: GEA, 2006).
106. Green and Nix, *Geothermal.*
107. Oak Ridge National Laboratory and U.S. Department of Energy, *Biomass as Feedstock for a Bioenergy and Bioproducts Industry: The Technical Feasibility of a Billion-Ton Annual Supply* (Washington, DC: U.S. Department of Energy, 2005, DOE/GO-102995-2135).
108. Jack Barkenbus et al., *Resource and Employment Impact of a Renewable Portfolio Standard in the Tennessee Valley Authority Region* (Knoxville, TN: Institute for a Secure and Stable Environment at the University of Tennessee, 2006).
109. David Pimentel, Megan Herz, Michele Glickstein, Matthew Zimmerman, Richard Allen, Kratina Becker, Jeff Evans, Benita Hussan, Ryan Sarsfeld, Anat Grosfeld, and Thomas Seidel, "Renewable Energy: Current and Potential Issues," *Bioscience* 52, no. 12 (December 2002): 1111–1118.
110. DOE, *Water Resources.*
111. Shimon Awerbuch, J. C. Jansen, and Luuk W. Beruskens, *Building Capacity for Portfolio-Based Energy Planning in Developing Countries: Shifting the Grounds for Debate* (Paris: United Nations Environment Program, August 7, 2004); Shimon Awerbuch and Martin Berger, *Applying Portfolio Theory to EU Electricity Planning and Policymaking* (Paris: International Energy Agency, February 2003, EET/2003/03).
112. Brent Gale, "Renewable Energy Requirements and Net Billing—Lessons Learned from Iowa" (presentation to the IOWA PUC, 2000).
113. Benjamin K. Sovacool, "The Costs of Failure: A Preliminary Assessment of Major Energy Accidents, 1907 to 2007," *Energy Policy* 36, no. 5 (May 2008), 1802–1820.
114. Stefan Hirschberg and Andrzej Strupczewski, "Comparison of Accident Risks in Different Energy Systems: How Acceptable?" *IAEA Bulletin* 41, no. 1 (1999): 25–31.
115. Stefan Hirschberg, Peter Burgherr, Gerald Spikerman, and Roberto Dones, "Severe Accidents in the Energy Sector: Comparative Perspective," *Journal of Hazardous Materials* 111 (2004): 57–65.
116. A. F. Fritzsche, "Severe Accidents: Can They Only Occur in the Nuclear Production of Electricity?" *Risk Analysis* 12 (1992): 327–329.
117. Peter Fontaine, "The Gathering Storm," *Public Utilities Fortnightly* 8, no. 142 (2004): 50–61.
118. Ross Gelbspan, *Boiling Point: How Politicians, Big Oil and Coal, Journalists, and Activists Are Fueling the Climate Crisis—And What We Can Do to Avert Disaster* (New York: Basic Books, 2004), 94–97.
119. These examples are all taken from Ted Williams, "Smoke on the Water," *Audubon Magazine* (January/February 2008): 55–60; and Felicity Barringer, "States' Battles over Energy Grow Fiercer with U.S. in a Policy Gridlock," *New York Times,* March 20, 2008, 3.
120. "Tide May Be Shifting Versus Coal," *Electricity Journal* 21, no. 2 (March 2008): 3–7.
121. U.S. Department of Energy, *Annual Report on U.S. Wind.*
122. Lovins, "Energy Myth Nine."
123. See C. Chen, R. Wiser, and M. Bolinger, "Weighing the Costs and Benefits of Renewable Portfolio Standards: A Comparative Analysis of State-Level Policy Impact Projections" (LBNL, January 2007).

124. C. Herig, *Using Photovoltaics to Preserve California's Electricity Capacity Reserves* (Golden, CO: National Renewable Energy Laboratory, 2002).
125. Carolyn Fischer, "How Can Renewable Portfolio Standards Lower Electricity Prices?" (Resources for the Future Discussion Paper, May 2006).
126. R. Wiser et al., "Easing the Natural Gas Crisis: Reducing Natural Gas Prices through Increased Deployment of Renewable Energy and Energy Efficiency" (LBNL, 2005).
127. Alan Nogee, Jeff Deyette, and Steve Clemmer, "The Projected Impacts of a National Renewable Portfolio Standard," *Electricity Journal* 20, no. 4 (May 2007): 7.
128. Ryan Pletka, "Economic Impact of Renewable Energy in Pennsylvania," *Black & Veatch Project 135401,* March 2004, 56pp.
129. Bruce Hannon, "Energy and Labor Demand in a Conserver Society," *Technology Review* 79, no. 5 (1977): 47–53; Bruce Hannon, Richard G. Stein, B. Z. Segal, and Diane Serber, "Energy and Labor in the Construction Sector," *Science* 202, no. 4370 (1978): 837–847; Stephen Casler and Bruce Hannon, "Updating Energy and Labor Intensities for Non-I/O Years," *Energy Systems and Policy* 9, no. 1 (1985): 27–48.
130. Ralph Cavanagh, "Least-Cost Planning Imperatives for Electric Utilities and Their Regulators," *Harvard Environmental Law Review* 10 (1986): 299–344.
131. United Nations Environment Program, *Natural Selection: Evolving Choices for Renewable Energy Technology and Policy* (Geneva: United Nations, 2000).
132. D. Kammen et al., "Putting Renewables to Work: How Many Jobs Can the Clean Power Industry Create?" *Rael Report* (January 2004).
133. Barkenbus et al., *Resource and Employment Impact.*
134. John Grieco, "How Much in Job Years?" *Power & Energy* 1, no. 3 (October 2004).
135. Union of Concerned Scientists, *The Southeastern United States Can Benefit from a National Renewable Portfolio Standard* (Washington, DC: UCS, 2007), 1.
136. Black and Veatch, *Economic, Energy, and Environmental Benefits of Concentrating Solar Power in California* (Los Angeles: Black and Veatch, April 2006).
137. Arizona Department of Commerce Energy Office, *Energy Dollar Flow Analysis for the State of Arizona* (Phoenix: State of Arizona, 2004).
138. Antonio Herzog, Timothy Lipman, and Jennifer Edwards, "Renewable Energy: A Viable Choice," *Environment* 43, no. 10 (December 2001), 8–20.
139. Martin Schweitzer and Bruce E. Tonn, *An Evaluation of State Energy Program Accomplishments: 2002 Program Year* (Washington, DC: U.S. Department of Energy, June 2005, ORNL/CON-492).
140. U.S. Department of Energy, *Energy Demands on Water Resources: Report to Congress on the Interdependency of Energy and Water* (Washington, DC: U.S. DOE, December 2006).
141. U.S. Department of Energy, *The Wind/Water Nexus* (Washington, DC: U.S. Department of Energy, April 2006, DOE/GO-102006-2218).
142. Ed Brown, "Renewable Energy Brings Water to the World," *Renewable Energy Access* (August 23, 2005).
143. American Wind Energy Association, *Wind Web Tutorial* (Washington, DC: AWEA, 2007).
144. Ari Reeves and Fredric Becker, *Wind Energy for Electric Power: A REEP Issue Brief* (Washington, DC: REEP, 2003).
145. Pacific Winds, *The Health Benefits of Altamont Pass Wind Power* (Tracy, CA: PowerWorks, 2005).

146. Ibid.
147. J. Spadaro, L. Langlois, and B. Hamilton, "Greenhouse Gas Emissions of Electricity Generation Chains—Assessing the Difference," *IAEA Bulletin* 42 (2000): 19–21.
148. U.S. Department of Energy, *Guide to Purchasing Green Power* (Washington, DC: U.S. Department of Energy, Energy Efficiency and Renewable Energy Federal Energy Management Program, 2006).
149. Union of Concerned Scientists, *Clean Power Blueprint* (Washington, DC: UCS, 2001).
150. Luc Gagnon, Camille Belanger, and Yohji Uchiyama, "Life-Cycle Assessment of Electricity Generation Options: The Status of Research in Year 2001," *Energy Policy* 30 (2002): 1267–1278.
151. Marin Pehnt, "Dynamic Life Cycle Assessment of Renewable Energy Technologies," *Renewable Energy* 31 (2006): 55–71.
152. V. M. Fthenakis, H. M. Kim, and M. Alsema, "Emissions from Photovoltaic Life Cycles," *Environmental Science and Technology* 42 (2008): 2168–2174.
153. Amory Lovins, *Nuclear Power: Economics and Climate-Protection Potential* (Snowmass, CO: Rocky Mountain Institute, September 11, 2005).
154. Bill Keepin and Gregory Kats, "Greenhouse Warming: Comparative Analysis of Nuclear and Efficiency Abatement Strategies," *Energy Policy* (December 1988): 538–561.
155. McKinsey and Company, *Reducing U.S. Greenhouse Gas Emissions: How Much at What Cost?* (Washington, DC: McKinsey, 2008).
156. Wind and coal estimates come from Mark Diesendorf, *Refuting Fallacies About Wind Power* (Sydney: University of New South Wales, August 27, 2006). PV and solar thermal estimates come from National Renewable Energy Laboratory, *Renewable Energy and Land Use* (Golden, CO: NREL, 2007).
157. Paul Brophy, "Environmental Advantages to the Utilization of Geothermal Energy," *Renewable Energy* 10, no. 2 (1997): 373–374.
158. AWEA, *Wind Web.*
159. Alan Nogee, *Responses to Senate Questions* (Washington, DC: Union of Concerned Scientists, March 25, 2005).
160. Mike Roth, "Could Wind Power Replace MTR Coal?" *Alternatives* (2006).
161. National Renewable Energy Laboratory, *PV FAQs* (2007).
162. Scott Anders, *Technical Potential for Rooftop Photovoltaics in the San Diego Region* (San Diego: Energy Policy Initiatives Center, 2003).
163. Flavin et al., *American Energy.*
164. Daniel Kammen and Sergio Pacca, "Assessing the Costs of Electricity," *Annual Review of Environment and Resources* 29 (2004): 301–344.
165. Hugh Nash, "Foreword," in *The Energy Controversy: Soft Path Questions and Answers,* ed. Hugh Nash (San Francisco: Friends of the Earth, 1979), 1–6.
166. The United Nations reported these findings in [REN21] Renewable Energy Policy Network for the 21st Century, *Renewables 2007: Global Status Report* (Washington, DC: REN21, 2008). The study reported that the LCOE for new renewable energy resources, excluding subsidies but adjusted to $2007, was 3 to 7 ¢/kWh for hydroelectric power, 4 to 7 ¢/kWh for geothermal, 5 to 12 ¢/kWh for wind, 5 to 12 ¢/kWh for biomass, 12 to 18 ¢/kWh for solar thermal, and 20 to 80 ¢/kWh for solar PV.
167. Paraphrase of definition in John Carlin, *Environmental Externalities in Electric Power Markets: Acid Rain, Urban Ozone, and Climate Change* (Washington, DC: EIA,

2002). See also NARUC, *Environmental Externalities and Electric Utility Regulation* (Washington, DC: NARUC, 1993), 3.

168. Russell Lee, "Externalities and Electric Power: An Integrated Assessment Approach," *Oak Ridge National Laboratory CONF-9507-206–2* (1995), 2–4.
169. U.S. Department of Energy and the Commission of the European Communities, "U.S.-EC Fuel Cycle Study: Background Document to the Approach and Issues," *External Costs and Benefits of Fuel Cycles,* ORNL/M-2500 (November 1992).
170. David J. Bjornstad and Marilyn A. Brown, *A Market Failures Framework for Defining the Government's Role in Energy Efficiency* (Knoxville, TN: Joint Institute for Energy and Environment, June 2004, JIEE 2004-02); Marilyn A. Brown, "Market Failures and Barriers as a Basis for Clean power Policies," *Energy Policy* 29 (2001): 1197–1207.
171. J. E. Pater, *A Framework for Evaluating the Total Value Proposition of Clean Power Technologies* (Golden, CO: National Renewable Energy Laboratory, Technical Report NREL/TP-620-38597, February 2006).
172. Shimon Awerbuch, *How Wind and Other Renewables* Really *Affect Generating Costs: A Portfolio Risk Approach* (Dublin: Irish Parliament, Oireachtas Joint Committee on Communications, Marine and Natural Resources, December 6, 2006).
173. See Godfrey Boyle, "Assessing the Environmental and Health Impacts of Energy Use," *Energy Systems and Sustainability* (Oxford: Oxford University Press, 2003), 519–566; H. Scott Matthews and Lester B. Lave, "Applications of Environmental Valuation for Determining Externality Costs," *Environmental Science & Technology* 34 (2000): 1390–1395; John P. Holdren and Kirk R. Smith, "Energy, the Environment, and Health," in *World Energy Assessment: Energy and the Challenge of Sustainability,* ed. Tord Kjellstrom, David Streets, and Xiadong Wang (New York: United Nations Development Programme, 2000), 61–110; W. Krewitt, P. Mayerhofer, R. Friedrich, A. Trukenmuller, N. Eyre, and M. Holland, "External Costs of Fossil Fuel Cycles," in *Social Costs and Sustainability: Valuation and Implementation in the Energy and Transport Sector,* ed. Olav Hohmeyer, Richard L. Ottinger, and Klaus Rennings (New York: Springer, 1997), 127–135; National Association of Regulatory Utility Commissioners, *Environmental Externalities and Electric Regulation* (Washington, DC: NARUC, September 1993, ORNL/Sub/95X-SH985C); Olav Hohmeyer, "Renewables and the Full Costs of Energy," *Energy Policy* (1992): 365–375; John P. Holdren, "Energy and Human Environment: The Generation and Definition of Environmental Problems," in *The European Transition from Oil: Societal Impacts and Constraints on Energy Policy,* ed. G. T. Goodman, L. Kristoferson, and J. Hollander (London: Academic Press, 1981), 100–101; Robert Stobaugh and Daniel Yergin, "Toward a Balanced Energy Program," in *Energy Future: Report of the Energy Project at the Harvard Business School,* ed. Robert Stobaugh and Daniel Yergin (New York: Random House, 1979), 216–227; William Ramsay, *Unpaid Costs of Electrical Energy: Health and Environmental Impacts from Coal and Nuclear Power* (Baltimore: Johns Hopkins University Press, 1979); Sam H. Schurr, Joel Darmstadter, Harry Perry, William Ramsay, and Milton Russell, "Chapter One: Health Impacts of Energy Technologies," *Energy in America's Future: The Choices Before Us* (Baltimore: Johns Hopkins University Press, 1979), 343–369; Robert J. Budnitz and John Holdren, "Social and Environmental Costs of Energy Systems," *Annual Review of Energy* 1 (1976): 553–580.
174. Thomas Sundqvist and Patrik Soderholm, "Valuing the Environmental Impacts of Electricity Generation: A Critical Survey," *Journal of Energy Literature* 8, no. 2 (2002):

1–18; Thomas Sundqvist, "What Causes the Disparity of Electricity Externality Estimates?" *Energy Policy* 32 (2004): 1753–1766.
175. See U.S. Department of Energy and the Commission of the European Communities, *U.S.-EC Fuel Cycle Study: Background Document to the Approach and Issues* (Knoxville, TN: Oak Ridge National Laboratory, November 1992, ORNL/M-2500); and U.S. Department of Energy and the Commission of the European Communities, *Estimating Externalities of Coal Fuel Cycles* (Knoxville, TN: Oak Ridge National Laboratory, September 1994, UDI-5119-94); Russell Lee, *Externalities and Electric Power: An Integrated Assessment Approach* (Oak Ridge, TN: Oak Ridge National Laboratory, 1995).
176. U.S. Office of Technology Assessment, *Energy Efficiency: Challenges and Opportunities for Electric Utilities* (Washington, DC: U.S. Government Printing Office, September 1993), 31–32.
177. Ian F. Roth and Lawrence L. Ambs, "Incorporating Externalities into a Full Cost Approach to Electric Power Generation Life-cycle Costing," *Energy* 29 (2004): 2125–2144.
178. Douglas L. Norland and Kim Y. Ninassi, *Price It Right: Energy Pricing and Fundamental Tax Reform* (Washington, DC: Alliance to Save Energy, 1998).
179. Kammen and Pacca, *Assessing the Costs.*
180. Ari Rabl and Joseph V. Spadaro, "Public Health Impact of Air Pollution and Implications for the Energy System," *Annual Review of Energy and Environment* 25 (2000): 601–627.
181. Paul Watkiss, "Electricity's Hidden Costs," *IEE Review* (November 2002): 27–31; Wolfram Krewitt, Thomas Heck, Alfred Trukenmuller, and Rainer Friedrich, "Environmental Damage Costs from Fossil Fuel Electricity Generation in Germany and Europe," *Energy Policy* 27 (1999): 173–183.
182. Sovacool, *Power Production Paradox.*
183. Ibid.

Chapter 3

1. Guido Calabresi and A. Douglas Melamed, "Property Rules, Liability Rules, and Inalienability: One View of the Cathedral," *Harvard Law Review* 85, no. 6 (April 1972): 1089–1092.
2. Douglas R. Bohi and Michael A. Toman, "Energy Security: Externalities and Policies," *Energy Policy* 30 (1993): 1093–1109.
3. Noah D. Hall, "Political Externalities, Federalism, and a Proposal for an Interstate Environmental Impact Assessment Policy," *Harvard Environmental Law Review* 32 (2007/2008): 50–94.
4. Ibid., 55.
5. Ibid.
6. "Mining and Pollution: National Treasure," *The Economist* 387, no. 8582 (May 31, 2008): 47.
7. My thinking on markets has been greatly influenced by Robert U. Ayres and Thomas R. Casten, *Recycling Energy: Growing Income While Mitigating Climate Change* (Chicago, IL: Primary Energy LLC, September 28, 2005); Jonathan G. Koomey, "Energy Efficiency in New Office Buildings: An Investigation of Market Failures and Corrective Policies" (doctoral dissertation, University of California Berkeley, Berkeley, CA, 1990);

Allen V. Kneese, "Natural Resources Policy, 1975–1985," *Journal of Environmental Economics and Management* 3 (1976): 253–288.

8. Alan H. Sanstad and Richard B. Howarth, "'Normal' Markets, Market Imperfections and Energy Efficiency," *Energy Policy* 22, no. 10 (1994): 813.
9. Severin Borenstein, James B. Bushnell, and Frank A. Wolak, "Measuring Market Inefficiencies in California's Restructured Wholesale Electricity Market," *The American Economic Review* 92, no. 5 (December 2002): 1376–1405; Severin Borenstein, "The Trouble With Electricity Markets: Understanding California's Restructuring Disaster," *Journal of Economic Perspectives* 16, no. 1 (2002): 191–211.
10. Richard B. Howarth and Alan H. Sanstad, "Discount Rates and Energy Efficiency," *Contemporary Economic Policy* 13, no. 3 (July 1995): 101–103.
11. Amory B. Lovins, "Energy Myth Nine—Energy Efficiency Improvements Have Already Reached Their Potential," in *Energy and American Society—Thirteen Myths,* ed. Benjamin K. Sovacool and Marilyn A. Brown (New York: Springer, 2007).
12. Carl Blumstein, Betsy Krieg, Lee Schipper, and Carl York, "Overcoming Social and Institutional Barriers to Energy Conservation," *Energy* 5 (1980): 355–371.
13. Jayant Sathaye and Scott Murtishaw, *Market Failures, Consumer Preferences, and Transaction Costs in Energy Efficiency Purchase Decisions* (Berkeley, CA: Lawrence Berkeley National Laboratory and California Energy Commission, 2004, CEC-500-2005-020).
14. Quoted in Joseph E. Stiglitz, "Globalization, Technology, and Asian Development," *Asian Development Review* 20, no. 2 (2003): 3.
15. Quoted in Rosemary Chalk, "AAAS Project on Secrecy and Openness in Science and Technology," *Science, Technology, & Human Values* 10, no. 2 (1985): 28.
16. William H. Golove and Joseph H. Eto, *Market Barriers to Energy Efficiency: A Critical Reappraisal of the Rationale for Public Policies to Promote Energy Efficiency* (Berkeley, CA: Lawrence Berkeley National Laboratory, March 1996, LBL-38059).
17. George A. Akerlof, "The Market for 'Lemons': Quality Uncertainty and the Market Mechanism," *The Quarterly Journal of Economics* 84, no. 3 (August 1970): 488–500.
18. George J. Stigler, "The Economics of Information," *The Journal of Political Economy* 69, no. 3 (June 1961): 213–225.
19. Herbert A. Simon, "A Behavioral Model of Rational Choice," *Quarterly Journal of Economics* 69, no. 1 (February 1955), 99–118; Herbert A. Simon, "Rational Decision Making in Business Organizations," *The American Economic Review* 69, no. 4 (September 1979): 493–513.
20. Mary Douglas and Baron Isherwood, *The World of Goods* (New York: Basic Books, 1979), 5.
21. Howard P. Marvel, "The Economics of Information and Retail Gasoline Price Behavior: An Empirical Analysis," *The Journal of Political Economy* 84, no. 5 (October 1976): 1033–1060.
22. Marilyn A. Brown, "Energy-Efficient Buildings: Does the Marketplace Work?" *Proceedings of the Annual Illinois Energy Conference* 24 (1997): 233–255; Marilyn A. Brown, "The Effectiveness of Codes and Marketing in Promoting Energy-Efficient Home Construction," *Energy Policy* (April 1993): 391–402; Dennis Anderson, "Roundtable on Energy Efficiency and the Economists—An Assessment," *Annual Review of Energy and Environment* 20 (1995): 562–573.
23. Southern States Energy Board, *Distributed Generation in the Southern States—Final Report* (Knoxville, TN: Energy Resources International, April 2003).

24. U.S. Department of Energy, *The Potential Benefits of Distributed Generation and Rate-Related Issues That May Impede Their Expansion* (Washington, DC: DOE, February 2007), iii.
25. Carolyn S. Kaplan, "Coastal Wind Energy Generation: Conflict and Capacity," *Boston College Environmental Affairs Law Review* 31 (2004): 170.
26. Marilyn A. Brown, "Obstacles to Energy Efficiency," *Encyclopedia of Energy* 4 (2004): 465–475.
27. Ernst Worrell, Rene van Berkel, Zhou Fengqi, Christoph Menke, Roberto Schaeffer, and Robert O. Williams, "Technology Transfer of Energy-efficient Technologies in Industry: A Review of Trends and Policy Issues," *Energy Policy* 29 (2001): 29–43.
28. Joseph O. Wilson, "The Answer, My Friends, Is in the Wind Rights Contract: Proposed Legislation Governing Wind Rights Contracts," *Iowa Law Review* 89 (2004): 1775–1800.
29. Allen M. Weiss, "The Effects of Expectations on Technology Adoption: Some Empirical Evidence," *Journal of Industrial Economics* 42, no. 4 (December 1994): 341–360.
30. Gary L. Nakarado, "A Marketing Orientation Is the Key to a Sustainable Energy Future," *Energy Policy* 24, no. 2 (1996): 188.
31. American Wind Energy Association, *The Economics of Wind Energy* (Washington, DC: AWEA, March 2002), 4.
32. U.S. Department of Energy, "Wind Power: Today & Tomorrow," *Report from the U.S. Department of Energy Energy Efficiency and Renewable Energy Wind and Hydropower Technologies Program* (Washington, DC: Department of Energy, 2004), 7–11.
33. IEA, *Distributed Generation,* 29–30.
34. Public Renewables Partnership, *PV Cost Factors* (Berkeley, CA: Lawrence Berkeley National Laboratory, 2005).
35. Galen Barbose, Ryan Wiser, and Mark Bolinger, *Supporting Photovoltaics in Market-Rate Residential New Construction: A Summary of Programmatic Experience to Date and Lessons Learned* (Berkeley, CA: Lawrence Berkeley National Laboratory, February 2006).
36. International Rivers Network, *Dammed Rivers, Damned Lives: The Case Against Large Dams* (Berkeley, CA: IRN, 2004), 1.
37. *The Report of the World Commission on Dams* (November 16, 2000).
38. Wolfram Jorb et al., *Decentralized Power Generation in the Liberalized EU Energy Markets: Results from the DECENT Research Project* (New York: Springer, 2003), 27–29.
39. Benjamin K. Sovacool, *The Power Production Paradox: Revealing the Socio-technical Impediments to Distributed Generation Technologies* (doctoral dissertation, Virginia Polytechnic Institute & State University, Blacksburg, VA, April 2006).
40. Ibid.
41. Ibid.
42. Ibid.
43. Shel Feldman, "Why Is It So Hard to Sell Savings as a Reason for Energy Conservation?" in *Energy Efficiency: Perspectives on Individual Behavior,* ed. Willet Kempton and Max Neiman (New York: American Council for an Energy-Efficient Economy, 1986), 27–40.
44. Gilbert E. Metcalf, "Economics and Rational Conservation Policy," *Energy Policy* 22 (1994): 819–825.

45. See Howard Geller and Sophie Attali, *The Experience with Energy Efficiency Policies and Programs in IEA Countries* (Paris: International Energy Agency, August 2005); Alan H. Sanstad, Carl Blumstein, and Steven E. Soft, "How High Are Option Values in Energy-Efficiency Investments," *Energy Policy* 23, no. 9 (1995): 739–743; Dermot Gately, "Individual Discount Rates: Comment," *The Bell Journal of Economics* 11, no. 1 (Spring 1980): 373–374.
46. Jonathan G. Koomey, *Energy Efficiency in New Office Buildings: An Investigation of Market Failures and Corrective Policies* (doctoral dissertation, University of California Berkeley, 1990), 2.
47. Alan Meier and J. Whittier, "Consumer Discount Rates Implied By Purchases of Energy-Efficient Refrigerators," *Energy* 8, no. 12 (1983): 957–963.
48. Mark D. Levine, Jonathan G. Koomey, James E. McMahon, Alan H. Sanstad, and Eric Hirst, "Energy Efficiency Policy and Market Failures," *Annual Review of Energy* 20 (1995): 535–555.
49. Jonathan G. Koomey, Alan H. Sanstad, and Leslie J. Shown, "Energy-efficient Lighting: Market Data, Market Imperfections, and Policy Success," *Contemporary Economic Policy* 14, no. 3 (July 1996): 98–109; Stephen J. DeCanio, "The Efficiency Paradox: Bureaucratic and Organizational Barriers to Profitable Energy-Saving Investments," *Energy Policy* 26, no. 5 (1998): 441–454.
50. Kenneth Train, "Discount Rates in Consumers' Energy Related Decisions: A Review of the Literature," *Energy* 10, no. 12 (1985): 1243–1253.
51. Jerry A. Hausman, "Individual Discount Rates and the Purchase and Utilization of Energy-Using Durables," *The Bell Journal of Economics* 10, no. 1 (Spring 1979): 33–54.
52. Kevin A. Hassett and Gilbert E. Metcalf, "Energy Conservation Investment: Do Consumers Discount the Future Correctly?" *Energy Policy* (1993): 710–716.
53. James Rendon, "In Search of Savings, Companies Turn to the Sun," *The New York Times,* October 12, 2003, B12.
54. Koomey, *Energy Efficiency,* 2.
55. Ibid.
56. Sovacool, *Power Production Paradox.*
57. Theresa M. Jurotich, *Network Infiltration: Gaining Utility Acceptance of Alternative Energy Systems* (master's thesis, Virginia Tech, Blacksburg, VA, May 2003), 2–3.
58. Kenneth J. Arrow, "The Economics of Agency," *Principals and Agents: The Structure of Business* (Boston: Harvard Business School Press, 1991), 37–51.
59. David J. Bjornstad and Marilyn A. Brown, *A Market Failures Framework for Defining the Government's Role in Energy Efficiency* (Knoxville, TN: Joint Institute for Energy and Environment, June 2004, JIEE 2004-02).
60. Marilyn A. Brown, "Energy-Efficient Buildings: Does the Marketplace Work?" *Proceedings of the Annual Illinois Energy Conference* 24 (1997): 233–255; Marilyn A. Brown, "The Effectiveness of Codes and Marketing in Promoting Energy-Efficient Home Construction," *Energy Policy* (April 1993): 391–402; Dennis Anderson, "Roundtable on Energy Efficiency and the Economists—An Assessment," *Annual Review of Energy and Environment* 20 (1995): 562–573.
61. Amory Lovins, *Energy Efficient Buildings: Institutional Barriers and Opportunities* (Boulder, CO: E-Source, 1992), 16.
62. Ibid., 12.

63. U.S. Office of Technology Assessment, *Building Energy Efficiency* (Washington, DC: USOTA, May 1992, OTA-E-518).
64. John DeCicco, Rick Diamond, Sandy Nolden, and Tom Wilson, *Improving Energy Efficiency in Apartment Buildings* (Washington, DC: ACEEE, 1996).
65. Sovacool, *Power Production Paradox.*
66. Richard B. Howarth, Brent M. Haddad, and Bruce Paton, "The Economics of Energy Efficiency: Insights from Voluntary Participation Programs," *Energy Policy* 28 (2000): 477–486.
67. Sovacool, *Power Production Paradox.*
68. Ibid.
69. Ibid.
70. Stephen J. DeCanio, "Barriers Within Firms to Energy-Efficient Investments," *Energy Policy* 21 (September 1993): 906–914.
71. Sovacool, *Power Production Paradox.*
72. Christopher Russell, *Barriers to Industrial Energy Cost Control: The Competitor Within* (Washington, DC: Alliance to Save Energy, 2004).
73. Sovacool, *Power Production Paradox.*
74. Ibid.
75. Ibid.
76. Ibid.
77. Ibid.
78. Ibid.
79. Fereidoon P. Sioshansi, "The Myths and Facts of Energy Efficiency: Survey of Implementation Issues," *Energy Policy* 19, no. 3 (April 1991): 231–243.
80. Richard Cowart, *Efficient Reliability: The Critical Role of Demand-Side Resources in Power Systems and Markets* (Washington, DC: National Association of Regulatory Utility Commissioners, June 2001).
81. Fereidoon P. Sioshansi, "Restraining Energy Demand: The Stick, The Carrot, or the Market?" *Energy Policy* 22, no. 5 (1994): 378–392.
82. Lovins, "Energy Myth Nine," 261.
83. A. Clark, "Demand-Side Management in South Africa: Barriers and Possible Solutions for New Power Sector Contexts," *Energy for Sustainable Development* 4, no. 4 (2000): 27–35.
84. Charles D. Feinstein, Ren Orans, and Stephen W. Chapel, "The Distributed Utility: A New Electric Utility Planning and Pricing Paradigm," *Annual Review of Energy & Environment* 22 (1997): 178–179.
85. Sovacool, *Power Production Paradox.*
86. Valery Yakubovich, Mark Granovetter, and Patrick McGuire, "Electric Charges: The Social Construction of Rate Systems," *Theory and Society* 34 (2005): 579–612.
87. Sovacool, *Power Production Paradox.*
88. Ibid.
89. Ibid.
90. Ibid.
91. Ibid.
92. Ibid.
93. Anthony Allen, "The Legal Impediments to Distributed Generation," *Energy Law Journal* 23 (2002): 505–523.

94. Sovacool, *Power Production Paradox.*
95. Brent Alderfer and Thomas J. Starrs, "Making Connections: Case Studies of Interconnection Barriers and Their Impact on Distributed Power Projects," *National Renewable Energy Laboratory Report NREL/SR-200-28053* (2000).
96. Richard Stavros, "Distributed Generation: Last Big Battle for State Regulators?" *Public Utilities Fortnightly* 137 (October 15, 1999): 34–43.
97. T. E. McDermott and R. D. Dugan, "PQ, Reliability, and DG," *IEEE Industry Applications Magazine* (September/October 2003): 17–23.
98. Sovacool, *Power Production Paradox.*
99. Ibid.
100. Ibid.
101. Ibid.
102. Thomas G. Bourgeois, Bruce Hedman, and Fred Zalcman, "Creating Markets for Combined Heat and Power and Clean Distributed Generation in New York State," *Environmental Pollution* 123, no. 3 (June 2003), 451–462.
103. Sovacool, *Power Production Paradox.*
104. The commissioners also observed that the coop's legal fees must have exceeded its cost to have entered into a net metering arrangement with the customer. For a summary of this case, see Interstate Renewable Energy Council, *Connecting to the Grid: FERC Rules PURPA Supports Net Metering;* FERC docket No. EL05-92-000, "Order Initiating Enforcement Proceeding and Requiring Midland Power Cooperative to Implement PURPA" (issued June 6, 2005).
105. Sovacool, *Power Production Paradox.*
106. Ibid.
107. Ibid.
108. Ibid.
109. International Energy Agency, *Variability of Wind Power and Other Renewables: Management Options and Strategies* (Paris: International Energy Agency, 2005), 8, 43–44.
110. European Wind Energy Association, *Large Scale Integration of Wind Energy in the European Power Supply: Analysis, Issues and Recommendations* (Paris: EWEA, December 2005), 7–9.
111. Richard Piwko, Dale Osborn, Robert Gramlich, Gary Jordan, David Hawkins, and Kevin Porter, "Wind Energy Delivery Issues: Transmission Planning and Competitive Electricity Market Operation," *IEEE Power & Energy Magazine* 3, no. 6 (November/December 2005): 47–56.
112. Edgar A. DeMeo, William Grant, Michael R. Milligan, and Matthew J. Schuerger, "Wind Plant Integration: Cost, Status, and Issues," *IEEE Power & Energy Magazine* 3, no. 6 (November/December 2005): 39–46.
113. EIA, *Impacts of a 10-Percent Renewable Portfolio Standard* (Washington, DC: U.S. Department of Energy, February 2002).
114. NYSERDA, "Report on Phase 2: System Performance Evaluation," *The Effects of Integrating Wind Power on Transmission System Planning, Reliability, and Operations* (Albany, NY: NYSERDA, 2005).
115. See Benjamin K. Sovacool and Christopher Cooper, "State Efforts to Promote Renewable Energy: Tripping the Horse with the Cart?" *Sustainable Development Law & Policy* 8, no. 1 (Fall 2007): 5–9, 78.
116. Sovacool and Cooper, "Tripping the Horse."

117. G. Czisch and B. Ernst, "High Wind Power Penetration by the Systematic Use of Smoothing Effects Within Huge Catchment Areas Shown in a European Example," *Proceedings of WINDPOWER 2001* (Washington, DC: American Wind Energy Association, 2001).
118. Cristina L Archer and Mark Z. Jacobson, "Supplying Baseload Power and Reducing Transmission Requirements by Interconnecting Wind Farms," *Journal of Applied Meteorology and Climatology* 46 (November 2007): 1701–1717.
119. Pumped hydro and compressed air storage systems are already commercially available and offer a combined 22.1 GW of installed capacity in the United States.
120. Paul Denholm, Gerald L. Kulcinski, and Tracey Holloway, "Emissions and Energy Efficiency Assessment of Baseload Wind Energy Systems," *Environmental Science & Technology* 39 (2005): 1903–1911.
121. Paul Denholm, "Improving the Technical, Environmental and Social Performance of Wind Energy Systems Using Biomass-Based Energy Storage," *Renewable Energy* 31 (2006): 1356.
122. Andrew Clark, "BP Accepts Blame for US Disasters and Agrees to Pay $373m in Fines," *The Guardian,* October 26, 2007.
123. Kurt M. Saunders, "Patent Nonuse and the Role of Public Interest as a Deterrent to Technology Suppression," *Harvard Journal of Law & Technology* 15 (2002): 391–392.
124. Ibid.
125. Ibid.
126. Joanna Lewis and Ryan Wiser, *Fostering a Renewable Energy Technology Industry: An International Comparison of Wind Industry Policy Support Mechanisms* (Berkeley, CA: Lawrence Berkeley National Laboratory, November 2005, LBNL-59116).

Chapter 4

1. Ronald C. Tobey, *Technology as Freedom: The New Deal and the Electrical Modernization of the American Home* (Los Angeles: University of California Press, 1996), 1.
2. See John G. Clark, *The Political Economy of World Energy: A Twentieth Century Perspective* (London: University of North Carolina Press, 1990), 6; as well as John G. Clark, *Energy and The Federal Government: Fossil Fuel Policies, 1900 to 1946* (Chicago: University of Illinois Press, 1987).
3. James Hershberg, *James B. Conant—Harvard to Hiroshima and the Making of the Nuclear Age* (New York: Alfred A. Knopf, 1963), 595.
4. Denis Hayes, *Rays of Hope: The Transition to a Post-Petroleum World* (New York: Norton & Norton, 1977).
5. Earl T. Hayes, "Energy Resources Available to the U.S., 1985–2000," in *Energy and the Way We Live,* ed. Melvin Kranzberg, Timothy A. Hall, and Jane L. Scheiber (San Francisco: Boyd & Fraser Publishing, 1980), 34–43; Michael C. Noland, "Solar Energy: Practice and Prognosis," in *Energy and the Way We Live,* ed. Kranzberg, Hall, and Scheiber, 372.
6. Sovacool, *Power Production Paradox.*
7. Ibid.
8. Ibid.
9. Ibid.
10. Ibid.

11. Ibid.
12. Ibid.
13. Ibid.
14. Ibid.
15. Ibid.
16. Ibid.
17. Ibid.
18. Peter Behr, "Solar Electric Industry Worried Reagan Might Pull the Plug," *Washington Post,* September 29, 1981, D7; Stephen Greene, "Solar Energy Industry Slips Into the Shadows; Fall in Oil Prices, Changes in Tax Rules Hurt Sales," *Washington Post,* November 9, 1986, B1; and M. K. Heiman, "Expectations for Renewable Energy Under Market Restructuring: The U.S. Experience," *Energy* 31 (2006): 1052–1066.
19. Sovacool, *Power Production Paradox.*
20. Ibid.
21. Ibid.
22. Ibid.
23. Ibid.
24. Ibid.
25. Ibid.
26. Ibid.
27. Ibid.
28. "Renewable Energy: Freezing the Sun," *The Economist* 387, no. 8584 (June 28, 2008): 42.
29. Sovacool, *Power Production Paradox.*
30. See Benjamin K. Sovacool and Christopher Cooper, "Big Is Beautiful: The Case for Federal Leadership on a Renewable Portfolio Standard," *Electricity Journal* 20, no. 4 (2007): 48–50.
31. Ibid.
32. Ibid.
33. Nancy Rader, "The Hazards of Implementing Renewables Portfolio Standards," *Energy & Environment* 11 (2000): 391–394.
34. Adam Segal, "Is America Losing Its Edge? Innovation in a Globalized World," *Foreign Affairs* 83, no. 6 (2004): 2–8.
35. Woodrow Clark and William Isherwood, "Distributed Generation: Remote Power Systems with Advanced Storage Technologies," *Energy Policy* 32 (2004): 1573–1589; S. Julio Friedman and Thomas Homer-Dixon, "Out of the Energy Box," *Foreign Affairs* 83, no. 6 (2004): 72–83.
36. Kurt E. Yeager, "Electricity Deregulation and Implications for R&D and Renewables," *Hearing Before the Subcommittee on Energy and Environment of the House Committee on Science* (March 31, 1998), 57.
37. John P. Holdren et al., *Powerful Partnerships: The Federal Role of International Cooperation on Energy Innovation* (Washington, DC: Office of Science and Technology Policy, 1999), ES-4.
38. Robert E. Burns and Michael Murphy, "Repeal of the Public Utility Holding Company Act of 1935: Implications and Options for State Commissions," *Electricity Journal* 19, no. 8 (October 2006): 32–41; Benjamin K. Sovacool, "PUHCA Repeal: Higher Prices, Less R&D, and More Market Abuses?" *Electricity Journal* 19, no. 1 (January/February

2006): 85–89; Steve Nadel and Marty Kushler, "Public Benefits Funds: A Key Strategy for Advancing Energy Efficiency," *Electricity Journal* 13, no. 10 (October 2000): 76.
39. Quoted in Yeager, "Electricity Deregulation," 97.
40. Ibid., 94–107.
41. Nancy L. Rose and Paul L. Joscow, "The Diffusion of New Technologies: Evidence from the Electric Utility Industry," *The RAND Journal of Economics* 21, no. 3 (Autumn 1990): 354–373.
42. Richard Munson, *From Edison to Enron: The Business of Power and What It Means for the Future of Electricity* (London: Praeger, 2005), 152.
43. Daniel M. Kammen and Gregory F. Nemet, "Reversing the Incredible Shrinking Energy R&D Budget," *Issues in Science & Technology* (Fall 2005): 84.
44. Sovacool, *Power Production Paradox.*
45. For three excellent articles exploring the difference between American and Danish approaches to wind R&D, see Ulrik Jorgensen and Peter Karnoe, "The Danish Wind-Turbine Story: Technical Solutions to Political Visions?" in *Managing Technology in Society: The Approach of Constructive Technology Assessment,* ed. Arie Rip, Thomas J. Misa, and Johan Schot (London: Pinter Publishers, 1995), 57–82; Matthias Heymann, "Signs of Hubris: The Shaping of Wind Technology Styles in Germany, Denmark, and the U.S., 1940–1990," *Technology & Culture* 39, no. 4 (October 1998): 641–670; Raghu Garud and Peter Karnoe, "Bricolage versus Breakthrough: Distributed and Embedded Agency in Technology Entrepreneurship," *Research Policy* 32 (2003): 277–300.
46. Janet Sawin, "The Role of Government in the Development and Diffusion of Renewable Energy Technologies: Wind Power in the U.S., California, Denmark, and Germany, 1970–2000" (Ph.D. dissertation, Tufts University, Boston, MA, 2001).
47. Allen L. Hammond and William D. Metz, "Solar Energy Research: Making Solar After the Nuclear Model?" *Science* 197, no. 4300 (July 15, 1977): 241–244.
48. Gail Greenberg, "Electric Vehicles: Lessons and the Legacy," *Energy* 8, no. 4 (Summer 1983), 6–9.

Chapter 5

1. Quoted in Lynn White, "The Historical Root of our Ecological Crisis," *Science* 155, no. 3767 (March 10, 1967): 1203–1207.
2. Richard R. Wilk, "Culture and Energy Consumption," in *Energy: Science, Policy, and the Pursuit of Sustainability,* ed. Robert Bent, Lloyd Orr, and Randall Baker (Washington, DC: Island Press, 2002), 110.
3. John A. Casazza and George C. Loehr, *The Evolution of Electric Power Transmission Under Deregulation: Selected Readings* (New York: IEEE, 2000), 17.
4. Quoted in Ralph Cavanagh, "Restructuring for Sustainability: Toward New Electric Service Industries," *Electricity Journal* (July 1996): 71.
5. Benjamin K. Sovacool, *The Power Production Paradox: Revealing the Socio-technical Impediments to Distributed Generation Technologies* (doctoral dissertation, Virginia Polytechnic Institute & State University, Blacksburg, VA, April 2006).
6. Eric Hirst, *Price-Responsive Demand as Reliability Resources* (Montpelier, VT: Regulatory Assistance Project, April 2002).

7. See Niels I. Meyer, "Distributed Generation and the Problematic Deregulation of Energy Markets in Europe," *International Journal of Solar Energy* 23, no. 4 (2003): 217–221; Rolf Kehlhofer et al., *Combined-Cycle Gas and Steam Turbine Power Plants* (Tulsa, OK: PennWell Publishing, 2003), 6; A. Douglas Melamed, "Electricity Competition: Volume 1," *Hearings Before the Subcommittee on Energy and Power of the Committee on Commerce, House of Representatives* (March 18–May 6, 1999), 257–264.
8. Martin J. Pasqualetti, "Morality, Space, and the Power of Wind-Energy Landscapes," *Geographical Review* 90, no. 3 (July 1, 2000): 384–386.
9. Paul C. Stern and Elliot Aronson, *Energy Use: The Human Dimension* (New York: Freeman & Company, 1984).
10. Bonnie Mass Morrison and Peter Gladhart, "Energy and Families: The Crisis and Response," *Journal of Home Economics* 68, no. 1 (1976): 15–18.
11. James C. Williams, "Strictly Business: Notes on Deregulating Electricity," *Technology & Culture* 42 (2001): 626–630.
12. Paul N. Edwards, "Infrastructure and Modernity: Force, Time, and Social Organization in the History of Sociotechnical Systems," in *Modernity and Technology*, ed. Thomas J. Misa, Philip Brey, and Andrew Feenberg (Cambridge, MA: MIT Press, 2003), 185–186.
13. "Special Report: America's Health-Care Crisis," *The Economist* (January 28, 2006), 24–26.
14. Sovacool, *Power Production Paradox*.
15. Marilyn A. Brown, "Obstacles to Energy Efficiency," *Encyclopedia of Energy* 4 (2004): 465–475.
16. Sovacool, *Power Production Paradox*.
17. Kentucky Environmental Education Council, *The 2004 Survey of Kentuckians' Environmental Knowledge, Attitudes and Behaviors* (Frankfort, KY: KEEC, January 2005).
18. Suzanne Crofts Shelton, *The Consumer Pulse Survey on Energy Conservation* (Knoxville, TN: Shelton Group, 2006).
19. Glenn Hess, "Bush Promotes Alternative Fuel," *Chemical and Engineering News* (March 6, 2006): 50–58.
20. Barbara C. Farhar, "Trends in US Public Perceptions and Preferences on Energy and Environmental Policy," *Annual Review of Energy and Environment* 9 (1994): 211–239.
21. Gordon Bultena, *Public Response to the Energy Crisis: A Study of Citizen's Attitudes and Adaptive Behaviors* (Ames, IA: Iowa State University, July 1976).
22. E. Riley Dunlap and Kenneth R. Tremblay, "Hard Times and Human Concerns: Assessing Probable Reactions to Scarcity" (paper presented at the Society for the Study of Social Problems, New York, 1976).
23. David Gottlieb and Marc Matre, *Sociological Dimensions of the Energy Crisis: A Follow-Up Study* (Houston, TX: University of Houston, The Energy Institute, April 30, 1976).
24. Sovacool, *Power Production Paradox*.
25. Ibid.
26. Ibid.
27. David E. Nye, *Consuming Power: A Social History of American Energies* (London: MIT Press, 1999).
28. David E. Nye, *Electrifying America: Social Meanings of a New Technology, 1880–1940* (London: MIT Press, 1990), 353.

29. Ruth S. Cowan, Mark H. Rose, and Marsha S. Rose, "Clean Homes and Large Utility Bills: 1900–1940," *Marriage and Family Review* 9, no. 1 (1985): 53–66.
30. These examples taken from Amory Lovins, "Soft Energy Technologies," *Annual Review of Energy* 3 (1978): 466–517, on 489; Norman Metzger, *Energy: The Continuing Crisis* (New York: Thomas Y. Crowell Company, 1984), 149–152; Bent Sorensen, "A History of Renewable Energy Technology," *Energy Policy* (January/February 1991): 8–12; Janet Sawin, "The Role of Government in the Development and Diffusion of Renewable Energy Technologies: Wind Power in the U.S., California, Denmark, and Germany, 1970–2000" (Ph.D. dissertation, Tufts University, Boston, MA, 2001), 94.
31. Bruno Latour, *Aramis or the Love of Technology* (Cambridge, MA: Harvard University Press, 1996).
32. Gabrielle Hecht, *The Radiance of France: Nuclear Power and National Identity After World War II* (Cambridge, MA: MIT Press, 1998).
33. Sovacool, *Power Production Paradox.*
34. Ibid.
35. Ibid.
36. Martin V. Melosi, *Coping with Abundance: Energy and Environment in Industrial America* (New York: Alfred A. Knopf, 1985), 8–10.
37. Most of these paragraphs are taken from the excellent history presented in Nye, *Consuming Power.*
38. Leo Marx, "American Institutions and Ecological Ideals," *Science* 170, no. 3961 (November 27, 1970): 945–952.
39. Roderick Nash, "The American Wilderness in Historical Perspective," *Forest History* 6 (1963): 3.
40. Nye, *Consuming Power,* 15.
41. White, *The Historical Root,* 1206.
42. Lewis W. Moncrief, "The Cultural Basis for our Environmental Crisis," *Science* 170, no. 3957 (October 30, 1970): 508–512.
43. Jan De Vries, "The Industrial Revolution and the Industriousness Revolution," *Journal of Economic History* 54, no. 2 (June 1994): 249–270.
44. Nye, *Consuming Power,* 167.
45. Metzger, *Energy,* 22–23.
46. John C. Fisher, *Energy Crisis in Perspective* (New York: Wiley & Sons, 1974).
47. Amory B. Lovins, L. Hunter Lovins, and Paul Hawken, "A Road Map for Natural Capitalism," *Harvard Business Review* (May/June 1999): 146, 152.
48. Langdon Winner, "Energy Regimes and the Ideology of Efficiency," in *Energy and Transport: Historical Perspectives on Policy Issues,* ed. George H. Daniels and Mark H. Rose (London: Sage Publications, 1982), 261–277.
49. Dorothy K. Newman and Don Day, *The American Energy Consumer* (Cambridge, MA: Ballinger Publishing Company, 1975).
50. Herman E. Daly, "Electric Power, Employment, and Economic Growth," in *Energy and Human Welfare—A Critical Analysis,* ed. Barry Commoner, Howard Boksenbaum, and Michael Corr (New York: Macmillan Publishing, 1975), 142.
51. Howard S. Geller, *Energy Revolution: Policies for a Sustainable Future* (New York: Island Press, 2002), 131.
52. Robert A. Ristinen and Jack P. Kraushaar, *Energy and the Environment* (New York: Wiley Press, 2005), 5.

53. Nye, *Consuming Power,* 202.
54. Economists have also shown that people will pay different amounts for the same item depending on who is providing it. Richard Thaler, for instance, showed that a thirsty bather would pay $2.65 for a beer delivered from a resort hotel but only $1.50 for the same beer if it came from a grocery store. See Steven D. Levitt and Stephen J. Dubner, *Freakonomics: A Rogue Economist Explores the Hidden Side of Everything* (New York: Penguin Books, 2006), 43.
55. Hillard G. Huntington, "Been Top Down So Long It Looks Like Bottom Up To Me," *Energy Policy* 22, no. 10 (1994): 853–839; Jonathan G. Koomey and Alan H. Sanstad, "Technical Evidence for Assessing the Performance of Markets Affecting Energy Efficiency," *Energy Policy* 22, no. 10 (1994): 826–832.
56. Lawrence J. Becker, Clive Seligman, Russell H. Fazio, and John M. Darley, "Relating Attitudes to Residential Energy Use," *Environment and Behavior* 13, no. 5 (September 1981): 590-609; L. J. Becker, C. Seligman, and J. M. Darley, *Psychological Strategies to Reduce Energy Consumption* (Princeton, NJ: Center for Energy and Environmental Studies, 1979).
57. Becker, Seligman, and Darley, *Psychological Strategies to Reduce Energy Consumption,* vii.
58. See Suzanne C. Thompson, "Will It Hurt Less If I Control It? A Complex Answer to a Simple Question," *Psychological Bulletin* 90, no. 1 (1981): 89–101; Sharon S. Brehm and Jack W. Brehm, *Psychological Reactance: A Theory of Freedom and Control* (New York: Academic Press, 1981); Michael B. Mazis, "Antipollution Measures and Psychological Reactance Theory: A Field Experiment," *Journal of Personality and Social Psychology* 31, no. 4 (1975): 654–660; Lawrence J. Becker, "Joint Effect of Feedback and Goal Setting on Performance: A Field Study of Residential Energy Conservation," *Journal of Applied Psychology* 63, no. 4 (1978): 428–433.
59. Quoted in Paul C. Stern and Elliot Aronson, *Energy Use: The Human Dimension* (New York: Freeman & Company, 1984).
60. Becker, Seligman, and Darley, *Psychological Strategies to Reduce Energy Consumption.*
61. John W. Reich and Jerie L. Robertson, "Reactance and Norm Appeal in Anti-Littering Messages," *Journal of Applied Social Psychology* 9, no. 1 (1979): 91–101.
62. Ellen J. Langer and Judith Rodin, "The Effects of Choice and Enhanced Personal Responsibility for the Aged: A Field Experiment in an Institutional Setting," *Journal of Personality and Social Psychology* 34, no. 2 (1978): 191.
63. Stern and Aronson, *Energy Use.*
64. Gottlieb and Matre, *Sociological Dimensions of the Energy Crisis.*
65. Ted Bartell, "Political Orientations and Public Response to the Energy Crisis," *Social Science Quarterly* 57 (1976): 430–436.
66. California Academy of Sciences, *National Survey of American public Reveals Profound Lack of Scientific Knowledge* (San Francisco: CAS, 2006).
67. Jeffrey S. Milstein, *Soft and Hard Energy Paths: What People on the Streets Think* (Washington, DC: U.S. Department of Energy, March 1978).
68. Sovacool, *Power Production Paradox.*
69. Jack W. Brehm, "Postdecision Challenges in the Desirability of Alternatives," *Journal of Abnormal and Social Psychology* 52 (1956): 384–389.
70. Stern and Aronson, *Energy Use,* 68–89.

71. Willett Kempton, Jeremy Firestone, Jonathan Lilley, Tracy Rouleau, and Phillip Whitaker, "The Offshore Wind Power Debate: Views from Cape Cod," *Coastal Management* 33 (2005): 119–149.
72. Jeremy Firestone and Willett Kempton, "Public Opinion About Large Offshore Wind Power: Underlying Factors," *Energy Policy* 35 (2007): 1584–1598.
73. Avraham Shama, "Energy Conservation in US Buildings: Solving the High Potential/Low Adoption Paradox from a Behavioral Perspective," *Energy Policy* 11, no. 2 (June 1983): 148–167.
74. Sovacool, *Power Production Paradox.*
75. Lynton K. Caldwell, "Energy and the Structure of Social Institutions," *Human Ecology* 4, no. 1 (1976): 31–45.
76. Nye, *Consuming Power,* 257–258.

Chapter 6

1. Walt Patterson, *Keeping the Lights On: Towards Sustainable Electricity* (London: Earthscan, 2007), 8.
2. Paul C. Stern and Elliot Aronson, *Energy Use: The Human Dimension* (New York: Freeman & Company, 1984).
3. Ian G. Barbour, *Technology, Environment, and Human Values* (New York: Praeger Scientific, 1980), 16.
4. See Roderick Frazier Nash, *American Environmentalism: Readings in Conservation History* (New York: McGraw-Hill, 1990), 1–8.
5. Barbour, *Technology,* 21.
6. See David Takacs, *The Idea of Biodiversity: Philosophies of Paradise* (Baltimore, MD: Johns Hopkins University Press, 1996).
7. Brian Czech, "A Potential Catch-22 for a Sustainable American Ideology," *Ecological Economics* 39 (2001): 4–5.
8. Aldo Leopold, *A Sand County Almanac* (Oxford: Oxford University Press, 1949), ix.
9. Nash, *American Environmentalism,* 257.
10. "The Environmental Movement: Endangered Species," *The Economist* (February 18, 2006): 32.
11. Ibid.
12. National Wind Energy Coordinating Committee, *Studying Wind/Bird Interactions: A Guidance Document* (Washington, DC: National Wind Coordinating Committee, December 1999).
13. Peter Asmusl, "Wind and Wings: The Environmental Impact of Windpower," *Electric Perspectives* 30, no. 3 (May/June 2005): 68–80.
14. Slightly more than 2,000 bats, for instance, were killed during a 7-month study at a wind location with 44 turbines in the Midwest. Another study examined wind resource areas in West Virginia and Pennsylvania and calculated that about 2,000 bats were killed during a much shorter 6-week study at 64 turbines. A third study conducted at Buffalo Mountain, Tennessee, when it was just 3 turbines, estimated that bat mortality was 21 bats per turbine per year. Several additional studies conducted in the Appalachian Mountains (focused on the region from Tennessee to Vermont) have found that large numbers of nocturnal migrants (including bats) are uniquely at risk of colliding with wind turbines. See U.S. Government Accountability Office, *Wind Power: Impacts*

on Wildlife and Government Responsibilities for Regulating Development and Protecting Wildlife (Washington, DC: U.S. GAO, September 2005, GAO-05-906); D. Daniel Boone et al., *Landscape Classification System: Addressing Environmental Issues Associated with Utility-Scale Wind Energy Development in Virginia* (Harrisonburg, VA: The Environmental Working Group of the Virginia Wind Energy Collaborative, 2005).

15. U.S. Department of Energy, *Guide to Purchasing Green Power: Renewable Energy Certificates and On-Site Renewable Generation* (Washington, DC: Department of Energy, 2004), 11
16. Carolyn S. Kaplan, "Coastal Wind Energy Generation: Conflict and Capacity," *Boston College Environmental Affairs Law Review* 31 (2004): 198.
17. Pimentel et al., "Renewable Energy," 29.
18. Calvin R. Trice, "Windmill Plan Good for County? Energy Plan Promises Revenue for Highland, but at Scenery's Cost," *Richmond Times-Dispatch,* July 10, 2005.
19. Peter A. Groothuis, Jana D. Groothuis, and John C. Whitehead, "Green vs. Green: Measuring the Compensation Required to Site Electrical Generation Windmills in a Viewshed," *Energy Policy* 36 (2008): 1545–1550.
20. Paula Berinstein, *Alternative Energy: Facts, Statistics, and Issues* (New York: Oryx Press, 2001).
21. World Commission on Dams, *Dams and Development: A New Framework for Decisonmaking* (London: Earthscan, 2000), 75–76.
22. E. Williams, C. James, and T. Tubiolo, *Distributed Generation and a Forecast of its Growth & Effects on the New York State Electric System from 2001 to 2020* (Washington, DC: Center for Clean Air Policy, 2003).
23. J. L. Silveira, J. A. de Carbalho, and A. D. C. Villela, "Combined Cycle versus One Thousand Diesel Power Plants: Pollutant Emissions, Ecological Efficiency and Economic Analysis," *Renewable and Sustainable Energy Reviews* 11 (2007): 524–535.
24. G. A. Heath et al., *Quantifying the Air Pollution Exposure Consequences of Distributed Electricity Generation* (Berkeley, CA: University of California Energy Institute, 2005).
25. Robert J. Budnitz and John Holdren, "Social and Environmental Costs of Energy Systems," *Annual Review of Energy* 1 (1976): 579.
26. Arpad Horvath, "Construction Materials and the Environment," *Annual Review of Environment & Resources* 29 (2004): 182.
27. K. S. Smallwood and C. G. Thelander, *Bird Mortality in the Altamont Pass Wind Resource Area* (Golden, CO: NREL/SR-500-36973, August 2005).
28. Michael Distefano, "The Truth About Wind Turbines and Avian Mortality," *Sustainable Development Law & Policy* (Fall 2007), 10–11.
29. M. D. Strickland, Gregory D. Johnson, Wallace P. Erickson, Sharon A. Sarappo, and Richard M. Halet, "Avian Use, Flight Behavior, and Mortality on the Buffalo Ridge, Minnesota, Wind Resource Area," *Proceedings of the National Avian Wind Power Planning Meeting III, San Diego, California, May 1998* (King City, Ontario: LGL Limited, June 2000), 70–79.
30. Stewart Lowther, "The European Perspective: Some Lessons from Case Studies," *Proceedings of the National Avian Wind Power Planning Meeting III, San Diego, California, May 1998* (King City, Ontario: LGL Limited, June 2000), 115–123.
31. Ibid.
32. Dong Energy, Vattenfall, Danish Energy Authority, Danish Forest and Nature Agency, *Danish Offshore Wind: Key Environmental Issues* (November 2006).

33. GAO, *Wind Power.*
34. Martin J. Pasqueletti, "Wind Power: Obstacles and Opportunities," *Environment* 46, no. 7 (September 2004): 22–31.
35. Ausilio Bauen, Jeremy Woods, and Rebecca Hailes, "Bioelectricity Vision: Achieving a 15% of Electricity from Biomass in OECD Countries by 2020," *Reported by the Centre for Energy Policy and Technology* (London: WWF International, 2004), 25.
36. Union of Concerned Scientists, *How Biomass Energy Works* (Washington, DC: Union of Concerned Scientists, 2005).
37. Lee R. Lynd, "Overview and Evaluation of Fuel Ethanol from Cellulosic Biomass: Technology, Economics, the Environment, and Policy," *Annual Review of Energy and the Environment* 21 (November 1996): 405–465.
38. Wendell A. Duffield and John H. Sass, *Geothermal Energy: Clean Power from the Earth's Heat* (Washington, DC: U.S. Department of Interior/U.S. Geological Survey, 2003).
39. Alyssa Kagel and Karl Gawell, "Promoting Geothermal Energy: Air Emissions Comparison and Externality Analysis," *Electricity Journal* 18, no. 7 (August/September 2005): 90–99.
40. Luc Gagnon and Joop F. van de Vate, "Greenhouse Gas Emissions from Hydropower: The State of Research in 1996," *Energy Policy* 25, no. 1 (1997): 7–13.
41. N. Strachan and A. Farrell, "Emissions from Distributed vs. Centralized Generation: The Importance of System Performance," *Energy Policy* 34, no. 17 (2006): 2688.
42. S. W. Hadley et al., "Quantitative Assessment of Distributed Energy Resources Benefits" (ORNL/TM-2003/20).
43. Benjamin K. Sovacool, "Distributed Generation (DG) and the American Electric Utility System: What Is Stopping It?" *Journal of Energy Resources Technology* 130, no. 1 (March 2008).
44. N. Strachan and H. Dowlatabadi, "Distributed Generation and Distribution Utilities," *Energy Policy* 30 (2002): 660.
45. International Energy Agency, *Distributed Generation in Liberalized Electricity Markets* (Paris: International Energy Agency, 2002), 92.
46. Noel Wise, "To Debate or to Rectify Environmental Injustice?" *Ecology Law Quarterly* 30 (2003): 353–370.
47. Elise Boulding and Kenneth E. Boulding, *The Future: Images and Processes* (London: Sage Publications, 1995).
48. Robert D. Bullard, *Unequal Protection: Environmental Justice and Communities of Color* (San Francisco: Sierra Club Books, 1994).
49. Eric M. Uslaner, *Shale Barrel Politics: Energy and Legislative Leadership* (Stanford, CA: Stanford University Press, 1989).
50. Dorothy K. Newman and Don Day, *The American Energy Consumer* (Cambridge, MA: Ballinger Publishing Company, 1975).
51. Kimberlianne Podlas, "A New Sword to Slay the Dragon: Using New York Law to Combat Environmental Racism," *Fordham Urban Law Journal* 23 (Summer 1996): 1283–1294.
52. Paul Mohai and Bunyan Bryant, *Environmental Racism: Reviewing the Evidence* (Boulder, CO: Westview Press, 1992), 174.
53. David W. Allen, "Social Class, Race, and Toxic Releases in American Counties," *The Social Science Journal* 38 (2001): 13–25.

54. United Nations Development Program, *Energy After Rio: Prospects and Challenges* (Geneva: United Nations, 1997).
55. Martha H. Keating and Felicia Davis, *Air of Injustice: African Americans and Power Plant Pollution* (Washington, DC: Clean the Air Task Force, October 2002).
56. Deanne M. Ottaviano, *Environmental Justice: New Clean Air Act Regulations and the Anticipated Impact on Minority Communities* (New York: Lawyer's Committee for Civil Rights Under Law, 2003).
57. Adam Swartz, "Environmental Justice: A Survey of the Ailments of Environmental Racism," *The Social Justice Law Review* 2 (Summer 1994): 35–37.
58. Ottaviano, *Environmental Justice,* 5–8.
59. U.S. Department of Energy and the Commission of the European Communities, *U.S.-EC Fuel Cycle Study: Background Document to the Approach and Issues* (Knoxville, TN: Oak Ridge National Laboratory, November 1992, ORNL/M-2500); and U.S. Department of Energy and the Commission of the European Communities, *Estimating Externalities of Coal Fuel Cycles* (Knoxville, TN: Oak Ridge National Laboratory, September 1994, UDI-5119-94); Russell Lee, *Externalities and Electric Power: An Integrated Assessment Approach* (Oak Ridge, TN: Oak Ridge National Laboratory, 1995).
60. Martin J. Pasqualetti, Paul Gipe, and Robert W. Righter, *Wind Power in View: Energy Landscapes in a Crowded World* (New York: Academic Press, 2002), 10–12.
61. Pasqueletti, "Wind Power: Obstacles and Opportunities."
62. M. Bisonnette, "Getting the Crop to Market: Siting and Permitting Transmission Lines on Buffalo Ridge, Minnesota" (presentation at the 2007 Power-Gen Renewable Energy & Fuels Conference).
63. Ibid.
64. Stern and Aronson, *Energy Use.*
65. In parallel, Stern and Aronson identified at least five different types of energy users. The *investor* regards energy as a cost that is carefully considered in making purchases such as equipment and capital, and views energy technologies as durable ways to recover costs over their useful life. The *consumer* thinks of their homes and automobiles as consumer goods that provide pleasures and necessities. The *conformer* sees energy technologies as a way to belong to a particular social group or attain status. The *crusader* sees energy use as an ethical issue and conserves energy as an expression of self-reliance and environmental stewardship. The *problem avoider* treats energy as no more than a potential source of annoyance or inconvenience, doing nothing about it until technologies break down and services cease.
66. Sam H. Schurr, Joel Darmstadter, Harry Perry, William Ramsay, and Milton Russell, *Energy in America's Future: The Choices Before Us* (Baltimore, MD: Johns Hopkins University Press, 1979).
67. Norman Metzger, *Energy: The Continuing Crisis* (New York: Thomas Y. Crowell Company, 1984), 181.
68. Walt Patterson, *Can Public Service Survive the Market? Issues for Liberalized Electricity* (London: Chatham House Briefing Paper, 1999).
69. Robert A. Bernstein and Stephen R. Horn, "Explaining House Voting on Energy Policy: Ideology and the Conditional Effects of Party and District Economic Interests," *The Western Political Quarterly* 34, no. 2 (June 1981): 235–245.
70. Gordon Bultena, *Public Response to the Energy Crisis: A Study of Citizen's Attitudes and Adaptive Behaviors* (Ames, IA: Iowa State University, July 1976).

71. Jim Powell, Program Manager for Energy Efficiency and Renewable Energy, U.S. Department of Energy, "Energy Efficiency and Renewable Energy: A Blueprint for Energy and Economic Growth" (presentation to the Appalachian Regional Commission, July 13, 2006, Huntsville, Alabama).
72. Energy Policy Project of the Ford Foundation, *A Time To Choose America's Energy Future* (Cambridge, MA: Ballinger Publishing, 1974).
73. Glenn Shippee, "Energy Consumption and Conservation Psychology: A Review and Conceptual Analysis," *Environmental Management* 4, no. 4 (1980): 297–314.
74. Amory B. Lovins, "Energy Myth Nine—Energy Efficiency Improvements Have Already Reached Their Potential," in *Energy and American Society—Thirteen Myths,* ed. Benjamin K. Sovacool and Marilyn A. Brown (New York: Springer, 2007), 240–241.

Chapter 7

1. J. K. Galbraith, *American Capitalism: The Concept of Countervailing Power* (Chicago, IL: University of Chicago, 1952), 42.
2. Philippe Menanteau, Dominique Finon, Marie-Laure Lamy, "Prices versus Quantities: Choosing Policies for Promoting the Development of Renewable Energy," *Energy Policy* 31 (2003): 799–812.
3. U.S. Energy Information Administration, *Policies to Promote Non-Hydro Renewable Energy in the U.S. and Selected Countries* (Washington, DC: U.S. Department of Energy, February 2005), 13; Marc Ringel, "Fostering the Use of Renewable Energies in the European Union: The Race Between Feed-in Tariffs and Green Certificates," *Renewable Energy* 31 (2006): 1–17.
4. Wilson H. Rickerson, Janet Sawin, and Robert Grace, "If the Shoe FITs: Using Feed-In Tariffs to Meet U.S. Renewable Electricity Targets," *Electricity Journal* 20, no. 4 (May 2007), 73–86.
5. Niels I. Meyer, "European Schemes for Promoting Renewables in Liberalized Markets," *Energy Policy* 31 (2003): 665–676.
6. Kornelis Blok, "Renewable Energy Policies in the European Union," *Energy Policy* 34 (2006): 251–255.
7. Volkmar Lauber, "REFIT and RPS: Options for a Harmonized Community Framework," *Energy Policy* 32 (2004): 1405–1414.
8. Jose Etcheverry, "Ontario's Experience with Renewable Energy Feed-in Tariffs" (presentation to the 5th Workshop of the International Feed-in Cooperation, Brussels, Belgium, April 8, 2008).
9. Stephen Lacey and Lily Riahi, "Will Renewables Trump Nuclear in Ontario?" *Renewable Energy Access* (Peterborough, NH: July 2, 2008).
10. "German Lessons," *The Economist,* April 3, 2008, 45.
11. Federal Ministry for the Environment, Nature Conservation, and Nuclear Safety, *Prospects for Renewable-Generated Electricity: Extract from Renewable Energy Sources Act (EEG) Progress Report 2007,* chap. 14 (Berlin: BMU, 2008).
12. Reinhard Kaiser, "Current Discussion on the RE Directive Proposal—Key Issues" (presentation at the 5th Workshop of the International Feed-in Cooperation, Brussels, Belgium, April 7–8, 2008).

13. Carlos Gasco, "Economic Impact of Renewable Energy Expansion" (presentation to the 5th Workshop of the International Feed-in Cooperation, Brussels, Belgium, April 8, 2008).
14. Marlene Kratzat, "Economic Costs and Benefits of Renewable Energy—Employment Effects" (presentation to the 5th Workshop of the International Feed-in Cooperation, Brussels, Belgium, April 8, 2008).
15. "German Lessons," *The Economist.*
16. Ibid.
17. Gasco, "Economic Impact."
18. United Nations Environment Program, *Energy Subsidies: Lessons Learned in Assessing Their Impact and Designing Policy Reforms* (Geneva: United Nations Foundation, 2004).
19. For an excellent introduction and overview to energy subsidies, see Douglas N. Koplow, *Federal Energy Subsidies: Energy, Environmental, and Fiscal Impacts* (Washington, DC: Alliance to Save Energy, April 1993); Doug Koplow and John Dernbach, "Federal Fossil Fuel Subsidies and Greenhouse Gas Emissions: A Case Study of Increasing Transparency for Fiscal Policy," *Annual Review of Energy & Environment* 26 (2001): 361–389; Doug Koplow, "Subsidies to Energy Industries," *Encyclopedia of Energy* 5 (2004): 749–765.
20. Estimates for DOE R&D for FY 2007 are taken from American Association for the Advancement of Science, *DOE R&D in FY 2007 Congressional Appropriations.* All other numbers are taken from Douglas N. Koplow, *Federal Energy Subsidies: Energy, Environmental, and Fiscal Impacts* (Washington, DC: Alliance to Save Energy, April 1993), and updated to 2007 dollars.
21. Koplow, *Federal Energy Subsidies.*
22. Irwin M. Stelzer, "It's Just as Well the World Ended, It Wasn't Working Anyway," *Electricity Journal* (December, 1990): 16.
23. Thomas R. Casten, *Turning Off the Heat: Why America Must Double Energy Efficiency to Save Money and Reduce Global Warming* (New York: Prometheus Books, 1998).
24. Janet Sawin, "The Role of Government in the Development and Diffusion of Renewable Energy Technologies: Wind Power in the U.S., California, Denmark, and Germany, 1970–2000" (Ph.D. dissertation, Tufts University, Boston, MA, 2001).
25. Koplow, *Federal Energy Subsidies.*
26. Marshall Goldberg, *Federal Energy Subsidies: Not All Technologies are Created Equal* (Washington, DC: Renewable Energy Policy Project, Report No. 11, July, 2000).
27. Ibid.
28. David W. Orr, "Problems, Dilemmas, and the Energy Crisis," in *Social and Political Perspectives on Energy Policy,* ed. Karen M. Gentemann (New York: Praeger, 1981), 1–17.
29. Goldberg, *Federal Energy Subsidies.*
30. U.S. Government Accountability Office, *Department of Energy: Key Challenges Remain for Developing and Deploying Advanced Energy Technologies to Meet Future Needs* (Washington, DC: U.S. GAO, December 2006, GAO-07-106).
31. U.S. Government Accountability Office, *Federal Electricity Subsidies: Information on Research Funding, Tax Expenditures, and Other Activities that Support Electricity Production* (Washington, DC: U.S. GAO-08-102, October 2007).
32. Koplow, *Federal Energy Subsidies.*
33. Data taken from Amory B. Lovins, "Energy Myth Nine—Energy Efficiency Improvements Have Already Reached Their Potential," in *Energy and American*

Society—Thirteen Myths, ed. Benjamin K. Sovacool and Marilyn A. Brown (New York: Springer, 2007), 259–260; and Dan Watkiss, "The Middle Ages."

34. Lovins, "Energy Myth Nine."
35. Anthony Heyes, *Determining the Price of Price-Anderson* (Washington, DC: CATO Institute, Winter 2002–2003).
36. Koplow, *Federal Energy Subsidies.*
37. Sadeq Z. Bigdeli, "Will the Friends of Climate Emerge in the WTO? The Prospects of Applying the Fisheries Subsidy Model to Energy Subsidies," *Carbon & Climate Law Review* 2, no. 1 (2008): 78–88.
38. Andrew Simms, *The Price of Power: Poverty, Climate Change, the Coming Energy Crisis, and the Renewable Revolution* (London: New Economics Foundation, 2004).
39. Tony Regan, "Relations Between Producers and Consumers of Energy" (seminar on Sustainable Development and Energy Security, Institute of Southeast Asian Studies, Singapore, April 22 and 23, 2008).
40. U.S. Office of Technology Assessment, *Energy Efficiency: Challenges and Opportunities for Electric Utilities* (Washington, DC: U.S. Government Printing Office, September 1993).
41. David Moskovitz, "Why Regulatory Reform for DSM?" in *Regulatory Incentives for Demand-Side Management,* ed. Steven Nadel, Michael W. Reid, and David R. Wolcott (Washington, DC: American Council for an Energy-Efficient Economy, 1992), 1–6.
42. Benjamin K. Sovacool, *The Power Production Paradox: Revealing the Socio-technical Impediments to Distributed Generation Technologies* (doctoral dissertation, Virginia Polytechnic Institute & State University, Blacksburg, VA, April 2006).
43. Ibid.
44. See Marilyn Showalter, *Mapping Electricity Policy* (Tacoma, WA: Power in the Public Interest, February 26, 2007); and Ken Rose, *The Impact of Competition on Electricity Prices: Can We Discern a Pattern* (Cambridge, MA: Harvard Electricity Policy Group, December 6, 2007). The number was even higher during electric utility restructuring. In 2000, for instance, at least 18 states implemented some sort of price cap or rate cap (Arizona, Connecticut, Delaware, District of Columbia, Illinois, Maine, Maryland, Massachusetts, Michigan, New Hampshire, New Jersey, New York, Ohio, Pennsylvania, Texas, and Virginia).
45. John Byrne and Daniel Rich, "Energy Markets and Energy Myths: The Political Economy of Energy Transitions," in *Technology and Energy Choice,* ed. John Byrne, Mary Callahan, and Daniel Rich (Newark, DE: University of Delaware, 1983), 124–160.
46. Richard Cowart, *Efficient Reliability: The Critical Role of Demand-Side Resources in Power Systems and Markets* (Washington, DC: National Association of Regulatory Utility Commissioners, June 2001).
47. PSCo Rule 40-2-124(1)(g)(I).
48. Richard Mignogna, "Implementing Colorado's Renewable Portfolio Standard" (presentation at the 3rd Annual Renewable Portfolio Standards Conference, Denver, Colorado, April 23, 2007).
49. According to the EIA, the 2003 industrial rate was 5.13 ¢/kWh while the residential rate was 8.70 ¢/kWh.
50. The EIA reports that Dominion Virginia Power, Virginia; Appalachian Power Co, Virginia; Indianapolis Power and Light Co., Indiana; Kentucky Power Co., Kentucky;

Cleveland Electric Illum Co., Ohio; Toledo Edison Co., Ohio; Rappahannock Electric Coop, Virginia; Lincoln Electric System, Nebraska; Cuivre River Electric Coop Inc., Missouri; Otter Tail Power Co., North Dakota; Wheeling Power Co., West Virginia; Matanuska Electric Association Inc., Alaska; Homer Electric Association Inc., Alaska; and Lower Valley Energy, Nebraska, have declining block electricity rates. See U.S. EIA, *National Action Plan for Energy Efficiency* (Washington, DC: EIA, July 2006), 5–2.

51. Sheryl Carter, "Breaking the Consumption Habit: Ratemaking for Efficient Resource Decisions," *Electricity Journal* (December 2001): 66–74.
52. Paul C. Stern and Elliot Aronson, *Energy Use: The Human Dimension* (New York: Freeman & Company, 1984).
53. Sovacool, *Power Production Paradox.*
54. Willett Kempton and Linda Layne, "The Consumer's Energy Analysis Environment," *Energy Policy* 22, no. 10 (1994): 857–866.
55. Thomas K. McCraw, *Prophets of Regulation: Charles Francis Adams, Louis D. Brandeis, James M. Landis, Alfred E. Kahn* (Cambridge, MA: Harvard University Press, 1984), 226.
56. John Dernbach, "Stabilizing and Then Reducing U.S. Energy Consumption: Legal and Policy Tools for Efficiency and Conservation," *Environmental Law Review* 37 (2007): 10003–100031.
57. LBNL Environmental Energy Technologies Division, *Tariff Analysis Project* (Berkeley, CA: LBNL, 2007).
58. Marilyn A. Brown, "Energy-Efficient Buildings: Does the Marketplace Work?" *Proceedings of the Annual Illinois Energy Conference* 24 (1997): 233–255; Marilyn A. Brown, "The Effectiveness of Codes and Marketing in Promoting Energy-Efficient Home Construction," *Energy Policy* (April 1993): 391–402; Dennis Anderson, "Roundtable on Energy Efficiency and the Economists—An Assessment," *Annual Review of Energy and Environment* 20 (1995): 562–573.
59. C. Goldman, N. Hopper, O. Sezgen, M. Moezzi, and R. Bharvirkar, *Customer Response to Day-Ahead Wholesale Market Electricity Prices: Case Study of RTP Program Experience in New York* (Berkeley, CA: Lawrence Berkeley National Laboratory, June 2004, LBNL-54761); C. Goldman, N. Hopper, O. Sezgen, M. Moezzi, and R. Bharvirkar, *Does Real-Time Pricing Deliver Demand Response? A Case Study of Niagara Mohawk's Large Customer RTP Tariff* (Berkeley, CA: Lawrence Berkeley National Laboratory, August 2004, LBNL-54974).
60. Justin Colledge et al., "Demand Side Management: Power by the Minute," *Power Economics,* February 28, 2002.
61. Bay Area Economic Forum, *Lightning Strikes Twice: California Faces the Real Risk of a Second Power Crisis* (San Francisco: BAEF, August 2004).
62. Jeffrey M. Fang and Paul S. Galen, *Issues and Methods in Incorporating Environmental Externalities into the Integrated Resource Planning Process* (Golden, CO: National Renewable Energy Laboratory, November 1994, NREL/TP-461-6684).
63. Sovacool, *Power Production Paradox.*
64. Ibid.
65. Ibid.
66. Willett Kempton and Linda Layne, "The Consumer's Energy Analysis Environment," *Energy Policy* 22, no. 10 (1994): 857–866.

67. Robin C. Winkler and Richard A. Winnett, "Behavioral Interventions in Resource Conservation," *American Psychologist* 37, no. 4 (April 1982): 421–435.
68. Robert H. Socolow, *Saving Energy in the Home: Princeton's Experiment at Twin Rivers* (Cambridge, MA: Ballinger Publishing, 1978).
69. Stern and Aronson, *Energy Use.*
70. John Byrne, Daniel Rich, Francis T. Tanniang, and Young-Doo Wang, "Rethinking the Household Energy Crisis: The Role of Information in Household Energy Conservation," *Marriage & Family Review* 9 (September 1985): 83–101.
71. L. J. Becker, C. Seligman, and J. M. Darley, *Psychological Strategies to Reduce Energy Consumption* (Princeton, NJ: Center for Energy and Environmental Studies, 1979).
72. Mark Bolinger, Ryan Wiser, and Garrett Fitzgerald, "An Overview of Investments by State Renewable Energy Funds in Large-Scale Renewable Generation Projects," *The Electricity Journal* (January/February 2005): at 78, 78–84; Brent Haddad and Paul Jefferiss, "Forging Consensus on National Renewables Policy: The Renewables Portfolio Standard and the National Public Benefits Trust Fund," *The Electricity Journal* (March 1999): 68, 70.
73. California Energy Commission, *Public Interest Energy Program* (Sacramento, CA: CEC, 2007).
74. Stern and Aronson, *Energy Use.*
75. Martin Schweitzer and Bruce E. Tonn, *An Evaluation of State Energy Program Accomplishments: 2002 Program Year* (Washington, DC: U.S. Department of Energy, June 2005, ORNL/CON-492).
76. OTA, *Energy Efficiency.*
77. Don Schultz and Joseph Eto, "Carrots and Sticks: Shared-Savings Incentive Programs for Energy Efficiency," *The Electricity Journal* (December 1990): 32–46.
78. U.S. Department of Energy Office of Policy and International Affairs and Oak Ridge National Laboratory, *Behavioral Research Workshop: Residential Buildings Energy Efficiency Draft Summary Report* (February 26, 2008), 8.
79. Bigdeli, "Will the Friends of Climate Emerge?"
80. Ibid.
81. Ibid.
82. Ibid., 82.
83. Eugene A. Rosa, Gary E. Machlis, and Kenneth M. Keating, "Energy and Society," *Annual Review of Sociology* 14 (1988): 149–172.
84. Kempton and Layne, *The Consumer's Energy.*
85. Ralph Cavanagh, "Energy-Efficiency Solutions: What Commodity Prices Can't Deliver," *Annual Review of Energy and Environment* 20 (1995): 519–525.
86. Marvin E. Olsen, "Public Acceptance of Energy Conservation," in *Energy Policy in the U.S.: Social and Behavioral Dimensions,* ed. Seymour Warkov (New York: Praeger, 1978), 91–109.
87. Lorie Huggins and Loren Lutzenhiser, "Ceremonial Equity: Low-Income Energy Assistance and the Failure of Socio-Environmental Policy," *Social Problems* 42, no. 4 (November 1995): 468–492.
88. Olsen, "Public Acceptance of Energy Conservation."
89. Alan M. Schneider, "Elasticity of Demand for Gasoline," *Energy Systems and Policy* 1, no. 3 (1975): 277–284.

90. Steven C. Hayes and John D. Cone, "Reducing Residential Electrical Energy Use: Payments, Information, and Feedback," *Journal of Applied Behavior Analysis* 10 (1977): 425–435.
91. R. B. Cialdini, "Basic Social Influence Is Underestimated," *Psychological Inquiry* 16, no. 4 (2005): 158–161.
92. P. W. Schultz, J. M. Nolan, R. B. Cialdini, N. J. Goldstein, and V. Griskevicius, "The Constructive, Destructive, and Reconstructive Power of Social Norms," *Psychological Science* 18, no. 5 (May 2007), 429–434.
93. Lovins, "Energy Myth Nine," 240.
94. L. J. Becker, C. Seligman, and J. M. Darley, *Psychological Strategies to Reduce Energy Consumption* (Princeton, NJ: Center for Energy and Environmental Studies, 1979).
95. Kempton and Layne, *The Consumer's Energy.*
96. Everett Shorey and Tom Eckman, *Appliances & Global Climate Change: Increasing Consumer Participation in Reducing Greenhouse Gases* (Washington, DC: Pew Center on Global Climate Change, October 2000).
97. Thomas A. Heberlein, "Conservation Information: The Energy Crisis and Electricity Consumption in an Apartment Complex," *Energy Systems and Policy* 1, no. 2 (1975): 105–118.
98. Stern and Aronson, *Energy Use.*
99. Thomas Petersik, "State Renewable Energy Requirements and Goals: Status Through 2003," *Report for the Energy Information Administration* (Washington, DC: Department of Energy, 2004).
100. Southern States Energy Board, *Distributed Generation in the Southern States: Barriers to Development and Potential Solutions* (Jackson, MS: Mississippi Development Authority, April 2003).
101. Sovacool, *Power Production Paradox.*
102. IEA, *Renewable Energy,* 37–55; Mirjam Harmelink, Monique Voogt, and Clemens Cremer, "Analyzing the Effectiveness of Renewable Energy Supporting Policies in the European Union," *Energy Policy* 34 (2006): 343–351; Richard Golob and Eric Brus, *The Almanac of Renewable Energy: The Complete Guide to Emerging Energy Technologies* (New York: Holt and Company, 1993), 139–148; Paul Gipe, *Wind Power Comes of Age* (New York: Wiley & Sons, 1995), 35–41.
103. For more on these issues relating to scale, readers are invited to see Benjamin K. Sovacool, "The Best of Both Worlds: Environmental Federalism and the Need for Federal Action on Renewable Energy and Climate Change," *Stanford Environmental Law Journal* 27, no. 2 (June 2008): 397–476.
104. OTA, *Energy Efficiency.*

Chapter 8

1. Lynn White, *Medieval Technology and Social Change* (Oxford: Oxford University Press, 1966), 28.
2. Marilyn A. Brown and Benjamin K. Sovacool, "Developing an 'Energy Sustainability Index' to Evaluate Energy Policy," *Interdisciplinary Science Reviews* 32, no. 4 (December 2007): 335–349.
3. Renewable Energy Policy Network for the 21st Century, *Renewables 2007: Global Status Report* (Washington, DC: REN21, 2008).

4. Kate Connolly, "Endless Possibility," *The Guardian* (April 16, 2008).
5. David A. Stockman, "The Political Process and Energy," in *Future American Energy Policy* (Lexington, MA: DC Heath & Company, 1982), 9–20.
6. These thoughts are heavily borrowed from David Orr and Langdon Winner. See David W. Orr, "Problems, Dilemmas, and the Energy Crisis," in *Social and Political Perspectives on Energy Policy,* ed. Karen M. Gentemann (New York: Praeger, 1981), 1–17; and Langdon Winner, "Energy Regimes and the Ideology of Efficiency," in *Energy and Transport: Historical Perspectives on Policy Issues,* ed. George H. Daniels and Mark H. Rose (London: Sage Publications, 1982), 261–277.
7. Wiebe Bijker, "Do Not Despair: There Is Life After Constructivism," *Science, Technology, & Human Values* 18, no. 1 (1993): 124–125.
8. A caveat must be introduced in advancing this conclusion, however. The way that the modern electric utility system has acquired momentum lends some legitimacy to those who adhere to notions of technological determinism. Of course, few seriously believe that the technologies used to generate, distribute, and consume electricity have any real chance of becoming autonomous or self-sustaining. However, the existing technological system in favor of fossil fuels and large power plants does condition the way that people—system builders, managers, and users—conceive electricity. In this way, the existing system constrains what individuals see as *possible* when they think about energy. This is not to say that the existing system controls society, but it certainly exerts influence on it, a sort of technological somnambulism instead of a technological determinism.
9. Orr, "Problems, Dilemmas, and the Energy Crisis."
10. Peter Lund, "Market Penetration Rates of New Energy Technologies," *Energy Policy* 34 (2006): 3317–3326; Peter Lund, "Effectiveness of Policy Measures in Transforming the Energy System," *Energy Policy* 35 (2007): 627–639.
11. John Byrne and Daniel Rich, "Energy Markets and Energy Myths: The Political Economy of Energy Transitions," in *Technology and Energy Choice,* ed. John Byrne, Mary Callahan, and Daniel Rich (Newark, DE: University of Delaware, 1983), 124–160.
12. Lee H. Hamilton, "Foreword," in *Energy: Science, Policy, and the Pursuit of Sustainability,* ed. Robert Bent, Lloyd Orr, and Randall Baker (Washington, DC: Island Press, 2002), xiv.
13. Lee Schipper, *Energy Conservation: Its Nature, Hidden Benefits, and Hidden Barriers* (Berkeley, CA: Lawrence Berkeley National Laboratory, June 1, 1975, UCID 3725 ERG 2), 60.
14. Carl Blumstein, Betsy Krieg, Lee Schipper, and Carl York, "Overcoming Social and Institutional Barriers to Energy Conservation," *Energy* 5 (1980): 358.
15. Greenpeace, *Oiling the Machine: Fossil Fuel Dollars Funneled into the U.S. Political Process* (Washington, DC: Greenpeace, 1997).
16. Megan Moore, *Energy & Environmental Giving in the States* (New York: National Institute on Money in State Politics, 2007).
17. John G. Clark, *The Political Economy of World Energy: A Twentieth Century Perspective* (London: University of North Carolina Press, 1990), 8.
18. Raymond Vernon, "An Interpretation," *Daedalus: Journal of the American Academy of Arts and Sciences* 104, no. 4 (Fall 1975), 1–14.

19. Riley E. Dunlap and Marvin E. Olsen, "Hard-Path Versus Soft-Path Advocates: A Study of Energy Activists," *Policy Studies Journal* 13, no. 2 (December 1984), 413–428.
20. David Gottlieb and Marc Matre, *Sociological Dimensions of the Energy Crisis: A Follow-Up Study* (Houston, TX: University of Houston, The Energy Institute, April 30, 1976).
21. Felicity Barringer, "States' Battles over Energy Grow Fiercer with U.S. in a Policy Gridlock," *New York Times,* March 20, 2008, 3.
22. Crystal Yednak, "ComEd is Behind 'Consumer' Warning: Critics Blast the Utility for Its Obscured Role," *Chicago Tribune,* January 5, 2007.
23. American Coalition for Clean Coal Electricity, *Behind the Plug: The Latest News on FutureGen* (Washington, DC: ACCCE, July 6, 2008).
24. Sadeq Z. Bigdeli, "Will the Friends of Climate Emerge in the WTO? The Prospects of Applying the Fisheries Subsidy Model to Energy Subsidies," *Carbon & Climate Law Review* 2, no. 1 (2008): 78–88.
25. Hugh Nash, "Foreword," in *The Energy Controversy: Soft Path Questions and Answers,* ed. Hugh Nash (San Francisco: Friends of the Earth, 1979), 1–6.
26. Lacey and Riahi, "Will Renewables Trump?"
27. Amory Lovins, "Soft Energy Technologies," *Annual Review of Energy* 3 (1978): 508.
28. For an excellent summary of this type of thinking, see Herman E. Daly, "On Thinking About Future Energy Requirements," in *Sociopolitical Effects of Energy Use and Policy,* ed. Charles T. Unseld, Denton E. Morrison, David L. Sills, and C.P. Wolf (Washington, DC: National Academy of Sciences, 1979), 232–240.
29. Neil Postman, *Amusing Ourselves to Death: Public Discourse in the Age of Show Business* (New York: Penguin Books, 1985), 6–8.
30. Catherine Mitchell and Bridget Woodman, *New Nuclear Power: Implications for a Sustainable Energy System* (University of Warwick: Warwick Business School, 2006).
31. Max Planck, *Scientific Autobiography and Other Papers* (London: Williams & Norgate, 1950), 33–34.
32. John A. Yager, "Energy in America's Future: The Difficult Transition," *Energy Policy in Perspective* (Washington, DC: Brookings Institute, 1981), 637–664.
33. Harvey Brooks, "Energy: A Summary of the CONAES Report," *Bulletin of the Atomic Scientists* (February 1980): 23.
34. Lovins, "Soft Energy Technologies."
35. S. David Freeman, "Is There an Energy Crisis? An Overview," *Annals of the American Academy of Political and Social Science* 410 (November 1973): 2.
36. Clark, *The Political Economy of World Energy,* 368.
37. Daly, "On Thinking About Future Energy Requirements."

Index

About the Author

Dr. Benjamin K. Sovacool is currently a research fellow in the Energy Governance Program at the Centre on Asia and Globalisation, part of the Lee Kuan Yew School of Public Policy at the National University of Singapore. He is an adjunct assistant professor at the Virginia Polytechnic Institute and State University in Blacksburg, Virginia, where he has taught for the Government and International Affairs Program and the Department of History. Dr. Sovacool recently completed work on a grant from the National Science Foundation's Electric Power Networks Efficiency and Security Program investigating the social impediments to distributed and renewable energy systems. He has also worked in research and advisory capacities with the Virginia Center for Coal and Energy Research, New York State Energy Research and Development Authority, Oak Ridge National Laboratory, and U.S. Department of Energy's Climate Change Technology Program. Apart from this book, his most recent work is an edited volume entitled *Energy and American Society—Thirteen Myths,* published by Springer in 2007. His e-mail address is bsovacool@nus.edu.sg